Synthese Library

Volume 497

Studies in Epistemology, Logic, Methodology, and Philosophy of Science

The aim of *Synthese Library* is to provide a forum for the best current work in the methodology and philosophy of science and in epistemology, all broadly understood. A wide variety of different approaches have traditionally been represented in the Library, and every effort is made to maintain this variety, not for its own sake, but because we believe that there are many fruitful and illuminating approaches to the philosophy of science and related disciplines.

Special attention is paid to methodological studies which illustrate the interplay of empirical and philosophical viewpoints and to contributions to the formal (logical, set-theoretical, mathematical, information-theoretical, decision-theoretical, etc.) methodology of empirical sciences. Likewise, the applications of logical methods to epistemology as well as philosophically and methodologically relevant studies in logic are strongly encouraged. The emphasis on logic will be tempered by interest in the psychological, historical, and sociological aspects of science. In addition to monographs *Synthese Library* publishes thematically unified anthologies and edited volumes with a well-defined topical focus inside the aim and scope of the book series. The contributions in the volumes are expected to be focused and structurally organized in accordance with the central theme(s), and should be tied together by an extensive editorial introduction or set of introductions if the volume is divided into parts. An extensive bibliography and index are mandatory.

Javier Pérez-Jara • Íñigo Ongay
Editors

Beyond Nature and Nurture

Perspectives on Human Multidimensionality

Springer

Editors
Javier Pérez-Jara
Department of Philosophy
University of Seville
Seville, Spain

Íñigo Ongay
Faculty of Social and Human Sciences
University of Deusto
Bilbao, Spain

ISSN 0166-6991 ISSN 2542-8292 (electronic)
Synthese Library
ISBN 978-3-031-87714-8 ISBN 978-3-031-87715-5 (eBook)
https://doi.org/10.1007/978-3-031-87715-5

This Springer imprint is published by the registered company Springer Nature Switzerland AG
The registered company address is: Gewerbestrasse 11, 6330 Cham, Switzerland

Contents

Chapter 1
Introduction: From the Ancient Greek Sophists' Binary Between Physis and Nomos to the Present

Javier Pérez-Jara and Íñigo Ongay

Abstract This introduction revisits the enduring binary of nature and nurture, tracing its roots from ancient Greek sophistic thought to contemporary scientific and philosophical debates. Using the Sphinx's riddle as a metaphorical point of departure, it poses the central enigma of human nature and interrogates what it truly means to be human. Across cultures and historical periods, humanity has been interpreted through binary oppositions—sacred and profane, body and soul, reality and illusion—yet few have remained as persistent or divisive as the nature/nurture divide. This volume examines how human traits such as aggression, love, and morality have been explained through both biological determinism and radical social constructionism, and how both perspectives have been alternately sacralized or challenged by different ideological camps. While the volume does not claim to definitively resolve the deeper riddle posed by the Sphinx, it offers a rigorous, pluralistic, and multidisciplinary approach to the question of human nature—understood as neither fixed nor entirely malleable, but shaped by a complex and evolving interplay of dimensions and variables.

An ancient Greek tradition tells us that, thousands of years ago, a treacherous and merciless creature called the Sphinx guarded the entrance to the city of Thebes. The Sphinx enjoyed strangling and devouring those who could not answer her riddle. Despair spread over the population, for no one could answer such an enigma. The monster asked all passersby what might be the most famous riddle in history: "Which creature has one voice and yet becomes four-footed and two-footed and three-footed?" Sophocles tells us that the legendary Oedipus, to the surprise of the

J. Pérez-Jara (✉)
Department of Philosophy, University of Seville, Seville, Spain

Center for Cultural Sociology, Yale University, New Haven, CT, USA
e-mail: jperez16@us.es

Í. Ongay
Faculty of Social and Human Sciences, University of Deusto, Bilbao, Spain
e-mail: Inigo.ongay@deusto.es

J. Pérez-Jara, Í. Ongay (eds.), *Beyond Nature and Nurture*, Synthese Library 497,
https://doi.org/10.1007/978-3-031-87715-5_1

Sphinx itself, solved the riddle. He answered: "Man, who crawls on all fours as a baby, then walks on two feet as an adult, and then uses a walking stick in old age." Devastated by defeat, the Sphinx killed herself. In other accounts, Oedipus killed her. Whatever the case, Oedipus saved the Thoebans by answering the Sphinx's riddle. This is how the myth usually ends. However, it was not the end of the story. Although Oedipus had solved the riddle, another even deeper mystery remained. Yes, the human being was the answer to the Sphinx's riddle. *But what is the human being?* For Kant and many other thinkers, this is the most important question we can ask.

We are humans, but *who* or *what*, if something, are we? Can we even talk about a unified "we" and a *universal essence* for all of us? While not entirely opaque, human nature is far from being transparent to itself. It seems like some Rorschach test where different individuals and societies see divergent things depending on their interpretative filters and perspectives. As such, there have been few consensuses about human nature throughout millennia. And yet, many societies do interpret human nature through clear-cut binaries, such as matter/spirit, illusion/reality, yin/yang, and nature/nurture. This book focuses on the last binary. The nature vs nurture duality has opened a debate that, as a spider web, has trapped thinkers since, at least, the Greek sophists and their binary between *nomos* and *physis*. The Sphinx's witty appeal to an artificial third leg was an early acknowledgment of the place of nurture in any definition of the essence of the human creature. Is aggression a product of nature or nurture? And what about selfishness, monogamous love, religious devotion, and homosexuality? The list of human themes analyzed through the lens of the nature/nurture debate seems never-ending.

Like the gods of Antiquity, cultural meanings often fiercely compete with one another. Along with harmony, stability, and compatibility, we constantly find contradictions, disarrangement, and impermanence in human cultures. Undoubtedly, the cultural side is one of the key elements that define our nature, whatever it is. However, are not *nature* and *culture* frequently interpreted as different realities, often even as antagonistic ones? This binary is usually translated as nature vs nurture (Hannon & Lewens, 2018). The tendency to contemplate humanity as the result of two opposing forces has made some thinkers privilege one pole and others the opposite. The binary is often interpreted as the polarity between the fixed and the flexible, malleable aspects of our identity. In the Western tradition, the idea that humans are not fixed creatures of any kind can be found, in one way or another, in several Greek sophists, Nietzsche, Foucault, Ortega y Gasset, Sartre, du Beauvoir and many radical social constructionists and postmodernists. Some of these thinkers held that human beings do not have nature but history; human nature is a radical open project whose "essence" is its plasticity. More clearly: The human *essence* is its *lack* of any essence.

On the other hand, *essentialism* (or the hypostatization of human nature) counts with equally famous defenders, from Confucius, Aristotle, Cicero, Aquinas, Spinoza, Hobbes, and Schopenhauer to twentieth century human ethologists and sociobiologists or more recently current radical evolutionary psychologists. Cicero famously stated that: *numquam naturam mos vinceret; est enim ea semper invicta,* that is, "custom will never conquer nature; for it is always invincible." (a statement that many radical conservatives and essentialists in the twentieth century and today would agree with, see for instance: Goldberg, 1973, 1993; Foster & Tennant, 2022).

Authors quarrel about the antiquity of the concepts of nature and culture. While some hold that these concepts are very old, others defend that they are modern inventions. According to some, "nature", in its modern and overwhelming used sense (distinct from the Greek's *physis* or the Medieval created *natura*), is mainly an invention from Humboldt's times (Wulf, 2016). Other authors hold in a similar way that "culture", in its objective sense (i.e. distinct from Cicero's *cultura animi*), is also a German invention of the nineteenth century (Latour, 1993). The binary culture/nature would be the product, through the work of Herder and others, of secularizing the medieval binary between *gratia* and *natura* (Bueno, 2002).

Aside from their antiquity, in axiological terms, there is also a lack of historical and social consensus regarding whether one of the poles (nature or nurture) should be regarded as good and the other as bad or both good or bad. This transcends the classic debate between followers of Rousseau and Hobbes in the Western tradition or between Confucius and Xunzi in Chinese culture. It is fascinating to see how both *nature* and *culture/nurture* have been, and still are, commonly used by both progressives and conservatives to support their moral claims (Bueno, 2002; Daston, 2019). Both nature and culture are constantly coded as one of the main sources of sacrality. From polygamy, homosexuality, and heteronormative marriage to cosmopolitanism, tribalism, atheism and religious devotion, all kinds of diverse and often contradictory human traits and behaviors are justified for being "natural" ("We have to respect it! It is in our nature!"). The same goes for *culture*. It is difficult to find a human institution or behavior that has not been justified by someone for being "cultural" ("We have to respect it! It is their culture!"). This is indeed astonishing. How can both nature and culture/nurture be mobilized to justify all kinds of things under the pretext of their presumed sacrality? What is going on here? Who is right, if anyone?

Despite its long and venerable history, many today regard the nature-nurture contest as obsolete, given recent developments in epigenetics, evo-devo, niche-construction and ecological inheritance or gene-culture co-evolutionary paradigms. Other authors think, though, that these theories have updated the debate in interesting new ways rather than overcoming it.

There is also a lack of consensus about the metaphysical side of the debate (Camprubí & Pérez-Jara, 2023). For some, human nature appears as a *reality* described by social, biosocial and human sciences; for others, it is just a pernicious *fiction*. Furthermore, even as a reality, human nature is sometimes presented in substantial terms, while in others, as a contingent aggregate of multiple processes, as many branches of Buddhism and some current philosophers defend (Dupré, 2018; Pérez-Jara et al., 2022; Romero et al., 2022). And if the human being should not be understood as the result of two opposing forces, but as the contingent asymmetrical intersection of many factors and dimensions, not necessarily in opposition, how many of these factors or dimensions would we have to distinguish?

The present book aims to answer these questions and many more. It gathers several leading authors in their respective fields related to the nature vs nurture debate. The book is divided into two main parts. Part I deals with the nature vs. nurture debate from a multidisciplinary approach, bringing together biology, psychology, sociology, and philosophy. Part II deals with case studies in the nature vs nurture debate, such as beauty standards in China, gender representations and laws in Thailand, territorial behavior, the microbiome-gut-brain system, breastfeeding, gender, masculinity, and childcare in Japan. Part I gathers chapters from Massimo Pigliucci, Liane Gabora and Isabel Gómez, Ullica Segerstrale, John Dupré, Javier Pérez-Jara, Gustavo E. Romero, Íñigo Ongay, and José Carlos Sánchez. Part II gathers chapters from Fujino Atsuko, Shinji Kajitani, Gabe Ignatow, Kimio Ito, Porranee Singpliam, Jun Nohara, and Renxue Wan.

We do not expect to have solved the Sphinx's meta-puzzle (i.e. *what is the human being?*). It seems this puzzle, like a shadow, will accompany us as long as humans exist. However, with this strong list of contributors and topics, we are sure the volume will be a significant contribution to the nature vs nurture debate. We hope the reader will enjoy these chapters as much as their authors did writing them!

References

Bueno, G. (2002). *Der Mythos der Kultur: Essay einer materialistischen Kulturphilosophie*. Bern: Peter Lang.

Camprubí, L., & Pérez-Jara, J. (2023). La arqueología como arqueología y la relativa independencia de la cultura. *Revista De Occidente, 510*, 36–53.

Daston, L. (2019). *Against nature*. The MIT Press.

Dupré, J. (2018). Human nature: A process perspective. In E. Hanon & T. Lewens (Eds.), *Why we disagree about human nature*. Oxford University Press.

Foster, M., & Tennant, D. B. (2022). *It's good to be a man: A handbook for godly masculinity*. Canon Press.

Goldberg, S. (1973). *The inevitability of patriarchy: Why the biological difference between men and women always produces male domination*. William Morrow and Company.

Goldberg, S. (1993). *Why men rule: A theory of male dominance*. Open Court.

Hannon, E., & Lewens, T. (Eds.). (2018). *Why we disagree about human nature*. Oxford University Press.

Latour, B. (1993). *We have never been modern*. Harvard University Press.

Pérez-Jara, J., Romero, G. E., & Camprubí, L. (2022). What is materialism? History and concepts. In G. E. Romero, J. Pérez-Jara, & L. Camprubí (Eds.), *Contemporary materialism: Its ontology and epistemology* (Synthese library) (Vol. 447). Springer. https://doi.org/10.1007/978-3-030-89488-7_1

Romero, G. E., Pérez-Jara, J., & Camprubí, L. (Eds.). (2022). *Contemporary materialism: Its ontology and epistemology* (Synthese library) (Vol. 447). Springer. https://doi.org/10.1007/978-3-030-89488-7_3

Wulf, A. (2016). *The invention of nature: Alexander von Humboldt's new world*. Vintage.

Part I
The Nature vs. Nurture Debate

Chapter 2
Nature vs Nurture: Time to Let It Go

Massimo Pigliucci and Jonathan Kaplan

Abstract The nature vs nurture debate is as old as recorded history. Apparently, during the uprising against Qin rule in 209 BCE, Chen Sheng—the leader of the uprising—asked the rhetorical question as a call to war: "Are kings, generals, and ministers merely born into their kind?" Modern approaches to developmental biology reveal the question to be somewhat misguided, and yet the debate continues. This essay argues that it is time to let go of an artificially dichotomous approach to the bases of phenotypic variation and accept the following conclusions of decades of research: (1) Measures of so-called heritability are not particularly informative, especially in humans; (2) Genes and environments interplay in complex and non-linear ways that simply do not allow the two components to be meaningfully disentangled; (3) Despite our best efforts, there is little we can say about gene-environment interactions in humans, except that genes do have pervasive effects and that such effects are sometimes radically modulated by the environment. Anything beyond that is likely to be ideologically driven and unsubstantiated by the available evidence.

The authors recognize that they are very much not the first to call for abandoning the nature vs nurture framework—indeed, a Google Scholar search for articles with "beyond," "nature," and "nurture" in their titles returns some 200 results, going back to 1979! We would of course be foolish to expect that our intervention will have any more impact than previous ones, but the point remains as true today as ever. Parts of this essay are based on and update "Heritability: a handy guide to what it means, what it doesn't mean, and that giant meta-analysis of twin studies," by Jonathan Kaplan, published in Scientia Salon, June 2015.

M. Pigliucci (✉)
Department of Philosophy, The City College of New York, New York, NY, USA
e-mail: mpigliucci@ccny.cuny.edu

J. Kaplan
School of History, Philosophy, and Religion, Oregon State University, Corvallis, OR, USA
e-mail: jonathan.kaplan@oregonstate.edu

J. Pérez-Jara, Í. Ongay (eds.), *Beyond Nature and Nurture*, Synthese Library 497,
https://doi.org/10.1007/978-3-031-87715-5_2

2.1 A Very Important Experiment

One of the least known and most interesting experiments ever carried out about the effect of nature and nurture on behavioral traits was published by Cooper and Zubek back in 1958. The experimental subjects were rats. The authors compared "intelligence," as measured by the ability to avoid mistakes in running through a maze, in two inbred, genetically distinct lines of rats.

One line had been selected for high performance in the maze ("bright" line), the other for particularly low performance ("dull" line). When reared under a standard environment, comparable to the one in which the selection process occurred, the two lines showed a highly significant difference in their abilities. Maze-running ability in these rats did, indeed, have a demonstrable genetic basis.

However, Cooper and Zubek then went on to rear individuals of the two lines in two other environments: a situation in which the cage was entirely devoid of visual and tactile stimuli ("poor" environment), and one in which the developing animals were exposed to brightly colored walls and toys ("enriched" environment). The results were simply stunning.

Under the poor conditions, the bright rats performed as badly as the dull ones; furthermore, under the enriched environment the dull rats did as well as the bright ones! In technical terms, this translates into a high heritability and marked genetic differences in the standard environment, but no heritability under either extreme environment. The inescapable conclusion is that the genetics of maze-running ability in rats are strongly affected by the environment, with different genotypes converging onto similar phenotypes when exposed to extreme environmental conditions.

This experiment was, of course, far from flawless. For one thing, only two genetic lines were tested, and they were certainly not a representative sample of natural populations of rats. Furthermore, the cognitive ability being tested was a relatively specific one, and its correlation to generalized intelligence (if there is such thing) remains to be ascertained. Finally, one could easily conceive of other (less artificial, more naturalistic) kinds of environments that might yield different results. Nevertheless, more recent work in mice suggests the same qualitative pattern, albeit with some important complications (Sauce et al., 2018).

This is exactly the kind of experiment we would have to do in order to begin to unravel the relative contributions of genes and environments on cognitive traits in humans. It is unfortunate, then, that not only this has not been done, but it cannot be done, for obvious ethical and logistical reasons. Logistically speaking, *Homo sapiens* has a much longer lifespan than rats (or mice), and it has a significantly lower fertility, so genetic experiments would be long and cumbersome. More obviously, such experiments would be ethically unacceptable, as they would involve mating people according to pre-determined experimental protocols and raising their offspring under strictly controlled environmental conditions.

Yet, experiments like the ones carried out by Cooper and Zubek and by Sauce et al. are the gold standard for studying genotype-environment interactions, and the impossibility of performing them on humans sets very strict limits to what can be meaningfully said on the topic, from a scientific perspective.

2.2 A Long and Controversial History

During the uprising against Qin rule in China in 209 BCE, Chen Sheng—the leader of the uprising—asked the rhetorical question as a call to war: "Are kings, generals, and ministers merely born into their kind?" (Qian/Dawson, 1994). It's one of the earliest recorded references to what modern scientists call the nature-nurture problem, or the issue of gene-environment interactions.

The debate on the relative importance of nature (genetics) and nurture (environment) in determining human traits has been prolonged, acrimonious, and—we would suggest—largely pointless. Great minds have engaged in it. Even just limiting ourselves to the last 300 years and what is called broadly 'the west,' we encounter everything from philosophers of the like of John Locke, Thomas Hobbes, and Jean-Jacques Rousseau, to natural scientists of the caliber of Stephen J. Gould, Richard Lewontin, and Edward O. Wilson, to social scientists like Arthur Jensen, R.J. Herrnstein, C. Murray, and Steven Pinker.

Part of what makes the debate so difficult and frustrating is the range of very disparate issues that get lumped together. Consider the following questions:

1. In this particular population, at this particular time, how much of the variation in some trait (e.g., height, realized performance on IQ exam, etc.) is associated with variation in genes, and how much with variation in environment?
2. In what way(s) are genes implicated developmentally in the production of a trait in an individual organism?
3. Is a difference between two populations the result of genetic differences between those populations, where those genetic differences are themselves doing the causal work?
4. Is there something about the genes common to a population (or to our species as a whole) that makes certain kinds of behaviors more or less likely, or even inevitable or impossible?
5. Is there something about the genes common to a population (or to our species as a whole) that makes certain kinds of social organization more or less likely, or even inevitable or impossible?

For reasons we will detail below, with a few interesting exceptions, none of these questions can be addressed rigorously in humans; certainly, the kinds of behaviors and traits on which people who want to argue about nature vs nurture are focused do not lend themselves to the sort of careful research necessary to reach reliable conclusions.

The few cases where research has pointed towards clear answers are instructive here. For example, the genes associated with lactase persistence are causally wrapped up in the average dairy consumption of populations, and indeed, the variation between populations in the availability of dairy during our species' recent evolution explains in large part the variation in the frequency of genes that provide lactase persistence. However, even for a straightforward physical trait such as this one, with a relatively straightforward genetic basis (as these things go), with clear implications for relatively simple behavioral differences, elucidating the nature-nurture relationship through rigorous research has been a challenge, and parts of the story are still very much debated today (Evershed et al., 2022).

There is little point in an allegedly scientific debate that labels, for example, geneticist Richard Lewontin a "blank slate-ist" (Pinker, 2002) because he believed that in the contemporary world genetic differences between individuals explain little of importance to our vastly different life-outcomes. The latter is a claim about the importance of particular social and political facts (systems that permit vast income inequality, the lasting impacts of colonialism, etc.), and Lewontin's well-known concerns about the unreliability of heritability estimates in humans are entirely orthogonal to those arguments. Nor does it make a lot of sense to argue about whether e.g. Charles Murray is a "genetic determinist" because he believes that the average difference in IQ test-taking performance between people identified as "black" and those identified as "white" in the U.S. are the result, at least in part, of genetic differences between the populations examined (see Herrnstein & Murray, 1994; Murray, 1998).

While we would argue that Lewontin's claim regarding life-outcomes is supported by significant evidence, and that Murray's claim regarding racial differences in IQ is largely unsupported by reliable evidence, in neither case is the available evidence even close to being of the sort that we have for the evolution and maintenance of lactase persistence. Still less is the evidence of the caliper available, say, in the case of different flowering times in local populations of the weed *Arabidopsis thaliana* (Banta & Pigliucci, 2010). Nor is either argument particularly relevant to what we would ordinarily consider nature vs nurture. Lewontin's claim might be impossible if the (social) environment were entirely irrelevant to development (a true "genetic determinist" position!), but few if any have ever made that claim. Murray's would be impossible if genetic differences between individuals had no impact at all on development (a true "blank slate" position!), but again, few if any have actually defended such a claim.

In fact, these debates can easily slip into the absurd. To what extent was Morton's classical statistical analysis of the brain sizes in his haphazard collection of skulls flawed (Gould, 1996)? To what extent was Gould's reanalysis of that data itself flawed (Lewis et al., 2011; Weisberg, 2014)? Was Gould right to think that the differences in results when Morton moved from using mustard seed to shot to carry out his measurements suggest that Morton was (unconsciously) manipulating the data, or is it more reasonable to attribute those differences to systematically divergent samples (Kaplan et al., 2015)? Part of this debate can, perhaps, be tied tangentially to arguments about human nature; part of it can be tied to arguments around the "Native American Graves Protection and Repatriation Act (NAGPRA)" and its impact on the collections of universities (see DiSanto, 2021). The material is *used* in arguments that *sound* like they might have something to do with debates about human nature, but without any great attention to the actual scientific issue of nature vs nurture.

2.3 The Problems with Heritability

Many of the problems highlighted so far stem, we believe, from too much attention paid in the literature to a measure that is not very useful, coupled with too little attention given to a fundamental concept that has revolutionized the study of gene-environment interactions in non-human species. The overvalued notion is that of heritability, the undervalued one that of reaction norms.

Let's start with heritability. In order to fix our ideas we consider an impactful paper published by Polderman et al. (2015), a comprehensive study in the prestigious journal *Nature Genetics* entitled "Meta-analysis of the heritability of human traits based on fifty years of twin studies." It garnered a lot of attention when it came out: blog posts and various news sites trumpeted its conclusions, putting their own ideological spins on the results.

The twin studies analyzed by Polderman and colleagues attempted to estimate the heritability of various human traits. While one can, for reasons we'll note briefly below, raise concerns about how accurately heritability can be estimated from twin studies in humans, the more important point is that heritability does not measure what most people think it measures. As a consequence, while twin studies may be answering some questions, they do not, and cannot, answer the questions that many people think they are answering—questions about the relative roles of nature and nurture in the development of important human traits.

To begin with, there are a few features that made the Polderman et al. study somewhat puzzling, starting with the study's design: a meta-analysis of twin-studies. Meta-analyses, as the term implies, use already published papers by combining the results they present in order to generate some overall measure of interest.

There are two sorts of situations in which researchers regularly run meta-analyses: (a) Where previous studies on the issue at hand have reached conflicting conclusions, or (b) Where previous studies record results that fail to reach statistical significance (or are "just barely" significant). In both sorts of cases, the hope is that an analysis of all the studies published so far will generate results that are more robust than any individual study.

In the case of the Polderman et al. paper, however, neither purpose applies. It is well established that twin studies estimating heritability of traits in humans generally find a substantial heritable component for essentially all traits measured (Turkheimer, 2000), nor is the statistical significance of these findings in doubt. Performing a large-scale meta-analysis of this sort (over 2500 studies) is a herculean effort, but in this case, the reason why it was done was never made clear.

Regardless, one finding that the authors chose to highlight in their abstract is that "across all traits the reported heritability is 49%." This fact got picked up by many of the aforementioned blogs and media outlets, using misleading headlines along the lines of "Nature versus Nurture a draw." But the heritability of the least heritable traits measured was only around 5%, and the heritability of the most heritable traits over 80%. What information, one wonders, is gained by averaging the heritabilities of traits as disparate as "adult height," "structure of the eye-ball," "cognition," and "social values"?

For comparison, imagine a similar meta-analysis of drug effectiveness that reported the average effectiveness of every randomized clinical trial of a drug over the past 50 years. Of what possible use would it be to know the average effectiveness of various drugs, designed to treat different diseases or disorders? All the different trials share is a vague area of interest ("drug effectiveness") and a basic methodology ("randomized clinical trials"). That is not enough to support the coherence of a meta-analysis. Certainly, no one is suggesting that the same genes underwrite the heritability of these different traits, nor that these wildly different traits are linked in any interesting ways.

It would also be a mistake to suppose, as many readers of the review apparently have, that about half of the variation in all human traits is genetic. There are a number of human traits that are not studied in this way, because the results of such a study are a foregone conclusion. Certainly, the language you speak is a "trait," and certainly it matters to who you are. But your primary language is determined more or less entirely by the household you grow up in. No one would be foolish enough to try to find the heritability of speaking, say, Japanese, taking a world-wide population as the relevant one. Similarly, while some food preferences are no doubt heritable, the cuisine you grow up with, again, surely plays a fundamental role in what foods you regard as "normal," what you eat regularly, and the like; and the foods we regard as normal and eat regularly are surely important traits.

The traits chosen for study matter, as do the traits not chosen. All one can say of a meta-analysis of all the studies done, is that if performed well, it really does include all the studies done that meet its criteria for inclusion. The studies not done, of course, can't be part of the analysis. But there are good reasons to think that the results for an arbitrarily chosen "trait" may not much resemble the traits that have historically been selected for study via these methods.

Let us now take a closer look at the concept of heritability itself. Put simply, it is the proportion of the phenotypic variation in a trait of interest, measured in a given studied population and in a given environment, that is statistically co-varying with genetic differences (however measured) among individuals in the same population.

Note that heritability is not a measure of "how genetic" a trait is. For heritability to make any sense at all as a statistic, the trait in question must vary in the population under study. So for humans, the heritability of "head number" is undefined—there is (almost) no variation in head-number for living humans (there are a vanishingly few conjoined twins that may count as exceptions; note that two-headedness is rather more common, albeit still rare, in for example snakes). Similarly, the ability to speak Jawaran among the Jawaran population also has an undefined heritability, because virtually all Jawarans speak the language. Also, since heritability is a measure of what is associated with variation in the trait, and not a measure of what causes the trait, the heritability of finger number in humans is essentially zero, and the vast majority of variation in finger number is environmental (traumatic amputations being the primary cause!).

Heritability can be calculated, in principle, by partitioning the variance (the squared deviation from the mean value) of the trait in question in the following way:

$$Vp = Vg + Ve + Vgxe + Verr$$

where Vp is the total phenotypic variance in the population, Vg is the *phenotypic* variance associated with genetic differences in the population (however measured), Ve is the *phenotypic* variance associated with shared environmental variation within the population, Vgxe is the *phenotypic* variance associated with gene by environment interactions, and Verr is everything else, which includes the *phenotypic* (again) variance associated with so-called "unique" environments, developmental 'noise,' independent epigenetic effects, and errors. Vg is generally broken down into two parts: the variance associated with additive genetic variation (Vg-additive), and the variance associated with non-additive genetic variation (Vg-nonadditive, sometimes written as Vgxg to suggest that it is associated with gene-gene interactions). Notice that *all* the above variances are calculated in units of phenotype, not in genetic units, even when the variance component is labelled as "genetic."

Given these definitions, so-called Broad-Sense Heritability, H^2, is then calculated as Vg/Vp while Narrow-Sense Heritability, h^2, is Vg-additive/Vp.

Narrow-Sense Heritability is a fantastically important measure, if one is interested in breeding for a particular trait, or in predicting the likely response of a trait to selection more generally. But the coarser Broad-Sense Heritability is the measure usually associated with studies of heritability in humans. The two should not be confused, though much of the literature does, speaking of "heritability" without specifying which type is meant.

In order to estimate (either version of) heritability, one needs to be able to distinguish variation associated with environmental differences from variation associated with genetic differences. For non-human animals raised under controlled experimental conditions (as was the case for Cooper and Zubek's rats), this can be achieved by distributing organisms with known genetic variations into known developmental environments. An overwhelming proportion of gene-environment studies are done in this fashion (Pigliucci, 2001). For reasons we have already highlighted, this is neither logistically nor ethically possible in humans.

The basic trick exploited by the sort of twin-studies reviewed in Polderman et al., then, is to compare monozygotic (aka "identical") twins ("MZ twins") to dizygotic (aka "fraternal") same-sex twins ("DZ twins"). The assumption is that the family environment experienced by MZ and DZ twins will be relevantly similar, and therefore that the difference between how similar MZ and DZ twins are can be used to estimate how heritable the trait is.

There are a number of quibbles one can raise regarding the reliability of the estimates generated this way. MZ twins usually share a placenta; DZ twins never do. It is not clear that this makes a difference, but it might. MZ twins are sometimes, and perhaps usually, treated differently than DZ twins—dressed more similarly, assumed by both their parents and the other people that they meet to be more similar in a variety of ways, etc., which may also have a measurable effect.

Further, we take the question of how we should think about twin-heritability (heritability estimates derived from doubling the difference between how similar MZ twins are to DZ twins) seriously—is it an estimate of the broad-sense heritability, of narrow-sense heritability, something between the two, or something else entirely? Should we expect DZ twins to share none of the gene-gene interactions that influence the trait, or some fraction of those shared by MZ twins? It is sometimes said that DZ twins share half the genes shared by MZ twins; this is true in a sense, but it is also possible (in fact, it seems rather likely) that the alternative alleles available from each parent are not representative of the alleles available in the population as a whole. Again, whether these kinds of issues make a difference is fiercely debated and Joseph (2014) provides a spirited attack on these assumptions.

The upshot is that people who think that twin-studies have serious limitations compared to the methods used in research on non-human animals and plants will be unimpressed with a meta-analysis of twin-studies. Several thousand studies, all with the exact same methodological problems and limitations, are no better than one such study.

Moreover, heritability is a local measure: the heritability of a trait is relative to a particular population and a particular range of developmental environments. Recall that heritability is the fraction of the phenotypic variance of a trait associated with genetic variation measured in phenotypic units. So anything that changes either the total variation associated with a trait, or changes the fraction of the variance associated with genetic variation, will change the heritability.

One way to increase heritability is to reduce the variation in the environment for that population; at the extreme, a trait that is mildly heritable under "ordinary" conditions can be made entirely heritable by reducing the environmental variation to near zero. If the environment doesn't vary in the relevant way, all phenotypic variation associated with environmental variation will be eliminated, and the only variation left will be that associated with genetic differences. Similarly, increasing the range of environments considered will often reduce heritability; if environments are added that are associated with differences in the trait, the heritability will decrease, as the overall variance is increased. All of this is not just theoretical as we have experimental evidence of the counterintuitive environmental-dependence of heritability (e.g., Mazer & Schick, 1991).

Changes in heritability due to changes in the population considered are also possible; if we reduce the amount of genetic variation in a population, the variation associated with the genetic component will decrease, and more of the variation left will be associated with whatever environmental variation exists. Similarly, if we increase the genetic variation that is associated with differences in the trait, the total variation will increase, and the total fraction associated with genetic variation will also increase.

Polderman et al.'s meta-analysis reports a heritability for "cognition" of around 0.57. Consider the narrower part of this category, "performance on standard IQ tests." Assume, for the sake of argument, that the heritability of performance on standard IQ tests in the U.S. populations is really around 0.6 (contemporary estimates range from around 0.3 to around 0.8, give or take). IQ tests are designed to yield a mean of 100 and a standard deviation of 15. The total population variance is therefore 225, of which 135 (60%) is associated with genetic variation.

But it is well established that IQ test-taking performance has increased steadily over time in many countries, the U.S. included. Reasonable estimates put the average IQ in the 1940s, as measured by 1990s tests, at perhaps 70 or 80 (see e.g. Pietschnig & Voracek, 2015). If we take 80 as our estimate, the adjusted standard deviation would be around 12. Assume, again for the sake of argument, that the heritability of IQ test-taking performance was 0.6 in 1940 as well. What would happen if we combined an equal number of individuals from our 1940 population with our 1990 population, and thought of them as a single larger population? Using 1990's tests as our standard, the first thing to note is that we no longer have a normal curve, but a curve with two 'humps' (one at 80 and one at 100). The second thing to note is that the standard deviation has increased (to around 16.6). The third thing to note is that there is a new "shared environment" factor that is responsible for a significant portion of the variance—whether the person is from the 1940s or the 1990s explains a significant chunk of the variance in our total population!

In fact, the shared environment associated with the year in this example will account for about 36% of the total variance (1/2 [(100–90)2 + (80–90)2])/16.62). Once the variance associated with year is partitioned out, 60% of the remaining variance will be associated with genetic variation; heritability has therefore been reduced to around 0.38 [0.6 * (16.62—(1/2 [(100–90)2 + (80–90)2])/16.62)].

So what, in the end, is the "actual" heritability of IQ? The question makes no sense. Heritability, as a measure, is always and only relative to particular populations at particular times in particular places. This problem is not merely hypothetical. Turkheimer et al. (2003) found that in the US populations that were studied, in relatively poor families, most of the variation in IQ test-taking ability (about 60%) was associated with the shared familial environment, and almost none with genetic variation (the rest was associated with "unique" environments). In relatively affluent families, the reverse held, with most of the variation in IQ test-taking ability (70%) being associated with genetic variation, and almost none associated with shared familial environment (these findings have been relatively robust in the US context; note that this pattern, heritability increasing in better environments, is qualitatively different then the pattern observed in mice and rats, discussed above, where heritability *decreased* in better environments). What, then, is the heritability of IQ in these populations in the U.S.? Should we take the average? Should we adjust for the frequency of the tested socio-economic status in the population? Does the question even make sense?

Another, slightly more fanciful scenario, adapted very loosely from an example of James Flynn (Flynn et al., 2012) may drive the point further home. Imagine a population in which no one plays basketball. It isn't a sport anyone is familiar with. Imagine that we then test every young adult in this population for basketball playing ability (after explaining the rules, etc.). Much of the variation in the measured ability between individuals will likely be broadly genetic (e.g., height will make a big difference, and within populations today, variation in height is mostly associated with genetic variation; note as well that between populations, however, variations in height can be largely environmental). If we compare MZ to DZ twins, we'll very likely find that the heritability of basketball playing ability is quite high in this population.

Now imagine that we take, randomly, half the kids from that population, and train them extensively in basketball, while leaving the other half untrained. If we consider them part of the same population, and measure the heritability of basketball playing ability across the population after some time, heritability will be very low—the vast majority of the difference in ability will come down to whether the people were in the highly trained group or not. (Identical twins, one of whom was put in the "high-training" regime, and the other in the "no training" regime, will not be particularly similar to one another in their realized basketball playing ability!)

What about a society that cares enough about basketball that to be "good"—good enough to play on a school team, etc.—you have to be really good, because there is a lot of competition and lots of people try out. Everyone plays a little when they are very young, because it is an important sport that most kids are interested in. And when young, most of the small differences in ability will be down to odd little differences, some of which will likely be heritable within that population—difference in body type, reaction speed, perhaps interest. But then, small differences in abilities and interest will get magnified—those who start out not very good and not very interested are unlikely to pursue it much, very unlikely to get specialized training, etc. Those who start out with some "natural" talent and some interest are likely to be recognized, rewarded, and eventually highly trained.

In such a society, any small differences in ability and interest that are related to genetic variation will be greatly magnified. But here is the wrinkle. Any of those differences that are related, however distantly, to those early differences in heritable traits, will show up in an analysis as "genetic." MZ twins will tend to share the same training regime (or lack thereof) rather more often than DZ twins, because they will tend, more often, to share those small variations that make them either more or less likely than average to pursue basketball. But on one plausible view, what's doing most of the work in creating differences in abilities is training and practice—not "genes"! The trait will be highly heritable, but differences in ability will be mostly down to environments—environments selected (self-selected or imposed by others) at least in part because of genetic endowments. (For more on the different ways in which environments and genes can covary and the influence this has on heritability estimates, see e.g. Lynch & Bourrat, 2017).

So, in this population, is basketball playing ability mostly genetic, or mostly environmental? The question makes no sense—or rather, depending on how one interprets it, one can defend either answer, or neither, equally well.

It is worth, at this point, stressing another oddity of twin studies. When "shared environment" is spoken of in this context, it means explicitly family environment. Other environments are not considered. So a twin-study would either identify the training regime (or lack of it) noted above as genetic—part of heritability—or if a portion of it was not associated with any genetic differences, it would be chalked up to "unique" environmental effects. But there is nothing "unique," in the usual sense of the word about basketball training, either in our hypothetical example or in the real world. Unique simply means, for our purposes here, not part of the familial environment that is shared equally by MZ and DZ twins—nothing more.

2.4 Intermezzo: The Promise and Perils of Genome Wide Association Studies

Another approach to the study of the genetic basis of human traits that has been gaining traction in recent years is known as Genome Wide Association Studies (GWAS). Very roughly, it works by comparing places on the genome where there are differences (within a particular population) with differences in the trait of interest (within the same population). If a variant at a particular genomic location is associated with a specific form of the trait of interest more often than one would expect by chance, that is taken to be some evidence that a gene near that location *may* be causally associated with the trait.

For human behavior genetics, which is perhaps the field most wrapped up in nature vs nurture debates, the deployment of this technique was a source of hope. The promise was that, finally, we could move beyond heritability to identifying actual genes that made a difference.

Unfortunately, while GWAS has proven to be a valuable tool in other areas, it has failed to live up to its promise in human behavior genetics. The hope that research would uncover a relatively small number of genes associated with much of the variation predicted to be linked to genetic variation by heritability studies were dashed when early studies failed to find much of anything of interest (Plomin, 2018).

What GWAS has revealed about the genetics of human behavior can best be summed by a very disappointing, albeit true, "it's complicated." For most traits studied, there are no single loci associated with more than a tiny fraction of the variation (no evidence of any genes with a big effect), and hundreds or thousands of loci are associated with the trait (see e.g. Chabris et al., 2015 and cites therein).

Enormous samples have failed to find loci that together account for any sizable fraction of the variation in the traits of interest, even when statistical standards are weakened to try to capture more of the possible "true positives" at the cost of increased false positives (the "missing heritability" problem; see: Zuk et al., 2012; Shen & Feldman, 2020; Matthews & Turkheimer, 2021). Population structure—the fact that different populations have slightly different frequencies of alleles—makes the interpretation of human GWAS results difficult at best.

The hope that GWAS would provide an entry into a mechanistic understanding of the relationship between genetic variation and variation in behavioral traits by elucidating the developmental pathways has been dashed (Kaplan & Turkheimer, 2021). Behavior genetics has retreated back into finding statistical associations, which are for all practical purposes no more reliable than heritability estimates more generally, and share the same conceptual problems (Turkheimer, 2012).

2.5 Reaction Norms and the Limits of Human Genetics

Despite being by far the most common way to talk about nature vs nurture, we have argued that the concept of heritability—especially in its broad form—is the least informative and most misleading way to do it. The prominent evolutionary biologist Theodosius Dobzhansky, one of the founders of modern population genetics, proposed that the correct way of thinking about the relationship between an organism's genotype and its phenotype is through the lens of the reaction norm. The partial reaction norm for a particular trait and a particular genotype is the way that that trait develops, given that genotype, over a defined range of environments.

The idea has been around ever since Wilhelm Johannsen (1911) invented the concepts of genotype and phenotype at the beginning of the twentieth century. More than a century later, thousands of papers have been published demonstrating variation in reaction norms for almost any conceivable trait across a huge range of living organisms, including bacteria, protozoa, fungi, plants, invertebrates, and vertebrates (for recent treatments see: Arnold et al., 2019; Reid & Acker, 2022; Napier et al., 2023).

From a reaction norm perspective (Schlichting & Pigliucci, 1998) genotype does not determine the phenotype; rather, we can think of it as determining how the organisms will develop given a specific environment. Of course, while it is possible to compare how one trait develops, given a set of genotypes, over a few different environments, it is impossible to determine the complete norm of reaction for even a single genotype. As Dobzhansky (1955) put it:

> The norm of reaction of a genotype is at best only incompletely known. Complete knowledge of a norm of reaction would require placing the carriers of a given genotype in all possible environments, and observing the phenotypes that develop. This is a practical impossibility. The existing variety of environments is immense, and new environments are constantly produced. Invention of a new drug, a new diet, a new type of housing, a new educational system, a new political regime introduces new environments (pp. 74–75).

If we switch from heritability to reaction norms thinking, questions about "nature" and "nurture" reveal themselves to be ill-formed. Rather, the right questions to ask are more of the form "how does this genotype respond to changes in this environmental variable?" and 'how does the response of this gene to changes in the environment depend on or vary with the rest of the organism's genes?" For some traits, the norm of reaction will be basically flat against most reasonable developmental environments—most human genotypes, in most developmental environments, produce humans with 10 digits on their hands. In other cases, the trait will develop differently in different environments (height, weight range, and so forth).

Strictly speaking, the only way to determine the reaction norm of a trait, given a particular genotype, for a particular range of environments, is to raise genetically identical clones in the variety of different environments in which one is interested. Partial norms of reaction, looking at the response of specific genetic variants averaged against varying genetic backgrounds need not use clones, but still require that organisms be able to be sorted into developmental environments based on their genetic endowments. The problem is that in the case of humans we can at best guess at the real shape of reaction norms for a number of traits, and the intellectually honest thing to do is therefore to acknowledge that we are indeed engaging in guesswork.

2.6 The Take Home Message

So, does it make sense to say that the nature vs nurture debate is over, and that it is about 50/50, or 60/40, or whatever percentages come out of heritability studies? Hardly. In good part, this is because heritability studies are simply ill-designed to answer nature vs nurture questions, however much some people would wish otherwise. Is our ability to read and write mostly "nature" or mostly "nurture"? The question, as stated, is too ambiguous to answer. Humans are the only animals we know of that can read and write, and our ability to perform those tasks clearly has something to do with our nature—with the kinds of creatures that we are. But, just as obviously, particular environmental contexts are necessary for those skills to develop—"nurture."

In the end, heritability doesn't tell us much that we ought to want to know. It doesn't tell us whether a trait will be easy or hard to change. It doesn't tell us what developmental resources are necessary for the trait to develop normally, nor how changes in those resources will change the development of the trait in question. It is simply, at best, a snap-shot of how much of the variation in the trait that there happens to be now is statistically associated with the genetic variation there is in this population, in this range of environments, with this particular distribution of genotypes into those environments. All of the above does not make heritability measures meaningless, or useless. But it does put severe limits on what can be deduced from such measures, and using them to try to answers questions about nature vs nurture is misguided.

A reaction norm perspective, by contrast, *would* provide us with answers to many of the questions we are interested in. Even better, it would actually lead to meaningfully and informatively revise the *kind* of questions we are asking.

Unfortunately, reaction norm studies in human beings are practically and ethically inconceivable.

Perhaps we should keep in mind Richard Lewontin's (1998) words, spoken about the related topic of the evolution of cognition. They are a salutary and refreshing reality check: "I must say that the best lesson our readers can learn is to give up the childish notion that everything that is interesting about nature can be understood. History, and evolution is a form of history, simply does not leave sufficient traces. . . . Form and even behavior may leave fossil remains, but forces like natural selection do not. It might be interesting to know how cognition (whatever that is) arose and spread and changed, but we cannot know. Tough luck."

References

Arnold, P. A., Kruuk, L. E. B., & Nicotra, A. B. (2019). How to analyse plant phenotypic plasticity in response to a changing climate. *New Phytologist, 222*(3), 1235–1241.

Banta, J. A., & Pigliucci, M. (2010). Genetic and environmental variation affect the ontogeny of reproductive traits in Arabidopsis thaliana. *International Journal of Plant Developmental Biology, 4*, 112–121.

Chabris, C. F., Lee, J. J., Cesarini, D., Benjamin, D. J., & Laibson, D. I. (2015). The fourth law of behavior genetics. *Current Directions in Psychological Science, 24*(4), 304–312.

Cooper, R. M., & Zubek, J. P. (1958). Effects of enriched and restricted early environments on the learning ability of bright and dull rats. *Canadian Journal of Psychology, 12*, 159–164.

DiSanto, J. (2021). Penn Museum announces the repatriation of the Morton Cranial Collection. *Penn Today*. Retrieved January 27, 2023, from https://penntoday.upenn.edu/news/penn-museum-announces-repatriation-morton-cranial-collection

Dobzhansky, T. (1955). *Evolution, genetics, and man*. Wiley.

Evershed, R. P., Davey Smith, G., Roffet-Salque, M., Timpson, A., Diekmann, Y., Lyon, M. S., et al. (2022). Dairying, diseases and the evolution of lactase persistence in Europe. *Nature, 608*(7922), 336–345.

Flynn, J., Shaughnessy, M. F., & Fulgham, S. W. (2012). An interview with Jim Flynn about the Flynn effect. *North American Journal of Psychology, 14*(1), 25–38.

Gould, S. J. (1996). *The mismeasure of man*. W.W. Norton.

Herrnstein, R. J., & Murray, C. (1994). *The bell curve: Intelligence and class structure in American life*. Free Press.

Johannsen, W. (1911). The genotype conception of heredity. *American Naturalist, 45*, 129–159.

Joseph, J. (2014). *The trouble with twin studies: A reassessment of twin research in the social and behavioral sciences*. Routledge.

Kaplan, J. M., Pigliucci, M., & Banta, J. A. (2015). Gould on Morton, Redux: What can the debate reveal about the limits of data? *Studies in History and Philosophy of Science Part C: Studies in History and Philosophy of Biological and Biomedical Sciences, 52*, 22–31.

Kaplan, J. M., & Turkheimer, E. (2021). Galton's quincunx: Probabilistic causation in developmental behavior genetics. *Studies in History and Philosophy of Science, 88*, 60–69.

Lewis, J. E., DeGusta, D., Meyer, M. R., Monge, J. M., Mann, A. E., & Holloway, R. L. (2011). The mismeasure of science: Stephen Jay Gould versus Samuel George Morton on skulls and bias. *PLoS Biology, 9*(6), e1001071.

Lewontin, R. C. (1998). The evolution of cognition: Questions we will never answer. In D. Scarborough & S. Sternberg (Eds.), *Methods, models, and conceptual issues: An invitation to cognitive science* (Vol. 4, pp. 106–132). The MIT Press.

Lynch, K. E., & Bourrat, P. (2017). Interpreting heritability causally. *Philosophy of Science, 84*(1), 14–34.
Matthews, L. J., & Turkheimer, E. (2021). Across the great divide: Pluralism and the hunt for missing heritability. *Synthese, 198*, 2297–2311.
Mazer, S. J., & Schick, C. T. (1991). Constancy of population parameters for life history and floral traits in *Raphanus sativus* L. I. Norms of reaction and the nature of genotype by environment interactions. *Heredity, 67*, 143–156.
Murray, C. (1998). IQ will put you in your place. eugenics.net, since taken down.
Napier, J. D., Heckman, R. W., & Juenger, T. E. (2023). Gene-by-environment interactions in plants: Molecular mechanisms, environmental drivers, and adaptive plasticity. *The Plant Cell, 35*(1), 109–124.
Pietschnig, J., & Voracek, M. (2015). One century of global IQ gains: A formal meta-analysis of the Flynn effect (1909–2013). *Perspectives on Psychological Science, 10*(3), 282–306.
Pigliucci, M. (2001). *Phenotypic plasticity: Beyond nature and nurture*. Johns Hopkins University Press.
Pinker, S. (2002). *The blank slate: The modern denial of human nature*. Viking.
Plomin, R. (2018). *Blueprint, with a new afterword: How DNA makes us who we are*. The MIT Press.
Polderman, T. J. C., Benyamin, B., de Leeuw, C. A., Sullivan, P. F., van Bochoven, A., Visscher, P. M., & Posthuma, D. (2015). Meta-analysis of the heritability of human traits based on fifty years of twin studies. *Nature Genetics, 47*, 702–709.
Qian, S. (91 BCE/1994). *Records of the grand historian* (vol. 048). English version: Dawson, R.S., *Historical records*. Oxford University Press.
Reid, J. M., & Acker, P. (2022). Properties of phenotypic plasticity in discrete threshold traits. *Evolution, 76*(2), 190–206.
Sauce, B., Bendrath, S., Herzfeld, M., Siegel, D., Style, C., Rab, S., Korabelnikov, J., & Matzel, L. D. (2018). The impact of environmental interventions among mouse siblings on the heritability and malleability of general cognitive ability. *Philosophical Transactions of the Royal Society B: Biological Sciences, 373*(1756), 20170289.
Schlichting, C. D., & Pigliucci, M. (1998). *Phenotypic evolution: A reaction norm perspective*. Oxford University Press.
Shen, H., & Feldman, M. W. (2020). Genetic nurturing, missing heritability, and causal analysis in genetic statistics. *Proceedings of the National Academy of Sciences USA, 117*(41), 25646–25654.
Turkheimer, E. (2000). Three laws of behavior genetics and what they mean. *Current Directions in Psychological Science, 9*(5), 160–164.
Turkheimer, E. (2012). Genome wide association studies of behavior are social science. In K. S. Plaisance & T. A. C. Reydon (Eds.), *The philosophy of behavioral biology* (pp. 43–64).
Turkheimer, E., Haley, A., & Gottesman, I. I. (2003). Socioeconomic status modifies heritability of IQ in young children. *Psychological Science, 14*(6), 623–628.
Weisberg, M. (2014). Remeasuring man. *Evolution & Development, 16*(3), 166–178.
Zuk, O., Hechter, E., Sunyaev, S. R., & Lander, E. S. (2012). The mystery of missing heritability: Genetic interactions create phantom heritability. *Proceedings of the National Academy of Sciences USA, 109*(4), 1193–1198.

Chapter 3
Human Nature: A Process Perspective

John Dupré

Abstract This sketch of an account of human nature begins with the claim that we should see humans as a kind of process, a life cycle, rather than as a kind of substance or thing. A particular advantage of such a process perspective is that it readily accommodates the developmental plasticity that has been an increasingly important concept in recent biological theory. Human behaviour, on this account, should be understood as providing adaptive and flexible responses to an unpredictable environment. It is, therefore, generally misguided to provide a standard account of human nature in terms of behaviour or behavioural dispositions. If there is such a thing as human nature it is a uniquely refined propensity for novel and unpredictable behaviour.

3.1 Introduction

According to the ever-reliable Wikipedia, 'Human nature refers to the distinguishing characteristics—including ways of thinking, feeling and acting—which humans tend to have naturally, independently of the influence of culture.[1] This simple statement summarizes several of the central features of many mainstream views on this topic.

First is the proposal that human nature is a set of distinguishing characteristics. In its strongest form, this is the view that human nature is the human essence, a necessary and sufficient condition for something to be human. But weaker

[1] I am very grateful to Tim Lewens for comments on an earlier draft that led to a number of significant improvements. The research leading to this chapter has received funding from the European Research Council under the European Union's Seventh Framework Programme (FP7/2007-2013)/ERC grant agreement n° 324186.

J. Dupré (✉)
Egenis, the Center for the Study of Life Sciences, University of Exeter, Exeter, UK
e-mail: J.A.Dupre@exeter.ac.uk

J. Pérez-Jara, Í. Ongay (eds.), *Beyond Nature and Nurture*, Synthese Library 497,
https://doi.org/10.1007/978-3-031-87715-5_3

interpretations—for example, clusters of properties some sufficient number of which will qualify an entity as human—or even defeasible Wittgensteinian criteria for humanness, might suffice.

A second noteworthy point is the emphasis on the psychological. This is, indeed, what people almost always have in mind when talking about human nature, and this is further foregrounded by the following contrast with culture. I shall follow this tradition in much of what follows, but it is worth pointing out that this deviates quite significantly from more general uses of the term 'nature' with reference to the living world. With reference to complex animals, we might emphasize behaviour or action as central to their nature, but hardly thought or feeling. For plants or bacteria, we are likely to think of physiology or morphology.

Finally, there is the contrast with culture. This is very often exactly the point of appeals to human nature. If someone does something that is characteristic of human nature, it is otiose to appeal to more specific social, personal, or cultural explanations. This is just what people do, regardless of any such contingencies.

I think all these aspects of this simple definition are problematic. I shall begin by focusing on the first and apparently most innocuous. What can be wrong with distinguishing the (typical) characteristics of humans?[2] And if that's all that the concept of human nature means, how can it do any harm? A familiar response to these questions is that describing a set of characteristics as 'human nature' tends to distract attention from the diversity of humans and, worse, to make humans who lack such characteristics appear as abnormal or unnatural. I think this is a legitimate concern, but here I want to start in a slightly different place.

Supposing I say that it is part of human nature to speak a language. It is true that not everyone can speak a language, as people can lack some of the many physiological attributes that underpin this capacity. In view of the existence of deaf people, many of whom have a language but do not *speak* it, we might do better to write of using a language than of speaking it. For those who lack any language capacity, it is perhaps not controversial to attribute to them a serious deficiency, a lack of a normal part of human nature. Any reasonable concept of a biological nature will allow such exceptions.

But there is a more serious difficulty. Humans a few months old, or human foetuses, do not have language. Some people, in extreme old age, lose some or all of their ability to speak. This simple point is in fact of great importance. If there is a human nature, it does not consist just of a set of properties that humans possess, but of properties that humans possess, or typically possess, at particular stages of their lives. A human is not a thing with a fixed set of properties, but a life cycle. I suggest that, as such, we are better thought of as processes than as things or, in traditional philosophical language, substances.

---

[2] A prominent philosophical defender of the concept of human nature, Edouard Machery (e.g. 2008, and Chap. 1 this volume) follows this kind of line, requiring only that ingredients of human nature be sufficiently widespread, and widespread as a result of evolutionary processes.

The distinction between things and processes is an ancient one. The thesis of the ontological centrality of processes is generally traced back to Heraclitus, but his view on the whole lost out to the thing-centred atomism of Democritus, Parmenides, and others. While a number of subsequent philosophers have been sympathetic to a process ontology, again the majority are more or less loosely aligned with the substance view.[3] The growing influence of science, and of the atomism of leading contributors to the scientific revolution (Locke, Boyle, and so on), did much to embed this dominance.[4]

The well-known exception in the last century has been Alfred North Whitehead,[5] whose process philosophy has attracted a number of philosophical followers as well as—perhaps equally significantly for present purposes—the attention of several prominent theoretical biologists, including Conrad Waddington, Joseph Needham, and Ludwig von Bertalanffy. Though Whitehead's work undoubtedly encompasses crucial insights, I shall not discuss it here. First, Whitehead uses a good deal of esoteric terminology that many have found off-putting. And second, his philosophical system also includes theological and idealistic elements that I wish to avoid. I find it easier to risk reinventing the wheel than to articulate my differences from Whitehead in detail. But in common with Whitehead and other process philosophers, I believe that the assumption of a substance ontology has been a disaster for our thinking about human nature.

3.2 Humans as Processes

A process necessarily persists for some length of time. Within their temporal extension, many processes have an origin, sequence of states, and conclusion that are more or less similar to the corresponding stages of a larger class of processes, and this fact makes possible the allocation of processes to kinds, such as thunderstorms and battles. More important examples for present purpose are living systems. The human process is naturally identified with the typical human life cycle—typical, though of course life cycles are often cut short, and they may deviate from the norm in countless ways.

As I have noted, properties characteristic of humans will vary according to the stage of the life cycle. On the other hand, this variation is far from random. Human life cycles follow characteristic patterns. So, provided we index properties with stages of the life cycle, does it make much difference moving to the process perspective? To address this question, we may begin by attending to a further important difference between a substance and a process. The default condition for a substance is stasis: if it has a particular set of properties at some time, other things

[3] For a clear introduction to process philosophy, see Rescher (1996).

[4] Though it is perhaps surprising that the rise of quantum mechanics did not reverse this tendency.

[5] The *locus classicus* is Whitehead (1978).

being equal it will continue to do so at later times. Explanation of some kind is called for when its properties change. For a process, on the other hand, its very persistence will normally require explanation. For a thunderstorm, say, to persist, the matter that composes it must continue to move in specific ways if it is not just to cease or to dissipate. Properties of the constituent matter and features of the whole system combine to explain why the relevant movement persists. Arguably, the most important difference between a substance and a process is that whereas for the former the central problems, both empirically and philosophically, concern the explanation of change, for the latter the central problems concern understanding stability, or persistence.

Much of biology, in fact, is engaged in explaining the stability of biological systems. To understand this point properly, it is necessary to appreciate the hierarchical nature of biological systems—not so much (as is commonly considered) the spatial hierarchy, but rather the temporal one. Cell biologists describe sequences of events that contribute either to the persistence of the cell (e.g. RNA transcription or oxidative phosphorylation) or to the execution of the cell cycle (e.g. the various processes that constitute mitosis). While these are sequences of changes, they are also central elements of the explanation of the stability of the cell or the multicellular organism of which they are part. And indeed, on an evolutionary timescale, it is this contribution that explains their occurrence. The cell cycle, as well as serving the short-term persistence of the organism, in conjunction with processes of cell differentiation and death (apoptosis) is at the heart of the explanation of a cyclical process at a much longer timescale, namely, development. Finally, developmental cycles explain, in part, the stability of the lineages that constitute the evolutionary process.

One reason why it is so important to keep in mind these differences of temporal scale is that they allow us to explain something that a process perspective certainly had better come to terms with: the success of mechanistic explanations. Mechanism, a philosophical view currently undergoing a major resurgence, is a quintessentially substantialist position (Machamer et al., 2000; Glennan, 1996; for criticism, see Nicholson, 2012, 2013.). Mechanism prescribes the explanation of the behaviour of complex systems such as are addressed by the life sciences in terms of the properties and interactions of their parts. As a matter of fact, I am somewhat suspicious of the very idea of a unique and objective decomposition of biological systems into their parts.[6] Nonetheless, there are, typically, entities that are stable on the timescale of a particular process we are trying to understand. When, for example, we explain metabolic processes, it is often the case that important enzymes or cell membranes can be treated as stable things. But of course these may also be understood as stages in processes of construction and degradation. This perspective on mechanistic

[6] It is no surprise that the approach to human nature that gives least respect to the dynamic features of human development and human evolution, evolutionary psychology, sees the mind as divided into a quite specific and largely autonomous set of parts or modules.

explanation should not only account for its considerable successes but also point to its limits, the places where it can be expected to break down.

This kind of stabilization also applies to organisms. At much shorter timescales than the transitions between life cycle stages, humans undergo a wide variety of processes—for example, metabolic—that maintain their short- or medium-term stability. Basic processes such as oxidative phosphorylation occur at all stages of the life cycle, though perhaps less efficiently at very late stages. Again, it is essential to keep in mind a hierarchy of processes. An organism lies at the intersection of a very large number of subordinate processes that together maintain its short-term stability and longer-term developmental trajectory.

The stable cycle of stages through which human lives typically pass may be viewed from a different perspective, as involved in the stabilization of a process of much longer timescale, the human lineage. This is also one way of understanding the connection of human behaviour with evolution, though, as I shall suggest below, one that carries rather different implications from those of more familiar accounts of this relation. For now I note only the obvious points that human life cycles are the constituents of the human lineage, and the success of the former in persisting and launching new such cycles is the main determinant of the persistence of the latter.

As well as processes that pass through a sequence of stages, like human life cycles, many, like storms or eddies, have no particular trajectory. They may nonetheless be very stable. The Red Spot on Jupiter, a gigantic storm on the planet's surface, has persisted for centuries, albeit with some changes in shape and location. This persistence is not just a default, in the absence of any intervention, but requires a lot of activity—in the case of the Red Spot high-speed prevailing winds that maintain its boundaries and provide the energy of its rotating matter. Were these sustaining processes to stop, the Red Spot would instantly disintegrate. If, as process philosophers generally suppose, at the most fundamental level reality is flux, then some explanation is needed wherever we encounter persistent structure. Whereas mechanists are concerned only with explaining how structures enable the execution of functions, process philosophers see also the necessity of explaining how function maintains structure.

3.3 Process and Plasticity

Since living things exist in complex and changing environments, the best way to persist is not always determinable in advance; it will often be adaptive to have the capacity to respond flexibly to environmental contingencies. Successful living systems, in fact, display plasticity at multiple levels. The extent of this plasticity, right down to the molecular level, has been one of the most striking themes of the biology of the last few decades. However, and as already noted, discussions of human nature are generally directed especially to characteristic human behaviour, and from now on this will be the main topic here; certainly, nowhere are the benefits of plasticity more important than in the realm of behaviour.

It is tempting to suggest that distinctive modes of flexibility characterize the major different modes of organization of life. For microbes, plasticity is chemical or metabolic. Microbes appear to be able to extract energy from almost any chemical substance from which this is in principle possible. They do this, frequently, by forming consortia of different microbes providing different chemical resources. Some of these resources are packaged in forms, such as in plasmids, that enable them to be shuffled conveniently from one microbial cell to another. These chemical capacities of microbes are exploited by all or most multicellular eukaryotic organisms, most obviously (though by no means exclusively) in digestion. Humans, in particular, exist in symbiotic relations with trillions of microbes, which contribute crucially to processes including digestion, immune response, development, and more. These symbioses provide one dimension of plasticity, or responsiveness to changes in the environment, at a far shorter timescale than that envisaged in standard models of evolutionary adaptation.

Plants, while outsourcing many of their metabolic needs either to independent microbial consortia or to long-captive endosymbionts (plastids), have introduced a new form of plasticity in morphology. The growing tip of a plant process, or meristem, includes a collection of stem (*sic*) cells that allow differentiation into various organs (leaves, branches, flowers) in response to the affordances offered by the environment. As gardeners who engage in such practices as taking cuttings or layering will be aware, the extent of this plasticity is highly variable; but all plants exhibit such morphological plasticity to some degree. The different capacities of the meristem also allow rapid morphological evolution, as strikingly demonstrated in the artificial differentiation of the wild cabbage *Brassica oleracea* into such morphologically diverse forms as cabbage, broccoli, cauliflower, Brussels sprouts, and kohlrabi.

Multicellular animals, on the other hand, generally adopt relatively determinate morphologies. I should emphasize the word 'relatively'. As is demonstrated in Mary Jane West-Eberhard's (2003) encyclopedic study of developmental plasticity, animal morphology is by no means fully fixed. What is clear, at any rate, is that animals have added a whole new range of flexible response to the environment, through evolving nervous systems that allow the production of behaviour responsive to environmental contingencies.

It is useful to distinguish several grades of behavioural flexibility. At the very simplest, least plastic level are little more than reflexes, typically automatic responses to potentially harmful or beneficial stimuli, such as the chemotaxis found in many bacteria. Much more complex are the inflexible sequences of behaviour found, for instance, in many arthropods. Examples well known to philosophers are the provisioning behaviours of sphecid wasps discussed in several places by Daniel Dennett, or the nest-building behaviour of leaf-cutter bees of the genus *Megachile*. Despite their impressive complexity, simple experiments appear to demonstrate that these behaviours are produced in inflexible sequences that cannot respond appropriately to changes in the animal's circumstances.[7]

Behaviour that is composed of stereotypical responses to specific environmental cues may characterize a good deal of arthropod behaviour. But if so, such behaviour need be in no way simple. Think, for instance, of the well-known waggle dance of the honeybee. This quite elaborate performance is elicited by the discovery of a pollen source, and produces in a suitable audience a specific response: movement to the pollen source. Such behaviour becomes particularly impressive when it is integrated at the social level, as the waggle dance appears to be involved in controlling the distribution of pollen among members of the colony to maintain an optimal food supply. More striking still is the organization of stereotypic behaviour in termites to maintain the structure of a mound that, among other things, provides a constant near-optimal temperature. This is an instance of niche construction, to which I shall return below.

The preceding examples exhibit activities of the organism that modulate its interaction with the environment in ways that are beneficial to its survival and reproduction. They do not emphasize elements that are characteristic of the behaviour of 'higher' animals and that are carried to unprecedented levels in humans, namely, learning and intelligence. These are both, needless to say, enormously difficult and contentious topics. They are also, however, at the core of human nature, and of why a description of human nature in the standard sense is unattainable. The importance of learning is that it implies that behaviour of the organism at later stages in its life cycle cannot be predicted without some knowledge of the series of environments that it has encountered during earlier stages. One interpretation of the importance of intelligence is that it implies that even with this knowledge, the response of the organism to an environmentally posed problem cannot be reliably predicted. Intelligence, that is to say, involves ingredients such as insight or imagination; its conclusions may be emulated by a similar or superior intelligence, but cannot be inferred from fixed features of the agent.

Seeing human behaviour in this way, as a uniquely developed capacity for flexible response to the environment, makes it unsurprising that it is difficult to provide a definitive account of human behaviour (and hence of the human nature that is supposed to govern behaviour) and unsurprising that human populations in different environments exhibit significantly different behaviour. But there is a further point. Humans constantly transform their environments to meet their needs: they build cities, cultivate fields, dam rivers. And they construct institutions, such as hospitals and schools, to provide enriched environments for the development of their offspring. Applying intelligence to these forms of behaviour, improvements are often made to this constructed niche, and the environment in which humans must

[7] Actually, while this inflexibility has been widely asserted by philosophers and cognitive scientists, a careful study of the history of research on sphecid wasps by Fred Keijzer (2013) suggests that the story is really a lot more complicated, and that wasp behaviour shows a subtle mix of rigid and flexible aspects. For discussion, see Lewens (2015, pp. 233–237).

survive is changed by their own behaviour. The relation between human behaviour, on the one hand, and on the other hand the niche that humans construct in which to behave is, in short, dialectical.

We now have the basic ingredients to summarize an account of human nature. Humans are life cycles during which a great deal of learning—mainly socially mediated (training) but also individually acquired—provides a changing and partly variable range of knowledge and behavioural capacities. Moreover, a developing capacity to first analyse a problem situation in the light of this knowledge and experience and then to decide on an appropriate course of action means that even adding a full knowledge of the individual's experience to a full account of the individual's original biological endowment will not enable a deterministic prediction of behaviour. Human intelligence is both fallible and creative. Behaviour may not be optimal; and it may not lie within any predetermined range of options. Moreover, as just discussed, the environment in which humans behave is constantly created and recreated by their own actions on it. Humans are not blank slates; they have a particular biological inheritance. But neither are they machines with a predetermined programme. A crucial part of their inheritance is their capacity for learning and intelligence.

3.4 Individual Life Cycles and Lineages

How does this account of the human fit into the popular, if not notorious, project of relating human nature to evolution? I have said nothing, so far, about natural selection. Following John Reiss's (2009) excellent book *Not by Design: Retiring Darwin's Watchmaker*, I shall not start there, but rather with the concept familiar from Darwin, but more closely connected to Cuvier, of *conditions for existence*. It would be impossible to summarize in any detail Reiss's complex and challenging argument, but at its core is the simple point that all organisms must, of logical necessity, have the properties that enable them to survive in the conditions they encounter—what he describes as satisfying the conditions for existence.[8] The stabilization I talked about in the first sections of this chapter is very close to the satisfaction of the conditions for existence that Reiss discusses. Whichever way it is described, this is a stringent condition. For a lineage—which I take to be a process on a longer timescale—to persist, not only must individual life cycles satisfy

[8] Reiss develops the argument by pointing out what follows from this simple point, namely, that contrary to Darwin and many of his successors, there cannot be a gap between the organism and its adaptedness for natural selection to bridge. The natural selection that Darwin and Wallace (importantly) contributed to our understanding is what Reiss calls 'narrow-sense selection': differential survival and reproduction among classes to the extent that this is caused by the distinguishing characteristics of these classes. But this provides a relatively small part of the explanation of the characteristics of a species. For a more detailed synopsis of Reiss's book, see Dupré (2010).

these conditions, but they must also spin off new life cycles, namely, reproduce. Continuing to meet the condition for existence of a lineage requires meeting this additional condition.

I have said that the persistence of a process requires explanation, and much of biology, from the study of metabolism to the study of behaviour, is concerned with providing such explanations. This range of investigations describes the multiple, interacting processes that enable the human life cycle to complete its course and, on the way, to reproduce itself. The latter, as just noted, contributes to the explanation of the persistence of a lineage. There is, however, a further crucial contributor to this persistence: stabilizing selection. What this means is just that the individuals who make up the lineage produce enough new individuals for that lineage to persist, despite the production of a certain proportion that—for reasons ranging from bad luck (struck by lightning, ambushed by a predator, and so on) to all manner of developmental failures—do not survive. The production of offspring must, for the lineage to persist, be sufficient to compensate for these losses. How much excess this requires will vary massively. For many fish, the chances of getting eaten in early youth are so high that millions of offspring must be produced. For modern humans in developed nations, the chances of viability are so high that little more than two offspring per female are required to sustain the population.

Aficionados of evolutionary theory will immediately suspect that this talk of stabilizing (or purifying) selection reveals an enthusiasm for the neutral theory of evolution, and this would be correct. In fact, given Reiss's denial of the gap between existence and adaptedness, standard ways of distinguishing between changes due to selection and changes due to drift become inapplicable. Neither, at any rate, makes the population better adapted. A great deal of the power of the concept of natural selection can be seen in its ability to answer (in part at least) the question of how a process, such as a lineage, persists for hundreds of millions of years. Bearing in mind that natural selection ensures that the form that the lineage takes must satisfy the conditions for existence, it is entirely plausible that what strike us as adaptations will nonetheless be the outcome of no more than a random walk through the rich but constrained space of biological possibility.

I do not want to disagree with what Motoo Kimura, the founder of the neutral theory, made clear: that the theory is not supposed to rule out the generation of novelties through mutation and natural selection. I only claim that this is a much rarer phenomenon than genetic drift. I also want to suggest that he might have gone further by allowing that phenotypic change, as much as genetic change, might be sufficiently explained as a chance process. However, in the case of the topic at hand—namely, human nature—the important point is rather different: human behaviour is highly plastic, substantially unconstrained by human biology. It is not that changes in human behaviour come about by chance, but rather that they are open to a variety of other explanations, ranging from broad cultural forces to individual decision-making. And such explanations can also account for the adaptedness of such traits.

In earlier work critical of evolutionary psychology (Dupré, 2001, 2008), I have emphasized the importance of timescales, and the role of the indefensible assumption that at some important level of description (e.g. psychological modules), human behaviour can change only at the rate of natural selection of randomly generated alleles. One of the advantages of a process ontology is that it immediately draws attention to the importance of processes at different timescales. The persistence of lineages is facilitated by processes at a range of very different timescales. Even if, as is supposed by some enthusiasts for the Modern Synthesis, selection of genetic mutations is the fundamental process that adapts the lineage to a changing world, there are certainly much more rapid processes that enable much more agile adaptation.

To begin with, it is well known now that the same genome (in terms of its sequence) can be used in very different ways, depending on the different expression of its constituent genes. This is obvious from the diversity of cell types with sequentially identical genomes in multicellular organisms, and is now studied extensively under the rubric of epigenetics. It remains somewhat controversial whether epigenetic changes to an organism can be inherited across generations, and hence whether they can be part of an evolutionary process. I say 'somewhat controversial', but this is a rather curious controversy. The evidence that there is some inheritance of epigenetic marks seems increasingly incontrovertible (see e.g., Morgan et al., 1999: Tang et al., 2015), and most of the argument seems now to be about the robustness of such inheritance over multiple generations.

It is hard to avoid the suspicion that much of this controversy is ideological. Supporters of epigenetic inheritance occasionally suggest that this will provide an opportunity for 'Lamarckian' inheritance (Jablonka & Lamb, 1995), the ultimate prohibition in orthodox Darwinian thought. And while this is certainly possible, there is no obvious connection between epigenetic inheritance and Lamarckianism.[9] A more important point is that the inheritance of epigenetic characteristics need in no way depend on transmission through the germline. The best illustration of this is the paradigm for behavioural epigenetics, the work of Michael Meaney and collaborators on maternal care in rats (Champagne et al., 2006). Meaney showed that proper care affected the developing brains of rat pups, making them less liable to be fearful and also, crucially for female pups, more likely to provide proper care to their own offspring. Although there is some controversy surrounding Meaney's interpretation of his results, at the very least the example illustrates how it is entirely possible in principle for an epigenetic trait to be inherited via the epigenetic effect of its behavioural manifestation.

If this seems unlikely to be a widespread phenomenon, it is at least worth reflecting on the ubiquity of epigenetic expression in the brain. Any experience that affects an organism in any lasting way will, presumably, leave some physical change in their brain, and this, we generally suppose, will involve the expression of

[9] Understood here in the somewhat vulgar sense as any inheritance of acquired characteristics that are acquired because they enhance fitness. Epigenetic changes are indeed acquired, but whether they are acquired because they are adaptive is another matter. If the reason that an organism acquires adaptive characteristics is that it has evolved a capacity to do so by orthodox Darwinian means, this would not be a significant threat to the orthodoxy.

genes and the production of proteins. Triggering such a process is what epigenetic modifications to, or chemical alterations of, the genome do. In a reasonably broad sense of 'epigenetics', the brain is undergoing epigenetic modification moment by moment.[10] That these modifications should be involved in more or less elaborate processes that enable their passage to other organisms, including their offspring, should not be surprising. In fact, of course, we know well that it happens all the time, and most spectacularly in our own species, by means of socially mediated learning.

3.5 Learning

Evolutionary psychologists tend to become very exercised when it is suggested that learning presents a problem for their views. They suppose that learning is taken to be some universal, content-independent way of acquiring any arbitrarily selected bit of competence or information, whereas in fact it is the exercise of a very specifically directed capacity. Thus Cosmides and Tooby (n.d.) write: 'Students often ask whether a behavior was caused by "instinct" or "learning". A better question would be "which instincts caused the learning?" ' Later they continue:

> The cognitive architecture, like all aspects of the phenotype from molars to memory circuits, is the joint product of genes and environment. But the development of architecture is buffered against both genetic and environmental insults, such that it reliably develops across the (ancestrally) normal range of human environments. [Evolutionary Psychologists] do not assume that genes play a more important role in development than the environment does, or that 'innate factors' are more important than 'learning'. Instead, [Evolutionary Psychologists] reject these dichotomies as ill-conceived.

Indeed, in terms of learning, we cannot even sensibly ask which of genes or the environment is the more important. For:

> Any developmental biologist knows that this is a meaningless question. Every aspect of an organism's phenotype is the joint product of its genes and its environment. To ask which is more important is like asking, Which is more important in determining the area of a rectangle, the length or the width?

Learning, then, is a way in which the environment contributes to the production of the universal human cognitive architecture. Genes and environment are inextricably linked in this process, and the reason it makes no sense to separate their contributions is because they are both doing exactly the same thing: jointly producing the universal species-typical outcome. Now certainly I don't want to deny that there is a good deal of buffering in development. Without this, it is hard to imagine any successful development happening at all. But surely learning would be a very strange device to evolve merely to buffer development against environmental

[10] It is suggestive that, according to Tang et al. (2015), the parts of the genome that resist demethylation of germ cells are disproportionately expressed in the brain.

noise. In fact, Cosmides and Tooby do allow that learning may function as a kind of switch, determining which of two developmental pathways, both selected by past evolution, an organism will follow. No doubt they would allow a predetermined range of developmental trajectories. But even this takes us only to the level of plasticity I described with examples from social insects; it does not begin to capture the functional plasticity found in higher mammals and, especially, humans.

Here is where the universality of plasticity should lead us to suspect that we are going down a wrong path. Certainly development must produce a sufficiently constant product to reliably meet the conditions of existence. This is no mean feat. But successful lineages, it appears, manage to embed within this reliable reproduction of their kind a great deal of flexibility that enables at least fine-tuning to the particular exigencies of the ever-changing and always differentiated environment, and to do so on a timescale far more rapid than could result from the natural selection of randomly generated genetic mutations. Whether, in the spirit of the neutral theory, we should see the latter as mainly irrelevant to the processes of adaptation is a matter for another time. But what is important now is that organisms have a wide variety of resources—developmental, metabolic, behavioural, and no doubt others—that have evolved precisely to facilitate adaptation at more or less rapid timescales. The human mind is a spectacular example of such an evolved set of capacities.

3.6 Niche Construction

More or less intelligent behaviour is of obvious importance in sustaining the lives of individual humans, but it plays an equally important part in the stabilization of human lineages.[11] Most obviously, the rearing and educating of children is a domain of behaviour essential to the perpetuation of any human lineage. The processes by which behaviour is transmitted between generations has been extensively investigated by students of cultural evolution (see e.g. Richerson & Boyd, 2005). But perhaps even more intriguing than these vital processes is the production and maintenance of so-called developmental scaffolding (Bickhard, 1992; Sterelny, 2003) or developmental niche construction. The latter, according to Karola Stotz (2010), asks not 'what is inside the genes you inherited', but 'what the inherited genes are inside of' and with which they form a wider whole—their internal and external ontogenetic niche, understood as the set of epigenetic, social, ecological, epistemic and symbolic legacies inherited by the organism as necessary developmental resources.

[11] I shall not address here the interesting question of the relevant scope of a human lineage. My assumption is that there is no unique answer: humans are socially integrated at multiple scales, with variable of degrees of permeability between these integrated groups. Proper attention to this complication would, no doubt, add a significant layer of complexity to the discussion.

Developmental niche construction began with an idea that goes back at least to Darwin's (1881) treatise on earthworms, and that has been brought back into prominence by Odling-Smee et al. (2003). Organisms do not just respond to the environment that they find, but devote a great deal of energy to transforming the environment, generally in ways to which they are adapted or in ways that promote the development of their offspring (or other younger members of their species).

Of particular importance to the human case is scaffolding or niche construction that promotes cognitive development, and indeed the lives of juvenile humans are largely structured around their cognitive development. Schools and a host of associated learning technologies are the most obvious structures developed to this end, but many informal interactions between children and parents, other adults, and peers are more or less explicitly focused on cognitive development. The physical environment in which humans develop and live, moreover, is structured in many subtle ways to facilitate intelligent activity. Think of an everyday site of complex activity, the kitchen. Heights of surfaces, spatial relations between appliances and work surfaces, the sizes and dispensing devices of containers for ingredients, the design and positioning of utensils, and so on are all more or less well designed to guide and facilitate effective action in the execution of complex food-preparation tasks.

There is of course much more that could be said on this topic. My present point, however, is a relatively simple one. This active shaping of the environment to promote cognitive and hence behavioural effectiveness must be understood in terms of the organism as open-ended, developing in a partially unpredictable way, rather than as a thing with a fixed repertoire of species-universal dispositions.[12] Perhaps the best way of seeing this is to see how this claim locates the sound middle ground between the extremes of a blank slate and a mind programmed to produce predetermined responses to given situations. In terms of the latter, if that were the kind of cognitive development that were required, it could be innate, in the sense of appearing largely irrespective of contingencies of the environment, and this would save the adult phases of the organism an enormous amount of effort. But on the other hand, as evolutionary psychologists like to insist, a blank slate would require far too much learning; trial and error would far too often lead the developing organism into fatal error. The solution is obvious: the 'blank slate' is constrained by the structures that are provided by parents and other conspecifics, living or dead. Human children, to take a trivial example, do not learn to cross roads by randomly trying a variety of techniques. The cost of failure would be too high. They are taught to do this by older humans. And this is just as well, since there were no roads in the so-called environment of evolutionary adaptation, and if they relied on past evolution they would mostly be dead. The important point about this strategy is that it enables evolution of the niche, in this case the invention of roads. This requires both cognitive creativity—someone has to think of

[12] This remark of course reflects my broader theme: that the human organism should be seen as a process rather than as a substance.

roads—and cognitive adaptation—our children have to learn to survive in a road-rich environment. Neither a blank slate nor a universal human cognitive architecture will do these jobs.

3.7 Back to Human Nature

There is much to be said for a flat-out denial that there is any such thing as human nature. The concept suggests something fixed, whereas, as I have argued, human behaviour is an adaptation for permitting plastic responses to a changing environment—an environment that may even be changed by the adaptive activities of humans themselves.

There is nothing in principle objectionable about a concept of human nature that consists of no more than the description of what humans at a particular time and place typically do—a kind of localized natural history. This is presumably the kind of thing that anthropologists, sociologists, or historians often try to provide, and I certainly don't mean to object to their practices. Such social scientists, while not generally committed to what evolutionary psychologists refer to as the 'standard social science model'—the blank slate—are certainly aware that the local natural histories they provide would not apply precisely, or sometimes even closely, to humans at distant times and places. On the whole, it seems best not to risk confusion of such local descriptive ventures with the much more portentous traditional concept of human nature.

I also advocate a stronger reason for rejecting the concept of human nature: that I take humans to be better understood as processes than as things or substances. The nature of a thing serves, among other functions, to determine what it is for a thing to persist. It does so for as long as it continues to realize the nature of the kind of thing it is. But if only because organisms undergo developmental pathways that may lead them through radically different states, this cannot be the kind of nature they have. It is true that human nature is generally understood as referring rather specifically to the behavioural or the cognitive, and often to these in relation to the adult. But here there is a clear tendency for such language to mislead. Attributing the nature of an organism to one particular developmental stage inevitably distracts attention from the process by which it reached that developmental stage; and it is not news that popular theories of human nature are often derived from evolutionary theories rightly criticized for their neglect of development. If development is, as is surely the case, a substantially plastic process, then overlooking it and attributing an invariant state to the developmental outcome is clearly misguided. Seeing the organism itself as a process, the intersection of many constituent processes that together reliably produce a stable structure that satisfies the conditions of its existence, makes errors of this kind impossible.

C. H. Waddington, perhaps the most important of the mid-twentieth-century biologists strongly influenced by Whitehead, distinguished between homeostatic and homeorhetic processes. Whereas the former referred to processes that maintained a constant condition of a system, the latter referred to the processes constraining a system to a particular pathway. Clearly both kinds of processes are exhibited by organisms, development being a central example of a homeorhetic process.[13] Could one then perhaps say that humans were distinguished by a particular species-specific set of homeorhetic processes, and the typically reproduced stages thus generated constituted the nature of the human (process)? If to characterize the nature of a kind of entity is just to describe it as well as possible, then indeed this is what one could and should say. However, I suggest that there is a great deal of baggage connected to the concept of human nature that is strongly at odds with the kind of account just indicated. It is not merely that a developmental process need not exhibit any characteristics that are constant throughout its career, but also that the plasticity I have stressed implies that—especially with respect to the psychological traits most generally associated with human nature—there is a lot of variability in the features manifested at particular stages of the process. All this is so far from the common associations with the expression 'human nature' that I think it is safer to dispense with its use altogether.

I have referred at several points to processes that maintain a stable structure. But note that I am not setting up the structure as something fixed and opposed to the functional or processual. As J. S. Haldane (1931, p. 22) summarized this crucial feature of a processual understanding of biology: 'Structure and functional relation to environment cannot be separated in the serious scientific study of life, since structure expresses the maintenance of function, and function expresses the maintenance of structure.' This thought beautifully captures the decisive shift from the organism as a thing with a certain set of properties conducive to its adaptation, to a process constantly acting to maintain its integrity and persistence. And from the latter viewpoint, adaptive plasticity, or flexible adaptation, is what behaviour is always aspiring to achieve.

References

Bickhard, M. H. (1992). Scaffolding and self-scaffolding: Central aspects of development. In L. T. Winegar & J. Valsiner (Eds.), *Children's development within social contexts: Research and methodology* (pp. 33–52). Erlbaum.

Champagne, F. A., Weaver, I. C., Diorio, J., Dymov, S., Szyf, M., & Meaney, M. J. (2006). Maternal care associated with methylation of the estrogen receptor-alpha1b promoter and estrogen receptor-alpha expression in the medial preoptic area of female offspring. *Endocrinology, 147*, 2909–2915.

[13] Note, however, that this classification of processes is, in an important way, relative to perspective and temporal scale. From an evolutionary perspective, development might better be described as a homeostatic process in which cycles of development spin off new cycles through reproduction, thereby stabilizing the lineage. Similarly, metabolic cycles that homeostatically maintain features of the cell are themselves, in finer detail, homeorhetic sequences of chemical reactions.

Cosmides, L., & Tooby, J. (n.d.). Evolutionary psychology: A primer. www.cep.ucsb.edu/primer.html (unpublished).
Darwin, C. (1881). *The formation of vegetable mould through the action of worms, with observations on their habits*. John Murray.
Dupré, J. (2001). *Human nature and the limits of science*. Oxford University Press.
Dupré, J. (2008). Against maladaptationism: Or what's wrong with evolutionary psychology? In M. Mazzotti (Ed.), *Knowledge as social order: Rethinking the sociology of Barry Barnes* (pp. 165–180). Ashgate.
Dupré, J. (2010). The conditions for existence: Review of John O. Reiss. Not by design: Retiring Darwin's watchmaker. *American Scientist, 98*, 170–172.
Glennan, S. S. (1996). Mechanisms and the nature of causation. *Erkenntnis, 44*, 49–71.
Haldane, J. S. (1931). *The philosophical basis of biology*. Hodder & Stoughton.
Jablonka, E., & Lamb, M. J. (1995). *Epigenetic inheritance and evolution: The Lamarckian dimension*. Oxford University Press.
Keijzer, F. (2013). The Sphex story: How the cognitive sciences kept repeating an old and questionable anecdote. *Philosophical Psychology, 26*, 502–519.
Lewens, T. (2015). *The meaning of science*. Pelican.
Machamer, P., Darden, L., & Craver, C. F. (2000). Thinking about mechanisms. *Philosophy of Science, 67*, 1–25.
Machery, E. (2008). A plea for human nature. *Philosophical Psychology, 21*, 321–329.
Morgan, H. D., Sutherland, H. G. E., Martin, D. I. K., & Whitelaw, E. (1999). Epigenetic inheritance at the agouti locus in the mouse. *Nature Genetics, 23*, 314–318.
Nicholson, D. J. (2012). The concept of mechanism in biology. *Studies in History and Philosophy of Biological and Biomedical Sciences, 43*, 152–163.
Nicholson, D. J. (2013). Organisms ≠ machines. *Studies in History and Philosophy of Biological and Biomedical Sciences, 44*, 669–678.
Odling-Smee, F. J., Laland, K. N., & Feldman, M. W. (2003). *Niche construction: The neglected process in evolution*. Princeton University Press.
Reiss, J. (2009). *Not by design: Retiring Darwin's watchmaker*. University of California Press.
Rescher, N. (1996). *Process metaphysics: An introduction to process philosophy*. SUNY Press.
Richerson, P. J., & Boyd, R. (2005). *Not by genes alone: How culture transformed human evolution*. Chicago University Press.
Sterelny, K. (2003). *Thought in a hostile world*. Blackwell.
Stotz, K. (2010). Human nature and cognitive–developmental niche construction. *Phenomenology and the Cognitive Sciences, 9*, 483–501.
Tang, W. W. C., Dietmann, S., Irie, N., Leitch, H. G., Floros, V. I., Bradshaw, C. R., Hackett, J. A., Chinnery, P. F., & Surani, M. A. (2015). A unique gene regulatory network resets the human germline epigenome for development. *Cell, 161*, 1453–1467.
West-Eberhard, M.-J. (2003). *Developmental plasticity and evolution*. Oxford University Press.
Whitehead, A. N. (1978). *Process and reality: An essay in cosmology*. Macmillan.

Chapter 4
Positioning Sociology in the Nature vs. Nurture Debate: An Ongoing Project

Ullica Segerstrale

Abstract In this essay I will be addressing the various efforts of American sociologists to investigate and regulate their field's relationship to biology over a period of some hundred years. I will be looking at the field's internal development as well as how it has reacted to various bigger social and historical events. We will particularly look at episodes of "nature vs. nurture" controversy as we follow sociology's changing self-definition over time. This means we will be tracking the field from its beginning in the early twentieth century to the present, paying special attention to sociology's encounters with biology, including the arrival of the neo-Darwinian or Modern Synthesis in the 1940s, the challenge of sociobiology in the 1970s, and various later efforts to "bring in Darwin" as the answer to sociology's academic identity problem. But "biology" also relates to sociology's role in regard to such things as IQ testing, race differences, and eugenics, and more recently initiatives such as sociogenomics, at the precarious intersection between scientific and moral/political concerns. It is even unclear if one can separate a social science's academic identity from its sociopolitical role. What is clear, however, is that sociology (with other social sciences) has experienced a number of collectively felt crisis situations during the past hundred years.

4.1 Introduction

To get an overview, we will start with biology-friendly sociologist Pierre van den Berghe's polemical essay, "Why most sociologists don't (and won't) think evolutionarily" (1981, 1990), which has been the standard reference for "explaining" sociologists or inviting them to self-scrutiny. That essay is a rather over-the-top list of errors of van den Berghe's fellow sociologists, but if it is even remotely true, hints at the seriousness of the problem.

U. Segerstrale (✉)
Illinois Institute of Technology, Chicago, IL, USA
e-mail: segerstrale@iit.edu

J. Pérez-Jara, Í. Ongay (eds.), *Beyond Nature and Nurture*, Synthese Library 497,
https://doi.org/10.1007/978-3-031-87715-5_4

I see two main ways of proceeding. One is to look at the problem with sociology as "internally caused", which would mean that what happened with sociology at any particular time would be explained as due to internal developments in that field. An alternative approach would refer to "social forces" or "public effects"—the wider social and socio-historical context—explainable by factors that are either known to have affected the field of sociology, or can be thought to have done so, at different points in time. I will attempt both. Finally I will take a quick peek at some very recent developments in biology and consider to what extent these would contribute—or not—to a rapprochement between nature and nurture.

In an Appendix I will present a "nature vs. nurture" case with which I am very familiar: the famous sociobiology debate, especially in its early days. I think it is in many respects a rather atypical nature-nurture controversy.

4.2 Ten Sources of Sociological Biophobia

A summary of Pierre van den Berghe's famous article

Why most sociologists don't (and won't) think evolutionarily

In 1990 the *American Sociologist* published the brief article "Why most sociologists don't (and won't) think evolutionarily" by sociologist Pierre van den Berghe.

His piece, often abrasive sounding, is an attempt to answer a particular puzzle in this respect: "why most sociologists, who for the most part claim to be scientists, continue to reject or ignore the relevance of evolutionary theory to the study of human behavior and sociality". Here I am presenting a serviceable overview of the many different factors van den Berghe saw as responsible for what he regarded as an alarming state of affairs. I am here taking some small liberties with naming and numbering for the sake of clarity, but for the specific tone of his article the reader will have to go to the original (van den Berghe, 1990).

The author starts with what he sees as the overriding problem: the *education* (training) that sociologists receive. He notes that in comparison to anthropologists, for instance, sociologists are quite underexposed to biological considerations. While anthropology students learn about such things as human evolution and adaptation to habitat, about other primates, etc., and in this way naturally come to perceive culture as a complex adaptation to a complex environment, sociologists have "succeeded in purging biology" even from the two fields where biology would be obviously relevant—demography and ecology—which now deal rather with statistical abstractions and data aggregations. And even "the otherwise promising field" of macrosociological ecology is making a clear distinction between cultural and biological evolution, the author notes.

Van den Berghe now goes on to identify the various sources of sociological biophobia. He distinguishes between those features that are generally true for all humans (species-wide features) and those features that he sees as representing a type of "trained aversion", which is his real target of criticism. In fact, the first four of his sources have to do with general human biases, related to our species-wide natural anthropocentrism, which he calls "sources of untrained human naivete", but which makes us experience evolutionary explanations as alien—or actually, as "repugnant".

Problems with evolutionary thinking relating to species-wide anthropocentrism:

1. *Unfamiliar time scale.* The time scale of evolution is not on our human time scale.
2. *Ban on teleological thinking.* Humans act purposefully and project purpose on the world, but evolution is considered to have no purpose—life forms just came into being through blind forces acting on random mutations. Indeed, even biologists sometimes lapse into teleological thinking—it is so ingrained in us. "Properly understood, evolutionary theory is repugnant to most of us", van den Berghe concedes.
3. *Assault on human uniqueness.* The process of displacement of humans from the center, which started with Copernicus, is doubly acutely felt by social scientists, since their very livelihood depends on human exceptionalism. Human sociobiology is attacking this last stronghold of anthropocentric conceit. (Van den Berghe calls this "the vanity factor").
4. *Evolutionary perspectives are deeply cynical.* Sociobiology reveals the self-interest behind human behavior—it explains even altruism as self-serving (as nepotism or reciprocity), and ideology and religion as involving delusion and self-deception. Evolutionary biology becomes a threat to our everyday understandings and comfort.

   The issue so far has involved the "untrained naivite" of sociologists, but now we go on to van den Berghe's more serious charges.

   Trained aversion

   Trained aversion has to do with some *implicit premises* sustained by the profession, which have led many sociologists and social scientists to ignore the relevance of biology to human behavior. This is because of the existence of particular dogmas about both sociology and biology.
5. *Belief in environmentalism* (as a dogma). The need to defend environmentalism is upheld as the main reason for the exclusion of biology. Environmentalism is here considered as being a contrasted to a hereditarian position attributed to biologists (no biologist supports this kind of view).

   In general, sociologists still largely think in terms of nature versus nurture, instinct versus learning, genes versus environment, humans versus other animals, etc. Dualism is a universal propensity of the human mind, but biologists and psychologists have realized it leads to false dichotomies. The scientific way of thinking is monistic, not dualistic.
6. *Aversion to determinism.* For many sociologists it means "exclusive determinism", and since the environmental and culture already clearly determine human behavior, it follows that genes cannot. However, cultural determinism

is a comfortable ideological position to hold. It is seen as implying control, man-in-charge-of himself.

7. *Aversion to reductionism.* Most sociologists don't know what it means but they are firmly against it. Reductionism is an epithet first popularized by Durkheim, according to whom social phenomena are not reducible to individual motivations or actions. However, this combination of ignorance and dogma is incapacitating when it comes to understanding science, where reductionism is a method and strategy. But why should antireductionism be such a cornerstone of the ramshackle mental edifice of sociology, asks the author? The answer is: For sociologists antireductionism is a territorial display. van den Berghe goes on to say:

   > By asserting that Homo sapiens is absolutely unique and positively radiating emergent properties, social scientists can claim a turf permanently beyond the purview of biology. By further claiming that sociology and culture are realities that transcend the individual, sociology and anthropology have respectively created their special impermeable turf versus psychology . . . Evolutionary biology extended to human behavior constitutes the most serious challenge to their intellectual turf that sociologists have had to face since behaviorism. No wonder they raise their antireductionist hackles.

8. *Aversion to materialism.* Evolutionary theory is materialist, while sociology is idealist. "It holds that the cultural superstructure grows out of the process of individual interaction, that is, out of the social structure, and that the latter, in turn, evolved genetically from a biological infrastructure". Meanwhile, many sociologists and other social scientists follow an idealist tradition. Sociologists typically believe that people's beliefs, values and norms determine their actions (they do opinion polls) while human sociobiologists are more skeptical and cynical about human motivation and behavior.
9. *The methodological distance between sociology and sociobiology (behavioral ecology).* There is a big difference between sociology and sociobiology (behavioral ecology). Instead of directly observing animals in the wild, like behavioral ecologists, sociologists deal mostly with abstract categories: classes of people, statistics, aggregated data, secondary analysis, etc. They are not generally trained or expected to observe individuals in natural settings. The variables they correlate are many steps away from actual behavior. Sociologists practice a type of "abstracted empiricism". This means that sociology and sociobiology (behavioral ecology) are far apart both theoretically and methodologically.
10. *Evolutionary theory looks simplistic.* Evolutionary theory is simple and straightforward, it explains a lot with few assumptions, which is considered a virtue in science. Sociologists, on the other hand, "have a penchant for obfuscation," and argue against simple language. Because humans are complex sociological theory must also be complex; the best theory says that everything is related to everything else. To sociologists simplicity looks simple-minded.

In conclusion van den Berghe asks: can sociologists continue to ignore biology? According to him, this may work for some sociologists: students of ideology, culture, knowledge, the arts, etc., but not for students in central specialties such as kinship and marriage, small groups, class and ethnicity, demography, ecology, deviance, political sociology, social change, etc., it won't. And here comes his final verdict: sociology will have to become scientific!

> *Sociology's claim to scientific status....must be anchored in a theoretical paradigm that satisfies scientific canons.* Such a theoretical paradigm sociology signally failed to produce in over half a century of self-conscious existence. The main reason for that failure is that sociology turned its back on the one theory that was overwhelmingly successful in explaining change and variation in life forms on this planet: evolution by natural selection. If sociology fails to join the scientific mainstream, it might survive for another century as a scholastic discipline, but it will never be taken seriously as a science. (my emphasis)

4.3 Darwin and Social Science in Early Twentieth Century America

One way to start the overview is to address the issue of social Darwinism, a doctrine that was generally seen as a defense of the social and economic hierarchy of nineteenth century America. After all, it was a sociologist, Herbert Spencer, who was behind this ideology. Social Darwinism recommends not assisting "the weak" in society, in order not to prevent society's envisioned stage wise evolution to higher states. Spencer's vision was soon to be picked up by the early American sociologist William Graham Sumner, while being strongly opposed by another sociologist, Lester Ward, early on suggesting the split reactions to biological type arguments that would be typical for social scientists for a long time to come.

Meanwhile, it appears that those conservatives labeled social Darwinists themselves typically didn't refer at all to natural selection or Darwin—their heroes were rather Adam Smith or Thomas Malthus. What about Darwin himself? If Darwin was a social Darwinist he was not a gloomy one in the Spencer sense, according to the historian Carl Degler. In regard to the human future he told his readers: "There is no cause to fear that the social instincts will grow weaker, and we may expect that virtuous habit will grow stronger becoming fixed perhaps by inheritance. In this case, the struggle between our higher and lower impulses will be less severe, and virtue will be triumphant." (Degler, 1991, p. 12). In the following, I will be using simple pages only for references to Degler.

In any case, the Spencerian type reasoning did not appeal to the social scientists in the early twentieth century. These people were social reformers, who saw evolution as a synonym for progress, and wanted to help show that the idea of evolution and Darwinism stood for a view of cooperation and social cohesion that would help achieve the goal of a new and improved American society. They emphasized that there was a difference between biological fitness and economic success. The vision of many of these early sociologists was basically a Darwinian one: by changing the social environment, human behavior would adjust to it, and here the social scientists would be helping (13, 14).

But there was a problem having to do with race. Darwin in *Descent of Man* had refuted a view of races as incipient different species (as was then held by contemporary American anthropologists), however, he still implicitly accepted the idea of a hierarchy among human societies. Ironically, as a result the most typical form of social Darwinism around the turn of the century came to involve racism, collapsing the difference between Darwinism and social Darwinism (15, 16).

The new science of psychology emerging at the end of the nineteenth century brought in Darwin from the start, especially his idea of instinct, which was highlighted by William James, the author of the first psychology textbook (32, 33). Sociologist also picked up the concept of *instinct.* Instinct was being used rather broadly to explain behavior, and became the subject of at least 600 articles and books in the early twentieth century (35). It was a concept that could be used to explain both normal and anti-social behavior, and in different fields, such as anthropology and sociology. Also a connection was being drawn between man and animals. Sociologists now came to see criminality as based on inborn tendencies, causing some people to imitate models that led them to a life of crime.

This was also the time of the introduction of the infamous large scale intelligence testing by Herbert Goddard. Following Goddard's lead, various social authorities set out to test their inmates, resulting in an astonishing number of "feeble minded" among prisoners, delinquents, and prostitutes. Goodard's famous book on the Kallikak family aimed at showing how inherited feeblemindedness over time could turn into a great social problem. The Kallikaks and similar studies have been examined by Stephen J Gould in his *The Mismeasure of Man* (1981). The classification was later changed, and biologically feeblemindedness, insanity and epilepsy were classified as simple recessive traits. Still, as late as the 1940s social scientists still regarded feeblemindedness as a heritable trait (Degler, 1991, 40).

The concern about heredity as the cause of social pathologies in turn led to the consideration of various measures to control the looming danger. And here we meet the eugenics movement as a way to improve social order. Francis Galton had come up with the idea in the later nineteenth century, but it didn't become popular until after Lamarckism had been abandoned after the rediscovery of Mendelism in the early twentieth century. After that breeding was considered the only way to improve the human race.

Eugenics was considered a positive movement. Its growth was helped by the social reform movement of Progressivism under Theodore Roosevelt and Woodrow Wilson. Eugenics actually became a fashionable reform movement supported by important intellectuals and statesmen and attracted American social scientist, especially psychologists (43). Degler notes that the social scientists accepted sterilization as a means of improving the heredity of the nation (46).

Eugenics also had a social policy angle. There was a concern to do something to control heredity. This usually translated into inhibiting the reproduction of mentally defective or criminals, with sterilization as the most common method. By 1930 30 states in the US had sterilization laws. It was by no means unique, other countries, too, enacted sterilization laws, e.g., the Scandinavian nations. In Germany, Degler notes, the sterilization law preceded Hitler's rise to power (46).

The IQ tests had earlier been instrumental in identifying people as mentally deficient (e.g., the famous Buck vs. Bell case). In the 1920s the IQ tests were put to use in a completely new way. The idea was to use them to measure and compare the intelligence of immigrants from various ethnic and racial groups of the more than 13 million European immigrants that had arrived to the United States between 1900 and 1914.

Madison Grant in the book *The Passing of the Great Race* (1916) had claimed that these recent immigrants from southern and eastern Europe war clearly inferior to the earlier immigrants from western and northern Europe—mentally, morally as well as physically. This claim in turn had been picked up and echoed by social scientists in a series of articles and books (48, 49).

The Army tests were the biggest event taking place in regard to IQ testing. These were organized by several psychologists. The task was to help the army make rational decisions about how to best employ the new immigrants, and testing about two million of them seemed like a good idea. This then became a source of material for social scientists, and later a reference to use for urging the government to limit immigration of undesirable groups. It was believed that this material was later used to argue for the strict new Immigration act of 1924. But later research shows that the psychologists and the Army test results may actually have had rather little impact on this decision; it was rather the American people that directly contacted Congress with their worries..

Ever since the First World War there had been a movement to restrict immigration; restricting immigration as such was nothing new. The labor unions and patriotic groups feared the new immigrants. The perceived problem was that there were now so many new immigrants who also seemed culturally so different from both older Americans and earlier immigrants. Hundreds of letters were sent to Congress with concern for social integration, homogeneity, and national cohesion. The IQ tests were not mentioned in these letters (52). As for the possible role played by psychologists in this matter, Samelson (1979) and later researchers came to the conclusion that very *few* social scientists actually participated or testified in Congressional hearings, or invoked IQ testing (54) Meanwhile there was a bigger concern at the time: a growing doubt among social scientists that biology was responsible for human intelligence and behavior.

We will return to the intelligence testing in a while, but first we will need to cover a parallel development in the early twentieth century: the promotion of the concept of *culture.*

The man in charge here was the anthropologist Franz Boas, who first presented culture as an alternative to racial explanations of differences in behavior in his book The Mind of Primitive Man (2011), a summary of his existing ideas. The culture idea in that book is that "savages" are no different from "civilized" people. This overthrew the anthropological view that peoples pass through stages, and that primitive people are at an earlier stage. Boas denied any differences in intelligence between "savages" and "civilized" people. According to Boas their visible differences are products of different histories or cultures. Boas views were to become very popular among social scientists in the early decades of the twentieth century. He later envisioned a single human nature, not superior and inferior people (62, 63).

Later Alfred Kroeber was to pick up where Boas left off and further develop the concept of culture. For Kroeber culture should be free from any connection to biology; he explained this in articles for *The American Anthropologist* 1910–17 (90). He also emphasized the universality of morality in all societies in a way similar to Darwin in the *Descent of Man* (92). Later Kroeber argued for a total distinction between biology and culture (99).

4.4 The Nature-Nurture Controversy and the Debate Over

4.4.1 Race Differences in Intelligence

In what Degler calls "the great professional debate of the 1920s over *the validity of measurements of innate intelligence*" there was great disagreement about the actual meaning of the scores obtained through the intelligence tests, as well as serious doubt about the whole idea of measuring race differences in intelligence (172–73). Confident statements like "races differ in intellectual stature, just as they differ in physical stature" (Harvard psychologist William McDougall, the promoter of the instinct concept) were early on challenged by research showing that "there is more likeness between children of the same social status but of different race than between children of the same race of different social status" (Ada Arlitt, psychologist at Bryn Mawr college, 173). The Army tests had been a breakthrough in regard to measurement, but there was criticism of the anti-immigrant and racist interpretations that could be found in *A Study of American Intelligence,* the summarizing book by Princeton psychologist Carl Brigham (1923) of the massive Army tests of intelligence. Already before that book, questions had been raised about what was really being measured: innate intelligence or learning? (170) A review in *Science* called Brigham's arguments and evidence "outdated" (175).

It seemed that the concept of *culture* was having an effect on professional opinion (175).

In 1930 Brigham himself recanted, admitting that the Army tests had been so badly designed and administered, that comparative studies of national and racial groups could not be made with those tests. By then quite a lot of critical studies of intelligence testing and racial differences had accumulated in the professional journals (176).

However, Degler believes that it was not only professional concerns for rigor and truth that stood behind the drastic change in opinion. He asks what it was that caused the various reservations about the existence of race differences in the 1920s debate. He singles out one dedicated psychologist, Otto Klineberg, who, inspired by Boas, devised an unusual 5 year research project to respond to virtually every argument or piece of research brought forth as evidence for the racial inferiority of some group. By locating villages with "pure ancestry" in Europe, he hoped to test whether "racial" background was more influential than history or environment when it came to intelligence. What he found was that urban boys scored consistently better than rural ones. What about those "Nordics"—did they consistently achieve higher test scores than other groups independently of the country in which they lived? They didn't (180).

Klineberg also found that the longer Southern black children had lived in New York, the higher their scores, again pointing to the environment rather than heredity or "selective migration" (i.e. the notion that more intelligent blacks had moved north in the Great Migration). He also tried to figure out what a "comparable urban environment" would mean for blacks and whites. Klineberg is an example of an unusually dedicated individual who put in an enormous effort, just like Boas and Kroeber.

But why was there such a determined effort to switch to a cultural explanatory framework during the 1920s? There is a puzzle here, according to Degler. Because "we are not dealing here with a long-held, well-substantiated theory (that is, race), which new and conclusive evidence had unambiguously disproved and overturned. *Rather we see essentially the substitution of one unproved (though strongly held) assumption by another*"... Culture or history could account just as well, if not better, than race and biology for differences in human group behavior." (187, my emphasis).

Among what Degler sociologically calls "the forces operating upon" the social scientists, was prominently *the ideology of equality of opportunity*. There was also a feeling of guilt about the treatment of blacks and other racial minorities. Degler registers a predisposition of social scientists as well as Americans in general to resist biological explanations. This was noted by the early eugenicist Charles Davenport, who complained that people objected to eugenics because they didn't want to believe in heredity, which he called "a pessimistic and fatalistic doctrine" (Davenport, 1911). Klineberg, too, admitted that, whenever justified by the data, he early on preferred the environmental explanation "because it was more optimistic, holding out the hope of improvement". He took the position that "we have no right to accept the hypothesis of the innate superiority of any one race over any other" until all other explanations have been excluded (188, my emphasis).

As for ideological underpinnings, not much is revealed. Some social scientists stated hopeful visions, such as the suggestion to stimulate cultural integration instead of biological differentiation, others candidly gave their opinions about the different achievements of various races while advocating for a view of the unity of human nature. Others were irritated with the endless seeming discourse.

The principle of equality of opportunity appears to have been behind the various efforts to cast doubt on race based explanations. In 1921, the psychologist Thomas Garth, concerned about prejudice in race psychology studies, and the stigma of low ratings which might result from hasty conclusions, formulated a "canon" for students of racial differences. This is what it said:

> In no case may we interpret an action as the outcome of the exercise of an inferior psychological faculty, if it can be interpreted as an exercise of one which stands higher on the psychological scale, but is hindered by lack of training (190).

Bitter accusations were later made against proponents of racial differences in intelligence. They were labeled enemies of the American principle of equality of opportunity and supporting preconceived notions with their mental tests (190). Meanwhile a strong motivating force was the sympathy that many social scientists felt with the social aspirations of black Americans. To help them concretely, Klineberg and others started addressing black audiences to train them in countering various negative arguments (191).

Degler's general point here is to present a contrasting picture to the typical claims about this being a period of racism and ethnocentrism in regard to blacks and immigrants from southern and Eastern Europe. He here deliberately presents an "almost contradictory" picture of the spirit among the social scientists, among whom hereditarian explanations were on the defensive for most of the period. He explains that this was because many American social scientists already from the nineteenth century "were predisposed to favor change and progress, social improvement and reform... The natural tendency of their world view was to prefer an environmental or cultural explanation." (192).

Meanwhile, there was also a counter reaction from yet another side in this story—the immigrants themselves. Scholars from immigrant backgrounds typically fiercely criticized the characterizations of recent immigrant groups in books. Madison Grant's popular *The Passing of the Great Race*, and Carl Brigham's scholarly *A Study of American Intelligence* both received devastating reactions from scholars of immigrant background (200). A number of prominent social scientists who participated in the crusade against test of racial differences were in fact immigrants themselves, or children of immigrants (201).

The historian of science Franz Samelson actually suggests that a major reason for the shift in the key concern of social scientists in the 1920s was *a shift in the social origins* of many of these. He makes a rather sharp point as he states: "Empirical data certainly did not settle the issue one way or the other... *The vast majority of psychologists have never taken a close look at all the data, but accepted, by and large, received opinion supported by appropriate pieces of evidence...*" The historian Hamilton Cravens brought in ethnic and social differences as additional sources of the change of view (201, emphasis added).

A major shift could also be observed in people's general assessment of the situation. Before 1920 an overriding concern had been the perceived deterioration of American genetic potential and intelligence. But 10 years later, it was race relations that became the new central concern. Samelson explains that one reason for this shift was the recognition that the general Americanization movement of the early 1920s was not working. Therefore the only viable alternative option at this point was seen to be the adoption of cultural pluralism (Samelson as quoted in Degler, 201).

4.5 The Trumph Triumph of Culture Over Race

So Boas and Kroeber had early on won the struggle within anthropology in regard to the race question by introducing the idea of culture, and within psychology matters had been settled by the nature-nurture debate. But what about the social sciences in general? Here we have to look at a variety of "forces" at play, some personal, some field specific and some having to do with larger historical background events.

Carl Degler comments on the fact that the radical shift in explanation of human behavior (which he calls "the triumph of culture") was not caused by any clear refutation of a well-established theory. What happened was rather the substitution of one unproven assumption with another equally unproven one! (187) So, according to Degler, because no crucial piece of new evidence or overpowering argument had been brought forth on either side, we need to look beyond professionalism or standard science to fully account for "the great transformation" in the explanation of behavior. Degler sees the story as "a complex mixture of happenstance, ideology, individual commitment and broad social influences" (187). Let's follow his narrative.

An important factor was the ideology of equality. There was a feeling of guilt about how blacks and other minorities were being treated. There was also the unstated assumption that if biological inferiority was indeed established, this would result in loss of job opportunities, which would be unfair. By 1930 the numbers of social scientists who still held on to a hereditarian view had finally dwindled. Two interesting dissenters were sociologist Pitirim Sorokin at Harvard who just dismissed any environmental interpretations in regard to blacks, and a president of the American Sociological Society (ASS, later ASA), Frank Hankings who still in 1938 continued resisting cultural explanations. He was a serious hereditarian, who had realized that the cultural explanation as it was thought about at the time left no space for evolutionary considerations (193–195).

As far as scientific factors were concerned, by the end of the 1920s the idea of gene-environment interaction, already recognized by natural science, had now become regarded as a good solution by psychologists. The good news about the mutual interdependence of heredity and environment was even being covered in a couple of popular books for social scientists (196).

In regard to bigger background factors we have the Great Migration of Southern rural blacks to Northamerican cities (such as Chicago and New York). Hundreds of thousands individuals participated in this great relocation starting just before the First World War 2014. For social scientists, and for everybody else, too, this event provided an unusual opportunity to study the role of the environment for human behavior while observing people in a natural setting. It soon became clear that as the new environment created enhanced job and educational opportunities, and it brought out new individual capabilities as well Also, during this time African Americans began entering the professions and the educational system. In 1928 a Chicagoan was elected as the first black member of Congress (197–199).

There was an overall change in outlook. Black scholars presented a new image of African Americans and a new perspective on their situation. Scholars of Jewish descent who had been excluded form American colleges were now showing up. Chinese scholars raised issues about comparative intelligence tests.

How did the Great Depression in 1929 affect the prevailing views of nature vs. nurture? This event, another "natural sociological experiment" of sorts, which did not discriminate between rich and poor, provided serious lessons of a different kind and is thought to have helped generate some understanding for the environmentalist stance. Samelson believed that it pushed psychologists toward a leftist social outlook (202).

Later there were the shocking news about beginning Nazi efforts to put into practice a eugenic program aiming at race purity and improvement of the population. (Later this escalated into the horrors of the Holocaust). Franz Boas spearheaded a mobilization movement against the Nazis. A resolution denouncing racism was taken at the 1938 meeting of the American Sociological Association (203). In the 1930s and '40 s articles using biological concepts totally disappeared from journals; they were replaced by cultural ones. A telling example was the *Encyclopedia of the Social Sciences*, which deliberately invited critical young scholars instead of well-known experts to write on topics such as IQ testing. Degler sees this as an example of the wholesale adoption in the 1930s of the concept of culture or environment as the primary shaper of human behavior (205).

It was not only the social scientists that were taking measures. William Provine (1973) describes how the genetics community in the late 1930s settled for an "agnostic" position on the question of racial cross-breeding. No new information on this issue had been forthcoming at this time. However, according to Provine, "geneticists revised their biology to fit their feeling of revulsion" [to the Nazis] (204, 205).

An important book was Boas' student Ruth Benedict's *Patterns of Culture* (1934). It pointed to the importance and power of culture and cultural differences, and the weakness and irrelevance of biology. Benedict's position can be described as "cultural relativist". In contrast, a complete cultural determinist was anthropologist Leslie White, and (what he called) his "culturological interpretation", according to which an individual should be regarded as a mere part of a vast sociocultural system. "The culture process may be regarded as a thing *sui generis",* White wrote in 1949 There is no place for the individual in this system. "Human nature is not natural at all, it is cultural" White wrote in 1949. In 1954, the anthropologist Ashley Montagu argued that man has no instincts, because everything he has been and has become, he has learnt (208, 209). As an indication that the cultural approach succeeded in penetrating sociology as well, here is part of what the ASA president Robert Faris said in 1961:

> Ability was generally held to be fixed in biological inheritance... [But now] barriers in many fields of knowledge are falling before the new optimism which is that anybody can learn anything... "[N]ow we perceive that an important part of the relevant causation in ability is necessarily sociological in nature and that control is most likely to come through penetration of this aspect of the subject."

Here he was arguing along the same lines as the Swedish sociologist Gunnar Myrdal whose famous 1400 pages book *An American Dilemma* (1944) voiced great hope for the role of sociology. This sociological two volume study of American blacks, funded by the Rockefeller Foundation, assumed a thoroughly culturalist approach in addressing a number of problems involving race while it described how the social sciences might provide new promising tools for research on social improvement. "We have today in social science a greater trust in the improvability of man and society than we have had since the Enlightenment" were Myrdal's final hopeful words. This signaled a thorough transformation for the social sciences. Myrdal characterized the current social sciences as "militantly critical" and "truly revolutionary in the cherished American tradition" (Degler, 215).

Indeed the attitude to race and ethnic groups had radically changed over the past half century—from human differences being seen as completely unchangeable, to later being regarded instead as "a complex of societal conditioning, in which culture is considered to be a more dominant factor than biology" (Odum, 1951). This was half a century after Frans Boas had started it all by upholding *culture* as the crucial explanation for human differences, later backed by Kroeber.

Yet another sign of the triumph of culture at the time can be seen the rise to prominence of B F Skinner and his behaviorism at Harvard. His famous experiments with conditioned responses in pigeons were thought to scientifically demonstrate that behavior could be modified by controlling an individual's learning environment, with obvious implications for humans and child development. (He is famously said to have experimented with his own daughter). Skinner was lecturing to huge audiences at Harvard in the 1950s and soon published a book on his principles for rigorous science (Skinner, 1953).

4.6 Harvard Takes the Lead

In the 1940s the theoretical and practical leadership in sociology moved away from the University of Chicago, whose department had been successfully focusing on the theory of symbolic interactionism, following in the footsteps of Charles Horton Cooley and Georg Herbert Mead This happened in 1946 under sociologist Talcott Parsons and his newly created multidisciplinary Department of Human Relations, which consisted of the merged faculty and students from the departments of sociology, anthropology, psychology and social psychology working under a common theoretical framework of functionalism. Functionalism had already been connected to anthropology under Radcliffe Brown and Malinowski; both strongly influenced by Durkheim. As a theoretical approach it focused on the function of the various parts of society for the maintenance of the whole. This functionalist perspective was now integrated in the expanded conception of functionalism developed by Parsons and his students (Sanderson, 2001, p. 13).

Parsons' overriding concern (following Durkheim) was the problem of *social order*. He rejected the idea of society being composed of individuals pursuing their own interests, in other words, any kind of utilitarian thinking. For the social order to be stable, there had to be ways to restrain such interests. *One obvious solution was the existence of shared values and norms, which society's members would internalize through socialization.* The traditional functionalist image of society is that of an organism, whose every part performs some essential function for the whole. Indeed, sociologist Herbert Spencer already in the nineteenth century had described society itself as an organism. So, within a functionalist sociological explanation, any aspect or part of society would be assumed to exist because of its importance either for some other part of society, or for society as a whole (Sanderson, 2001, pp. 13–14). (Emphasis added).

Already in the 1950s, however, there were clouds on the horizon. Something more needed to be done with functionalism, more sociological discoveries and contributions were required. There were rising criticisms that the Parsonian paradigm was too complex and static; it did not easily lend itself to generating testable hypothesis. Moreover, there was a clear disjunction between theory and method. Later, the radical sociologist C. Wright Mills was to famously characterize this situation as "Grand Theory vs. Abstract Empiricism" (Mills, 1959).

One person who felt that things were not going well, warned his colleagues and tried to do something early on was Parsons' student Robert Merton. And this is how we have a number of important contributions from him in their own right. For instance, Merton elaborated the discussion of function by considering also the notion of dysfunction, and the difference between what he called "manifest" and "latent" functions, in his book *Social Theory and Social Structure* (Merton, 1957) (Sanderson, 2001, p. 15). Merton had earlier tried to explore deviance as a social phenomenon, trying first a biological explanation, but learning that biological explanations were anathema, which led to his alternative theorizing, resulting in his famous "structural strain" theory. That was a logical and pedagogical 2 by 2 typology correlating an individual's acceptance (or not) of existing social goals with his or her access (or not) of existing means. The combinations of these give four main types of behavioral outcomes, such as conformism, rebellion, innovation, and ritualism. At the same time it can be seen as an explanation of deviance and implicit criticism of the American Dream idea. However, Merton was later criticized for using the term 'function' to speak about effects or outcomes, that is, he did not use it in the proper Parsonian way. And here we have Randall Collins' mysterious verdict: "Merton rose to prominence in the dark ages of American sociology in the cold war years of the 1940s and 1950s. He helped guide sociology across those mindless and repressive years, even as he helped perpetuate the darkness" (Collins 1975, cited in Sanderson, 2001, p. 15). Sanderson thinks that Merton should have rather helped overthrow the functionalist paradigm!

Parsons' work came under increasing criticism in the 1960s and by the mid 1970s it had lost much of its influence. It was not only Parsons work but functionalism in general that got attacked; leading critics were among others Ralph Dahrendorf (1959) and Irwin Zeitlin (1973) (Sanderson, 2001).

But during this time, was the Parsonian theory the only game in town? Was it all dependent on him? Actually, it seems this way, since Parsons was extremely dominant. Some of his students later continued the functionalist reasoning, others ended up developing exactly opposite approaches. Here we have Harold Garfinkel, who developed the new approach of ethnomethodology (Garfinkel, 1976) to show us how social reality is actually constructed by negotiation among individual actors—not something that simply gets internalized by socialization from the "outside". And anthropologist Robin Fox, who finally settled for ethology as the needed addition to his anthropology, ending up founding "biosocial anthropology" together with Lionel Tiger (e.g. Fox & Tiger, 1971, Fox, 1975). Or Pierre van den Berghe, who is respected as an early spokesman for a biological perspective in social science (e.g., van den Berghe, 1978, 1989).

C. Wright Mills was not a Parsonian student, which might explain his hyper critical tone in his 1959 book.

Well, Parsons was not the only game in town. There was also George Homans, a very eminent sociologist who later became chair of the sociology department at Harvard in the mid 1970s. He pursued a completely different approach, one that one could characterize as diametrically opposite to Parsons. If Parsons insisted on society being an external force that actually exercised power over our values and behavior, Homans was interested in 'Bringing Men back In//, (Homans, 1964).

Homans' theory was grouned in everyday human interaction; later it took the form of exchange theory. One of the theory's main theses was that individuals will do what they find rewarding and avoid things they don't. But this sounds a little like conditioning? Actually, it does, and this was actually Homans' point. For him the Skinnerian view of behaviorism as the scientific way to truth was very appealing. More importantly, for Homans the individual, not society, came first. However, as pointed out by critics, a theory completely dependent on conditioning or learning brings with it a tabula rasa idea of human nature. There is nowhere to find any suggestion that it is an evolved human nature with certain characteristics, needs, biases etc. that takes part in the social interaction ...

Having these two luminaries, Parsons and Homans, in the same department at the same time, might have created some rivalry or tension. Lopreato and Crippen (1999) may be hinting a little at this, as they discuss the department and the responsibility lying on "weak shoulders". So this rivalry, or the mere fact of these completely contrasting paradigms, might have contributed to the undermining of the Parsonian paradigm.

But there was still another aspect, which actually affected all of the activities of the Department of Human Relations at the time (this composite department was founded in 1946 and discontinued in 1964, when it was decomposed into its original departments, which then continued with their own chairs). That was the explicit pressure to "be scientific"! Where did this come from? This is not clearly analyzed in the books I have consulted, but we can maybe make some guesses.

What did it mean to be scientific in the Human Relations department? Well, for anthropologists it meant doing comparative research looking for patterns of culture, and many such patterns were indeed identified and analyzed. Here we can consult Robin Fox who was a student there "at the end tail of the boring 1950s and continued well into the 60s (Fox, 1995). He writes about a constant need to produce a large amount of papers and see to that the statistics was acceptable. The anthropologists could always rely on the Human Relations Area File collected by George Murdock at Yale for cultural comparisons of various kinds, and many comparative studies were conducted. The sociologists were testing various hypotheses, largely left to themselves. As a remedy for that Merton had been suggesting various possible measures in 1948, such as trying to develop "theories of the middle range" (as he himself proceeded to do), but he did not get much response. Meanwhile the Parsonian model was too abstract to be really useful for hypothesis generation. In any case, Fox remembers a constant command to "be scientific" in the late 1950s, which could have had something to do with the overpowering competition from behaviorism, that paragon of social scientific scientificness at the time.

4.7 The Crisis in Sociology

This section discusses what some sociologists, writing in the late 1990s, have characterized as the then recent "crisis in sociology". It includes various explanations for the crisis offered by fellow sociologists and other social scientists and academics, as well as suggestions for what to do to remedy the situation. The upshot of the criticism appears to be that for as long as sociology has existed there have been repeated missed opportunities to make it into a "real" science. There is still a possibility for rectifying the situation—but there is only one way, which is to rapidly bring in Darwin. However, time is running out. This is the view of Joseph Lopreato and Timothy Crippen in their 1999 book *Crisis in Sociology: The Need for Darwin*. In this overview social scientists, old and current, are classified according to whether or not they stand for—or have appeared to stand for—an evolutionary perspective. Those sociologists are OK according to our authors—while some leading figures like Parsons and Merton, get subjected to rather harsh critique. Lopreato and Crippen's book is from 1999, which means we'd better also take a look at what has happened to sociology over the last 20 years or so. Is the crisis still on, is sociology getting Darwinized, or what is happening?

The authors point back to the founders of sociology who were all "children of the Enlightenment and of the 'evolutionary centuries'" and "euphoric about crashing disciplinary boundaries" (3, 4). Soon after sociology had been established as a respected academic discipline, however, the field "went astray" and has "long been a discipline adrift" (4). The authors go on to cite a number of symposia and debates on the crisis of sociology in sociological journals taking place in the mid 1990s. Serious questions were raised about the discipline's academic respectability and ability to compete for scarce resources.

According to Lopreato and Crippen, "the crisis in sociology became unavoidable when we strayed from the premise generally accepted by the Founders that there can be no scientific sociology without general laws." (4). Sociological analysis is almost completely environmentalist, which means that its explanations are at best ephemeral. Meanwhile fields like psychology and anthropology have joined the new Darwinian revolution. “To survive [sociology] must rediscover the brain, the circumstance of its evolution, and the evolution of its products. It needs to learn Darwinian lessons, and contribute to their ever-expanding scope”. This is the authors’ verdict, and they go on to outline how they later in the book plan to teach the reader some basics of Darwinian science, and apply these to crucial sociological problems (4).

In regard to laws, they have in mind especially Auguste Comte, who advocated for applying the scientific method to study society and “invariable natural laws” “similar to the doctrine of Gravitation. He formulated as a scientific principle a law of three consecutive states through which the human mind needs to pass, ending up with a final positive state. However, the authors note, “a major example of Comtean error is the idealist philosophy that, his positivism notwithstanding, inhered in the law of the three stages” (6). They continue with Marx, whose vision was directly indebted to Darwin (9). Herbert Spencer had a Malthusian approach, seeing population increase as the mother of invention and progress, but adding that “not all persons and peoples are endowed by nature to attain the higher levels of skill and inventiveness”. He went on to explain the conquest of one people over another as the more adapted over the less adapted, and concluding that “the whole effort of nature is to rid itself of ‘feeble elements’ (10, 11). Crippen and Lopreato note that Spencer has been typically either reviled or disregarded, but inform us that “there are many capital lessons in Spencer’s evolutionary sociology” (12).

Durkheim, however, is getting energetically defended against the claim that he did later generations of sociologists a disservice with his statement that social and cultural facts are phenomena sui generis and hence having autonomous explications, i.e. not in terms of psychology and biology. The authors have also found a quote showing that Durkheim actually supported evolutionism. Durkheim did not try to replace evolution with some kind of functionalism—on the contrary, “To show how a fact is useful is not to explain how it originated or why it is what it is . . . .But if the usefulness of a fact is not the cause of its existence it is generally necessary that it be useful in order that it may maintain itself. . . .. (Else) the budget of the organism would have a deficit and social life would be impossible” (14).

The authors take this to mean that “culture must be useful; else it ends being acted upon unfavorably by natural selection (by having a ‘budget deficit’ – by being unadaptive)” (p. 14) Moreover, they demonstrate how Durkheim’s argument in *The Division of Labor in Society* can be “quite relevant to one of the most productive theories in evolutionary behavioral science, (Trivers, 1971).” They show in detail how the change from mechanical to organic solidarity represents a causal evolutionary theory. They also cite Durkheim referring to Darwin and “the struggle for existence” in relation to the division of labor. (15–18) In summary, they conclude that “Durkheim’s study makes sense best as a study of the relative deintensification of nepotistic or kin altruism (mechanical solidarity) and a concomitant intensification or diffusion of “reciprocal altruism” (organic solidarity) within the context of a society evolving away from ties based primarily on genetic kinship.” (18).

This shows that there is in principle a lot to discover in the Founders, say Crippen and Lopreato, but sociologists didn’t mine the material, even if in the

1930s there existed thorough commentaries on a number of the Founders. According to our authors, especially Talcott Parsons' famous (1937) book *The Structure of Social Action* "could have acted as a powerful invitation to later generations of sociologists to carefully examine the classics, retrieve those aspects that would help define the foundations of theoretical sociology, amend others, and of course discard the remainder." (19) Unfortunately, we are told, such a development never took place. Instead, sociology started moving away to issues considered relevant to the immediate problems of an ever increasing number of academic parishes" (19). The concrete problem with Parsons big book—expected to be based on an analysis of the classics—was that it had experienced an unexpected development, giving rise to a byproduct which Parsons called 'a voluntaristic theory of action'. According to one unfriendly commentator, this theory was "destined to become a major force in the theoretical trivialization of sociology" (19).

4.8 Ten Explanations for the Crisis in Sociology

This list is based on the various types of explanations and criticisms of the crisis in sociology during the 1990s especially, when sociological journals devoted space to symposia about this issue, and even *Newsweek* got involved at one point, as it asked for the relevance of sociology to the real world. That was also a time of increased activism among evolutionarily oriented sociologists to reform sociology by calling in Darwin. Meanwhile, the avid reformers may not have appreciated just how little interest or comprehension mainstream sociologists had at the time, largely because of simple underexposure to biological reasoning of any sort, and, of course, to discipline related "doctrines". However, among the many critiques there had also been the growing recognition that sociologists at large may actually need friendly basic education in biological reasoning, and should probably be following their sister disciplines anthropology and psychology at least in accepting at least the fact of genes and environment working together rather than representing two opposites (in fact, biologists routinely count culture too as part of he environment), as well as abandoning any tabula rasa ideas of human nature.

Here we have a list of ten explanations, boiled down from many more to be found scattered in the literature. Sometimes the criticism has been very harsh—especially from people on "the other side of the abyss"—but I am here typically paraphrasing. Among the criticisms have been many references to van den Berghe, in fact his 1990 article itself was part of one of the early symposia about what is wrong with sociology. I am here mostly focusing on explanations that are different and additional to van den Berghe's. Note that most of the critics—well, just like van den Berghe—are sociologists or social scientists themselves. The order is a (not totally systematic) attempt to go from very strongly held, or "dogmatic" beliefs of sociology or sociologists, to reflections on the basic nature of sociology, and various reasons why biological reasoning may feel uncomfortable or unnecessary to sociologists, to "political" reasons of changes in the ASA governance, or the advent of postmodernist and constructivist ideas, to basic beliefs about scientific laws, and finally a baffling blame game where the founding fathers themselves become targets! Some of the criticism is well formulated by Lopreato and Crippen themselves, they are worth quoting.

1. "The fundamental cause of the sociological crisis is based on the fact that too many sociologists adhere uncritically to the following methodological position:

   (a) human behavior is the result of *socialization*
   (b) the causes of human behavior are to be found in *social structure*

   There is an element of truth in both assertions. While humans are born with certain behavioral propensities, the latter are developed through processes of socialization." (Lopreato & Crippen, 1999, p. 72)
2. "We have culture to blame..... [C]ulture was a necessary concept for the emergence of a science of human behavior. But despite the participation in the development of evolutionary theory by 19th century sociologists and anthropologists, culture soon came to attain almost hypnotic powers." (Lopreato & Crippen, 1999, p. 110)
3. Unwillingness because of "ideology" (Lipset, 1994; Simpson & Simpson, 1994). Sociologists have always been reformers so this kind of ideology comes with the discipline, but in this case "ideology" is typically connected to aversion to biology because of certain political connotations. Horowitz (1993), devoting a whole book to the crisis in sociology suggests that sociology is often so involved in an ideology of self-righteousness and advocacy that it is unaware of developments in science (1993, p. 5) He notes that new feminist and other caucuses developed in the ASA to promote special group interests across the association have undermined the focus on promoting sociology as a discipline (35). Cole (1994, p. 148) observes that sociological problem choice often has to do with ideology or personal experiences.
4. Some sociologists believe that sociology actually is, or should be, about public policy, or a force for social criticism (Bellah et al., 1995, p. 297).
5. There is a complaint that sociology doesn't deal with important contemporary problems, which would need analysis. Cole (1994, p. 148) observes that sociological problem choice often are parochial, having to do with ideology or personal experiences. This was also observed in the *Newsweek* article. In Europe there may be nationalistic type sociologies competing with each other.
6. A 1992 *Newsweek* article "Is sociology relevant to the real world" could be interpreted as concluding that is actually isn't. The article was among other things featuring interviews with sociologists out there in "the real world". The challenge of this article became the focus of at least one symposium in an sociological journal. In his 1992 presidential address James Coleman, then President of ASA, referring to the *Newsweek* challenge, stated that sociology would need to show "utility" to be taken seriously. Meanwhile this was happening in a world where some of sociology's typical social initiatives and

mass education ambitions had been taken over by others, such as business and media organizations, which run successful professional programs. Society had become "sociologized" to some extent—which can perhaps be seen as the backside of sociology's earlier success. It was also noted that college students by now know about basic concepts of sociology and should rather learn to think comparatively in terms of systems (1993). Finally, sociology was threatened to be "cannibalized" not only by sociobiology (as E.O. Wilson famously declared in the early pages of his *Sociobiology,* 1975), but also by all the other bio-behavioral fields that had emerged, leaving it more narrow scope to function.

7. In the 1990s, various then recent academic developments such as postmodernism and social constructivism with their extreme conceptions of science had an undermining role on academia in general, and there was some participation of sociologists in the so-called Science Wars in the mid-1990s. See e.g., Segerstrale (2000b) on the Science Wars and the role of sociologists, especially sociologists of science.
8. The advocates for bringing in Darwinism or evolutionary theory as the rescue for sociology note that so far sociology has not been able to establish any law, which seriously undermines its scientific status and makes sociology being not taken seriously academically. They suggest bringing in at least one general law—and that is something that can actually be *borrowed* from a neighboring discipline, which is typically what happens in science (say between chemistry and physics). In this case the theory to borrow from would be evolutionary theory. The simple law that is suggested is *organisms tend to behave in a way that maximize their inclusive fitness.* (77) (This refers to Hamilton's extension of a basic Darwinian principle to include also the effects of social interaction among organisms ('kin"). This is all forcefully argued in Lopreato and Crippen's book, and they provide a few final chapters to pedagogically demonstrate how evolutionary explanations can become the ultimate explainers of many sociologically interesting phenomena of human behavior. Of course, beyond sociologists, one would expect criticism of this idea from people who dislike the idea of scientific borrowing of this type or deny the possibility of laws in sociology.
9. A contradictory point, involves the very Founders of sociology. Lopreato and Crippen seem to be of two minds about them. We saw earlier how it may indeed be possible to "discover" Durkheim to be an evolutionary sociologist (unwittingly perhaps, but demonstrable by textual exegesis). See e.g., this: "This is a causal, evolutionary theory, and the mechanism of evolution it proposes lies in the core of the current revolution in behavioral science." This refers to his theory about the changing division of labor (15). But Durkheim is also being criticized. Here it is Durkheim's *Suicide* which is being attacked for Durkheim not considering the individual, who commits the act. Meanwhile, our authors observe that "considerable progress toward a scientific grasp of suicide is being made in areas that Durkheim adamantly rejected as irrelevant to social facts, namely psychology and biology" (52).

10. And what about Weber? Although Crippen and Lopreato saved Durkheim, Weber is being slashed, especially for his method of Verstehen ("understanding", the idea of trying to understand the meaning that things have or had for people, and which therefore affected their decision-making). As a sociologist, one tries to grasp this meaning to get a fuller picture of a situation from the point of view of the actor, while Erklaren ("explanation") is what a sociologist does with the help of various "external" types of information she can collect about the situation and its broader context. Insightful sociological research uses both of these methods and have them inform each other. I believe I am myself a natural "Weberian" sociologist (see *Defenders of the Truth*, Segerstrale, 2000a, and *Nature's Oracle*, Segerstrale, 2000b, 2013). Moreover, there are ways of checking between the results of the two methods (see e.g. Schmaus et al., 1992; Segerstrale, 1993). These two methods of investigation in the broader human sciences are considered quite standard, and in my experience, the tension between them has been a well-known topic of academic lectures and discussions. So it is baffling to read what Lopreato and Crippen say, singling out Verstehen:

> This crucial aspect of Weber's sociology must share responsibility for the excesses of subjectivist approaches currently fashionable in sociology, despite, ironically, one of Weber's (1919) greatest contributions, namely his call for objective, dispassionate analysis. (55)

Weber responsible for excesses of subjectivism? He would have gotten a good laugh out of that.

4.9 Reconsidering Biology in a Changing Climate

More important, though, was the new information from biology that was accumulating at the time and reversing earlier conceptions. Especially Harry Harlow's famous monkey experiments (in which monkey infants clang to a fake terry cloth "mother" rather than go to the wire "mother" offering milk), and the surprise discovery that experimental animals refused to be conditioned for life, indicated the limits of the behaviorist paradigm (this was the observation by two behaviorist star trainers that experimental animals eventually started reverting to their natural selves, Breland & Breland, 1961). New types of empirical information from European ethologists, especially Niko Tinbergen and Konrad Lorenz, convinced readers that much of animal behavior really was instinctive, especially Tinbergen's carefully written *The Study of Instinct* (1951). This was further accentuated by the Nobel Prize that in 1972 was awarded to these two ethologists together with Karl v Frish (the discoverer of the famous "waggle dance" of bees). Psychologists who had been questioning the idea of using rats to conduct psychological experiments were now able to learn of the existence of different species genotypes.

Sociologists had a mixed response to Lorenz' book *On Aggression* (1966), but some were thrilled that ethology could shed new light on the connection between animal and human behavior. More influential were probably the mass of popular books now inspired by the new ethological approach; best sellers such as playwright Robert Ardrey's *African Genesis* (1961), and the *Territorial Imperative* (1966), and anthropologist Lionel Tiger and Robin Fox's *Men in Groups* (1969) and the *Imperial Animal* (1971). The book that beat them all was the one by the British ethologist Desmond Morris: *The Naked Ape: A Zoological Study of the Human Animal* (1967). Degler believes these book together "created a climate of thinking", which encouraged further exploration of a professional kind. (Degler does not mention that these books generated strong criticism in academic left and liberal circles. See more in the Appendix to this chapter and Segerstrale 2000a).

Many political scientists were especially taken with ethology, and the general possibility of rounding political science on a scientific basis. They saw natural parallels between ethologial concepts and human social organization. Especially Roger Masters, Albert Somit, and Thomas Wiegele were very active with publications and books and as organizers of interdisciplinary conferences (such as the Association for Politics and the Life Sciences, APLS). Progress in genetics and the study of identical twins impressed some sociologists and social psychologists in the 1960s (Lopreato and Crippen, 1999, p. 231).

Meanwhile the really big background event had been the Modern Synthesis, through which biology as a whole had reached a different stage of development, with its bringing together of Darwinism and Mendelism, or natural selection and genetics. The crucial bridging of the perceived unbridgeable gap had been R A Fisher's suggestion that small mutations provided the variety that natural selection needed for its work. Julian Huxley's famous explanatory book *Evolution: The Modern Synthesis* had been published already in 1942.

Meanwhile, many social scientists were disappointed with the progress in their own field, and was looking to biology for offering some kid of scientific certainty or lawfulness. An example of this is sociologist Walter Wallace's book *Principles of Scientific Sociology*, published in 1983, which argued that sociology is one of the natural sciences—at least much closer to biology than physics. New journals were founded, conferences were organized, various university programs were arranged to promote the integration of biology and social sciences. According to Degler, by the beginning of the 1980s it was becoming clear that biology was relevant to social science. He notes that an extra impetus was provided in the 1970s by Wilson's *Sociobiology* and the following controversy, but argues that, for social scientists, the most important aspect of that book was the understanding it provided of behavior made possible by a more recent generation of ethologists who studied animals in the wild (243).

However, the new interest in biology was different for different social scientists—while it for psychologists represented a return to an earlier interest, for political scientists it was a new and exciting opportunity to explore similarities between animals and humans. Political science particularly benefited enormously from various comparative studies and Roger Masters soon wrote a whole textbook on politics drawing ethological parallels of various sorts (Masters, 1989). An edited volume was ready already in 1981 (Wiegele, 1982) based on a lively conference bringing together political science and biology.

Toward the end of the twentieth century there were also various attempts to integration, such as Peter Weingart's 1 year collaborative workshop at the Zentrum fur Interdisziplinare Forschung (ZiF) in Bielefeld, Germany. (Actually, this could be called a meta-integration attempt, as will soon be apparent). In the early 1990s, Professor Weingart, the new director of this research center and an eminent sociologist, decided to make a serious effort to do something about the existing chasm between biology and social science—a phenomenon felt more globally especially after the start of the sociobiology controversy. Weingart's strategy was to bring together a number of leading researchers representing widely different approaches to address a common problem of their choice, the ultimate aim being an integrative edited book on a number of important topics. It was particularly valuable that the invited new residents at ZiF were able to meet each other in person on a daily basis and work on things together. This is how we ended up with the complex edited volume *Human by Nature* (Weingart et al., 1997), whose chapters were the outcomes of ongoing friendly debates within small cross-disciplinary teams, as well as the ongoing general discussion throughout the project. The volume contributors represented well-known scholars from a diverse number of fields: evolutionary psychology, anthropology, ecology, sociology, primatology, human ethology, biosocial science, history of science, history, linguistics, and medicine.

The ZiF project was also able to generate and fund two suggested cross disciplinary international conferences, one about nonverbal communication and another about the origin of language, both of which gave rise to rich multi-disciplinary edited books (Segerstrale & Molnar, 1997a, 1997b, 2016; Velichkovsky and Rumbaugh, 1996). The nonverbal conference brought together researchers from a wide number of fields from both biology and the human sciences. Many of the participants had been unaware of the research conducted by their cross-disciplinary colleagues.

This was also a time of a general changing attitude to "the gene"—it seemed that people in general had become more used to "gene talk", encouraged especially by the media which kept reporting on "genes for" such things as thrill seeking (Nelkin & Lindee, 1995). Meanwhile the intensity of the interchanges in the sociobiology controversy was diminishing—perhaps partly because E. O. Wilson together with Charles Lumsden with *Genes, Mind and Culture* (1981) had reinvented sociobiology so that it now encompassed culture as well (for discussion about that idea and its reception, see Segerstrale, 2000a). On the horizon, too, was the huge Human Genome Project—and of course, the Millennium.

This was also a time for some sociologists to explicitly voice the need for Darwin in social science journals, e.g., Maryanski (1994) and Lopreato (1984).

Meanwhile a shift was taking place in the general interest among evolutionary scholars—a move from altruism to cooperation. Cooperation was another one of Hamilton's four "basic" social behaviors, but he had mostly investigated it in one paper together with Robert Axelrod, identifying the conditions under which cooperation could come about and thrive—for humans as well as bacteria, most successfully when involving the strategy Tit for Tat (Axelrod & Hamilton, 1981). After Hamilton's death in 2000 his younger colleagues at Oxford picked up his mantle, deciding to look more closely into the workings of cooperation (e.g., West et al., 2007a, 2007b). For an interdisciplinary peek into the creativity of the evolutionary imagination, see Segerstrale (forthcoming), The Evolution of Cooperation and the Role of Hamilton's Rule.

Meanwhile one of those huge interdisciplinary and international Dahlem conferences, this time on the topic of cooperation, was arranged at the very beginning of the twenty-first century, bringing together a number of eminent evolutionary scholars in biology as well as the human sciences, and producing a very thoughtful edited book (Hammerstein, 2003). For a while there was quite a surge of interest in cooperation in the scientific world, including among economists who conducted various game theoretical experiments with surprising conclusions (Fehr & Gachter, 2002).

An issue that had meanwhile come to the fore among evolutionary scholars was the question how it had ever been possible for *large scale cooperation* to come about. Clearly the well-known mechanisms guiding kin selection and reciprocity were not sufficient in a world of strangers. A number of new concepts were suggested: "indirect reciprocity" (involving reputation for helpfulness) and "strong reciprocity" (punishment for non-cooperation). Also surprising was the general observation that among animals, reciprocity was not as common as people typically had imagined. Reciprocity worked mostly for humans (Clutton-Brock, 2009).

Another very fruitful development is the ongoing collaboration between Robert Boyd and Peter Richerson and others on what they call *cultural group selection*. They defend the position that our ancestors were indeed able to cope with large scale societies. Actually they are answering a bigger question about how human evolution actually happened—how did we survive in the first place?—and here the big point is that they point to culture as a driver. When humans needed a rapid change in behavior, culture came in handy, they were early on able to show this with examples (Boyd & Richerson, 1985; Richerson et al., 2003; Richerson and Boyd, 2005).

They have also been able to demonstrate this with mathematical models and provided plausible examples from studies of small scale societies and various current African tribes. One of their biggest points is that violence or group extinction is not necessarily involved—there are various reasons why groups dissolve and later regroup or get absorbed by other groups. Particularly their 2009 paper "Genes and Culture in a Time of Genomics" is full with potential ways to find out more, Richerson et al., 2009.

A huge paper in *Behavior and Brain Sciences* by Richerson and collaborators, presents the rationale of recent studies as well as a number of examples. Following the practice of BBS, Richerson et al. respond in detail to various queries, criticisms and challenges by an interdisciplinary group of peers. There seems to be backing from a number of fields, creating a sense of "consilience" (in the real Whewell sense of scientists from different fields coming to the same conclusion). (Richerson et al., 2016).We will return to the most recent developments at the end of this chapter.

4.10 Various Ways Forward (1)

The situation in regard to bringing evolutionary perspectives to bear on sociology is now much different from what it was around 1990 when Pierre van den Berghe wrote his famous verdict on sociological biophobia.

In the first place, ever since 2004, biology and evolution oriented sociological researchers have their own section in the American Sociological Association (#44), earlier called "Evolution, Biology and Sociology", but since 2023 going under the new name "Biosociology and Evolutionary Sociology". The section among other things sustains a newsletter and a website with useful information, sports a generously conceived online book display of relevant books from the interface between biology and the social sciences, as well as annual book and essay award competitions and various outreach activities. However, the section's membership number remains low, especially compared with long-standing and more typical fields, say, the sociology of culture, currently most popular of all 53 sections.

Since the 1990s, as we have seen, many developments have taken place, both within sociology and within biology, and in fields explicitly lying on the interface between these two. Biosocial science, founded by Tiger and Fox, for instance, can be traced back to the 1970s, as can various other hybrid fields with various hybrid names, all studying about the same things under their respective main field, such as anthropology (e.g. Chagnon and Irons, 1979).

What are the options currently available, then?

1. Neo-Darwinism in various forms, such as human sociobiology, human behavior ecology, evolutionary behavioral science, evolutionary sociology, and the like. The leading idea here is to sort social scientific concepts and theories under a broader neo-Darwinian framework, particularly invoking the explanatory force of inclusive fitness or kin selection.

   This is represented among others in books by Lopreato and Crippen (1999), Sanderson (2001), and Barkow (2006). Lopreato and Crippen already in their 1999 book (and in articles before that) advocated strongly for sociology to adopt as its overarching paradigm the neo-Darwinian idea that individuals strive to maximize their inclusive fitness. In this way, they suggested sociology would be safely linked to sociobiology, which in turn was linked to the Modern Synthesis, a stable scientific framework.

   Jonathan Turner and Alexandra Maryanski have tirelessly worked to bring biology closer to sociology, especially in the form of an evolutionary framework (Turner and Maryanski, 2008, 2018). It seems there has been special interest in launching the term "evolutionary sociology" (e.g. Maryanski, 1994, 1998). Most recently there is a massive launch of a 6 volume series edited by Turner and Machalek of a set of books with the aim of bringing biology closer to sociology (Turner and Machalek, 2018; Turner, 2020; Riley, 2021)

   One aspect of this move appears to be to patiently explicate the meaning of neo-Darwinism to underexposed sociologists, and present evolutionary explanations as actually useful for understanding some aspects of social life and especially human nature (Turner, 2020). Indeed, the collective recognition of an evolved human nature would go a long way toward accepting a role of biology for sociology, and maybe also toward bringing the social sciences closer to each other.

2. Evolutionary psychology. This can also be regarded as part of the neo-Darwinian paradigm, since this field can be seen as a version of E.O. Wilson style sociobiology (at least he thought so), while it later expanded and developed into a comprehensive explanatory framework of human behavior in its own right. According to evolutionary psychology, culture is not socially transmitted but rather evoked by various "cues" in different environments, which in turn are responded to by specially designed domain-specific mental mechanisms. Books here are among others Barkow et al. (1992), Buss (2005). (For an argument making the point that "much culture is not evoked", see Richerson & Boyd, 2005, p. 44).
3. Human ethology, which existed before sociobiology, has always paid special attention to aspects like development and learning. This field continued being broadly connected to traditional Darwinism. Important early representatives here were Patrick Bateson and Robert Hinde (1976), Irenaus Eibl-Eibesfeldt (1975), and Daniel Freedman (1979). Robin Dunbar writes popular books on the evolutionary origins of human behaviors (e.g. Dunbar, 1996). This inclusive field features the international organization ISHE, which among others organizes educational summer schools (For a peek into some interdsiciolinary evolution inspired research, see Segerstrale, forthcoming).
4. Culture-biology coevolution. This is a field which is a focal point for important cross-disciplinary research. After developing their original idea of dual inheritance, Boyd and Richerson (2006) have focused on the special role of *culture* as a driver of evolution, backing this up with anthropological data and different cross-disciplinary research (see e.g., the paper by Richerson et al., 2016, holding up well to a massive peer critique in the journal *Behavior and Brain Sciences*). One of their key points in favor of culture is that "little behavioral variation among groups is genetic" (Richerson & Boyd, 2005, p. 39). In fact, they devote their chapter two in their 2005 book to discussing false common beliefs and valid counterevidence. A central feature in Boyd and Richerson's approach is the idea of *cultural group selection*, which is theorized to come about in various ways, but—importantly—does not have to involve group extinction.

   Sociologist Brad Lovell Stone (2008) sees this approach as particularly promising for sociology. He suggests that this evolutionary theory of culture "nicely accommodates most of the traditional research interests of sociologists (83) and that "sociology is especially well positioned to contribute mightily to the empirical agenda of the evolutionary theory of culture. . . . The areas in which sociology can contribute uniquely include norm formation and internalization, the basic pattern of cultural variation, interaction rituals in the maintenance of group solidarity,.group assimilation, group differentiation and hierarchies . . . the institutional contexts for emotional expression . . . " [he adds a number of other things] (84). Of particular interest in the Boyd and Richerson approach is what they call "the tribal social instincts" which supplemented the "ancient social instincts" of kin selection and reciprocity. According to these authors, such social instincts arouse in the late Pleistocene through natural selection, because they made it possible to develop a cumulative culture which by being able to change faster than evolution made it possible for humans to adapt to the rapidly changing climate. Moreover, these instincts made it possible to create and sustain large scale human groupings (which was beyond the capability of kin selection and reciprocity). Later, complex societies found ways to create "work-around" institutions which mobilized those ancient tribal instincts for new purposes.

The tribal social instincts further facilitated conformist learning and "groupishness" or ingroup orientation, and even "moralistic punishment" for violating group norms (Richerson et al., 2003; Stone, 2008). Some of these ideas have got indirect support from experiments by behavioral economists as they discovered "altruistic punishment" (Fehr & Gachter, 2002).

Finally, how do we explain human intelligence, our most unique feature? It seems that our "costly" brains were selected for because they helped us cope with our social lives (Boyd & Richerson, 2006, pp. 72–73). Many have been trying to solve the mystery of the origin of our large brains, invoking some sudden "explosion" in the Pleistocene. Another approach has been to focus on the social function of the brain, some kind of "social brain" theory (e.g. Humphrey, 1976; Byrne & Whiten, 1988). Social ethologist Robin Dunbar finally famously found the answer (Dunbar's number). In primate species, the size of the cortex divided by bodyweight, correlates exactly with the typical group size of the species (Dunbar, 1996). For Boyd and Richerson the point with this is that there was no sudden intelligence explosion in the late Pleistocene. Instead, the evolution of human intelligence was a natural process.

5. There is an even strongly cultural evolution oriented approach, very capably handled by Mesoudi in his 2011 book Cultural Evolution. Mesoudi covers both the pioneering work of Feldman and Cavalli-Sforza, starting in 1981, and Boyd and Richerson's evolving projects ever since 1985, both cases involve models for various aspects of cultural behavior. Boyd and Richerson are meanwhile connected to researchers in various areas who can contribute valuable field observations of cultural change and cultural learning in small scale societies, and also studies of children's learning in this respect. They are developing various models of gene-culture coevolution—with culture rather than genes as a driver. More about this later.
6. The field of Biosociology has existed for some time already but not been much counted with by sociologists I believe. But currently is becoming more encompassing as it will include not only hormonal and physiological aspects but also genetic factors of various kinds (in the form, say, of behavioral genetics, or—see below!—the new field of sociogenomics). Probably also neural aspects belong here, as does nonverbal communication. One of the most steadfast and hardworking biosociologists is Allan Mazur, who already for a considerable time has conducted a number of interesting studies, e.g. on nonverbal communication, and written several books. He has done the kind of work which would seem to be

less controversial—dealing with our body's reactions to various situations, i.e. with proximate causes, which don't require any lofty evolutionary speculations (e.g., Mazur, 1981, 2004; Mazur and Robinson 1972). Another serious biosociologist is physical anthropologist Vernon Reynolds at Oxford who wrote a very useful book in this area as the sociobiology controversy was raging... But my favorite would be Robert Sapolsky, whose wonderful book *Behave!* (Sapolsky, 2017) I invite my introductory sociology students to flip around in to discover exciting things and then report to class. This is certainly a way to overcome biophobia before it has even started. Behave covers any imaginable aspect of biology from neurology to evolution with typical sociological situation as well as everyday happenings and sensational media stories thrown in and then explicated on a biological basis. He is a big believer in the explanatory role of hormones, especially testosterone...

4.11 Toward Peaceful Coexistence

As we see, within the first two decades of the twenty-first century, sociology became something of a "host' or coordinator of the various efforts to apply evolutionary theorizing to the social sciences. A good example is the Oxford Handbook of Evolution, Biology, and Society (2018) edited by Rosemary Hopcroft, then current chair of ASA's section of Evolution and Society. The title is intended to be "a catch-all phrase encompassing any scholarly work that utilizes evolutionary theory and/or biological or behavioral genetic methods in the study of the human social group", and ...contains a much needed overview of research in the area by sociologists and other social scientists". The summary of the book goes on to say that "the potential of theory and methods from biology for illuminating social phenomena is clear, and sociologists stand to gain from learning more about them and using them in their own work". To this end the Handbook included among others such relative novelties for social scientists as genetic association studies and hormonal level analyses. Meanwhile the topics covered by its authors seemed rather standard ones, except perhaps for the intriguing "determinants of human happiness".

It is of interest to see what happened at the 2019 ASA meeting in New York City in the session Author meets Critics. I will briefly cover a written version, later published as Rosemary Hopcroft's response to three critics' questions in the xxx section's *Newsletter*. Hopcroft starts off by saying she is pleased to be able to discuss the book, since—as one of her critics mentioned—such a discussion would not have been possible 30 years ago. She points out that "it has only been in the last two decades or so, that the use of evolutionary and biosocial theories of social behavior have begun to have some legitimacy in sociology." In the following I will present some of her key comments to her critics. This could be one possible indicator of the nature of the "rapprochement" between sociology and biology two decades into the twenty-first century.

One of her critics thought that the most useful chapters in the book addressed the interplay between biology and behavior, including gene environment interactions, but went to note that "many sociologists can keep doing what they are doing without ever learning more about evolution (or biology)".

The answer to this was as follows:

"It is true that sociologists whose work is far removed from the individual... do not need to explicitly use evolutionary theory in their work. Yet typically work in these areas, as in all areas of sociology, relies on assumptions about human nature and how people typically behave. For example, sociologists commonly assume that people value their families, follow group norms, and are concerned about fairness. Any such assumptions must be consistent with a human actor who is a product of evolution, just as work in chemistry must be consistent with what we know about individual atoms. Further, evolutionary theory can tell us WHY we have the human characteristics and tendencies we do. Why do we tend to value our families, why do we tend to follow group norms, and why do we tend to be concerned with fairness?" Hopcroft goes on to point out that an adequate explanation needs both types of explanation, both ultimate and proximate, answering both why and how something happens. Evolutionary answers deal with the why—especially involving kin selection. Why do people follow group norms? The answer is (according to sociologist and medical doctor Nicholas Christakis' book *Blueprint* (2012)?: *"We evolved to have a 'social suite' of predisposition that include evolved capacities for social learning and cooperation within the group that together promote the tendency to follow group norms"(emphasis added).*

Also, asked about criminal behavior, Hopcroft answers that an evolutionary approach can help give explanations for various ubiquitous aspects of criminal behavior, and especially why men tend to be disproportionately involved—this would not be possible using a Durkheimian functionalist theory. Another critic doubted that evolutionary biology was in line to become the new central paradigm in sociology. He got the following response:

1. Because evolutionary theory can give *ultimate explanations for many assumptions about human nature* that sociologists have made, it has the potential to *unify many disparate theories and areas of sociology.*
2. Meanwhile, it does not replace sociological theories, but rather gives *ultimate explanations for the proximate explanations that sociologists have developed (underlining added)*

She concludes by reassuring her critics that sociologists, with their knowledge of social structures and groups, will be essential in the twenty-first century. "So bringing in biology and an evolved human actor back in does not make sociology redundant in the slightest. But by adding biological variables and by adding ultimate explanations to the proximate explanations of sociologists, evolutionary theory and biosocial research can strengthen sociology." This I regard as a very responsible and rational handling of a situation where people typically have not quite internalized the idea of a natural complementarity between evolutionary and sociological approaches—exactly because sociologists know more about "how" things happen (or under what circumstances) than why they happen in the sense of a big why question.. On he other hand expecting every behavior to directly have something to do with fitness maximization is a too literal application—sociology—and social psychology—uses very serviceable observations and attributions of motives, etc. which makes us understand everyday life and others' motivations, as well as allowing for making general predictions . . .

4.12 Sociogenomics

Yet another new development is the field of Sociogenomics. This new field encourages collaboration between sociologists and genetics experts. The particular point with reaching out to sociologists is their survey research skills and training in regard to finding suitable measures and operationalizations for various sorts of queries. What is attempted here is nothing less than simultaneously getting hold of individual genomic data and the individual's psychological and social characteristics—information that can later be correlated and combined in various ways, even cross-culturally. A number of countries which are collecting national data or data on some groups, or on twins, are already participating in something called the SSGAC, social science genetic association consortium.

SSGAC organizes academic workshops, summer schools and graduate courses targeted at researchers from relevant backgrounds (economics, psychology, sociology, medicine etc), sometimes funded by foundations. While these events also strive to bridge knowledge gaps across disciplines, their specific aim is to introduce researchers to "key concepts, state-of-the-arts-methods, and computer tools in statistical genetics that can be applied in the social and medical sciences". Here can be found topics like: "the estimation and interpretation of heritability using molecular genetic data, genetic association studies, polygenic prediction, and identification strategies to isolate causal effects using genetic insights." (SSGAC website).

A recent job posting for a full-time 2-year research assistant position at UCLA says: "They're looking to hire students interested in social-science genomics: the identification of genetic variation associated with variation in behavior and other social-science-relevant phenotypes. Prior experience is not necessary. Research assistants should have an interest in applying to a quantitative PhD program (e.g., Economics, Statistics, Data Science, Genetics, Sociology) and pursuing an academic research career").(Job posting on the SSGAC website).

An article called "Sociology, Genetics, and the Coming Age of Sociogenomics" published in the authoritative *Annual Review of Sociology* in 2020 summarizes the advantages of this recently emerged new field. From chronicling "the post-2005 genome-wide association studies, the post-2015 Big Data era, and the emergence of polygenic scores" the authors go on to advocate the advantages of sociogenomics field for sociology:

"We argue that understanding polygenic scores, including their genetic correlations with each other, causation, and underlying biological architecture, is vital.

We show how genetics can be introduced to understand a myriad of topics such as fertility, educational attainment, intergenerational social mobility, well-being, addiction, risky behavior, and longevity. Although models of gene-environment interaction and correlation mirror agency and structure models in sociology, genetics is yet to be fully discovered by this discipline. .... We argue that sociogenomics can speak to long-standing sociological questions and that sociologists can offer innovative theoretical, measurement, and methodological innovations to genetic research" (Mills & Tropf, 2020).

However, as stated in the current SSGAC website:

> Because of the statistical power problem [see Sect. 4.6.4 in Benjamin et al., 2012], the SSGAC *only conducts studies on phenotypes that are available in large samples*. The SSGAC recognizes that there is a tradeoff between sample size and phenotypic quality and that when designing studies, it is important to assess this tradeoff quantitatively. *The first generation of studies conducted by the SSGAC are therefore conducted on phenotypes that are widely available.* We are currently prioritizing the following phenotypes (ul added):
>
> - Education
> - Subjective well-being
> - Fertility (number/timing of children)
> - Risk Aversion
> - Trust
> - Optimism
>
> (There is a note that the GWAS of fertility is being led by their partners, Sociogenome.)

What can be achieved? Here is a concrete example involving education. Daniel Benjamin and his international team were described as working on identifying variations in the human genome associated with how many years of education people get. This was the result of their 5 years of systematically increasing their database, as reported by *The Atlantic* in 2018:

> After scanning the genomes of 1,100,000 people of European descent—one of the largest studies of this kind—they have a much bigger list of 1,271 education-associated genetic variants. The team also identified hundreds of variants that are associated with math skills and performance on tests of mental abilities.

The obvious question is what such results are used for and who uses them? The *Atlantic's* attitude was optimistic, discussing various correlations, and various potential insights to derive from this study, some of these suggesting positive interventions or policy measures. But social scientists know about the nature-nurture debate of the 1920s and 30 s, the early 1970s controversy around Arthur Jensen, and the later 1984 Bell curve upheaval around Charles Murray and Richard Herrnstein. There have been at least three good reasons for resistance to purported measurements of IQ and educational ability: (1) they do not necessarily measure what they claim to measure, (2) they can easily be used for discriminatory purposes, (3) these kinds of "scientific" measures easily foster genetic determinism and the naturalistic fallacy, and the belief that "science" itself has spoken (see discussion of this in Segerstrale (2000a), *Defenders of the Truth*, Chap. 13).

4.13 Beyond the Genome

Meanwhile, there has been an initiative on another front, opposite seeming to the social genomics idea. A group of academics, keen on challenging the idea of the genome having privileged control over development and heredity has recently landed one of the biggest grants ever given to evolutionary research: a huge grant (5.7 million pounds) from the American Templeton Foundation for a project which involves a major rethink of how causation is understood in evolution.

This multidisciplinary, international team has been exploring what it calls "alternative perspectives" on the relationship between genes, organism, and environment, with special focus on the active role of the organism in evolution. The project has a complicated structure with major umbrella themes and lots of interlinked smaller projects; it involves colloquia and various opportunities for collaboration with additional institutions, a number of additional conferences, and the training of doctoral students. Its most unusual feature is probably the integrated collaboration with leading philosophers of science in regard to the understanding of evolution.

The team consists of 50 eminent experts from eight universities led by Kevin Laland (University of St. Andrews) and Tobias Uller (Lund University). They are supporters of what is called The Extended Evolutionary Synthesis (EES) (see e.g., Pigliucci & Muller, 2010), a notion that has been prominently up for scientific discussion e.g., in *Nature* (Laland et al., 2014).

At that time there were clearly two different views on how welcome an "extension" or modification of the Modern Synthesis would be. In a dialogue between proponents and critics of the EES in the *Natur*e issue, the group of "traditionalists" made several good counterpoints to the idea of radical reform. A later interview with one of the project leaders, however, made the whole thing sound less extreme:

> The extended evolutionary synthesis does not replace traditional thinking, but rather can be employed alongside it to stimulate research within evolutionary biology. The new perspective retains the fundamentals of evolutionary theory . . . but there are differences in how causation in biology is understood. As a result, some of he burden for adaptation and the diversity of life forms falls on the constructive processes of development. (Laland, interview in Uller group report).

An urgent task therefore has been to identify conceptual differences between the alternative and traditional interpretations, develop theory to fill the gaps in current understanding, and develop projects that could be regarded as critical tests for resolving points of disagreement. The first book published from this project was the book *Evolutionary Causation* (Uller & Laland, 2019).

4.14 Conclusions

We have here been exploring potential contexts around the pronouncements by Pierre van den Berghe, by looking into the conditions for and content of sociological research during the relevant time period. We started off by examining the early days of sociology in the United States and the kinds of expectations that were associated with being a sociologist. The sociologist was believed to be of help to society; identify its problems and help solve them; it was in many ways also a moral role. There continues being a lingering sense of this; sociologists want to be useful and promote social progress.

Being of use, at least for sociology's sister discipline psychology, landed that field straight in a massive project of IQ testing and IQ research, known as "the nature-nurture controversy"—for many this is THE nature-nurture controversy, lasting from about 1914 to 1940. That controversy involved the relative roles of inheritance and environment especially when it came to IQ, because that was a measurement tool when it came to streams of new European immigrants. There were quite fierce disputes about the importance of one or the other, psychologists took very fierce sides, and there were reports of even the same data being interpreted differently by the hereditarians and the environmentalists. In fact, Degler (1991) suggested that it often seemed that psychologists made up their minds less dependent on the data and more on the basis of personal belief.

Of course it was finally resolved that both nature and nurture are needed and that they are de facto inseparable. (But technically "finally" could draw out to 1940).

But there were other forces in operation. In the 1930s, the climate changed, with the Great Depression, and with a rising interest in eugenics. The Depression appears to have convinced many about the lack of a relationship between wealth and IQ—the disaster hit everybody. There was also the crucial influence of Franz Boas and his school in anthropology, and the launching of the concept of *culture*. Ruth Benedict emphasized the value of all cultures and the need to understand cultural relativism. Yale collected Human Relations Area Files. With Boas the emphasis was strong on culture, and Mead supported this with her Samoa study (1923).

Meanwhile we saw how the academic leadership in sociology moved from Chicago to Harvard in the 1930s and with that, a change in conception of sociology. University of Chicago's sociology had emphasized symbolic interactionism, a more "idealistic" type of sociology. Now Parsons was in charge at Harvard, founding the department of Human Relations which was operative from about 1946 to 1972, when it dissolved again into is various components: sociology, psychology, anthropology, and social psychology.

Looking at social scientists in general throughout this story, it would seem that the strongest voices have been carried by anthropologists and psychologists who have not shunned getting involved in veritable battles (I have in mind here especially the psychologists' intense "nature-nurture controversy" about IQ and race in the 1920s). Sociologists have been more of observers when it comes to major controversy (especially the sociobiology debate), but have impacted matters in other ways, for instance through critical commentaries and reanalyses of research results supporting hereditarian interpretations (Christopher Jencks' acclaimed reanalysis of the research data in the IQ controversy, 1975; Bellah et al.' (1995) critique of Herrnstein and Murray's (1994) book *The Bell Curve).* And we have seen various "internal" attempts to shake up or rectify the field, from Merton's call to action in

1957, to C Wright Mills' call for sociologists to fulfill "the promise" of their field (1959), to Pierre van den Berghe's frontal challenge of his fellow sociologists in regard to their "biophobia" (1981 and 1990).

And finally, of course, the various attempts over time of individual sociologists bravely pushing to make evolutionarily thinking into the ultimate umbrella for sociology's typical "proximate level" theorizing. Pioneers here were especially Lopreato and Crippen, individually and together (e.g., Lopreato & Crippen, 1999), Sanderson (e.g. 2001, 2014; Maryanski and Turner, e.g. 1993; Turner and Maryanski 2008). "Bringing in Darwin" was typically presented as the desired answer to what was widely experienced as a "crisis in sociology", an ongoing discussion in scholarly journals in which they eagerly participated themselves. A central figure in regard to supporting a more biology-friendly attitude is sociologist Rosemary Hopcroft who was instrumental in the original creation of the American Sociological Association section called "Evolution, Biology and Society," and also served as its president at crucial points.

Something that especially merits mentioning in this context is the bravery of the field of sociology itself. It should be congratulated for openly airing its serious internal disagreements in an ongoing discussion in journals and newsletters. The crisis has been discussed especially by the so-named book by Lopreato and Crippen (1999), who appear to have at least two candidate occasions for the crisis: one in the mid 1970s, and one in the 1990s, but it may simply have been one long crisis situation sitting on the back burner, which flared up from time to time.

The impression that sociology as a field was looking for a way to reinvent itself and gain greater scientific respectability can also be gathered from the ASA presidents' addresses at the association's yearly meetings. One example is Stanley Lieberson who protectively argued that the social science as a science needed to be probabilistic rather than deterministic, so as not to lend itself to easy falsifications (Lieberson, 1992; Lopreato and Crippen, 1999, 76). Then in 2007 there was radical ASA president Troy Duster, who kept the anti-racist torch lit by discouraging engagement in research that involved genetic aspects of behavior. He probably had in mind a particular recent issue of *American Journal of Sociology* which had been devoted to a discussion of the rapidly expanding new methods in genetics, which promised a wealth of data for behavioral researchers for possible correlations between genes and their "phenotypes' (behaviors or traits).

Much has been made of Talcott Parsons and his particular approach to sociology, which was often seen as being built on unrealistic assumptions of a goal of social order. C W Mills already made fun of the would-be "Parsonian paradigm" as Grand Theory combined with Abstract Empiricism, without these two necessarily connecting. According to Robin Fox, there was a lot of resistance by his students, but Fox says the debate was always intellectual, not ideological, in contrast to the later relativist and postmodernist activism in the discipline (Fox, 1995). Among Parsons' many illustrious students (which included Pierre van den Berghe) was also Harold Garfinkel, who pulled the ultimate trick on his teacher by questioning the very core of the Grand theory: social order.

Garfinkel just dismissed Parsons' well-socialized individuals as passive "culture dopes" and suggested instead that social order had to be constantly maintained by society's members, which gave it a precarious status. More important was surely the alternative theory by Parsons' colleague George Homans, which put the individual rather than society first, building forward towards an exchange theory. However, it retained the "environmental" or "cultural" tabula rasa view of the individual. In fact, Homans was a great admirer of his Harvard psychology colleague B F Skinner's theory of behavioral conditioning. In the 1950s and '60 s Skinner was regarded as the model for serious science in the social sciences at Harvard, since his approach allowed for prediction and control. Until, of course, it turned out that behaviorism's claims were wrong. Sustained experiments with animals started showing that their natural instincts sometimes overrode even careful conditioning.

Bigger background factors ("social issues" of various magnitudes) affected the theoretical situation for the social sciences in various ways—the massive immigration to the US, the Great Depression, the Nazi implementation of race theories and eugenics, the Holocaust. Among social scientists and many other academics there was a natural post World War II aversion to biology, and one can understand Parsons' own declaration of himself as a "cultural determinist" (following Boas and Kroeber). Skinner was still going strong upholding behaviorism, and Boas and his students were upholding culture, dismissing any biological origin of human universals. Sociology looked like the kind of field that E.O. Wilson had in mind with his explicit threat of "cannibalizing" the social sciences in his *Sociobiology: the New Synthesis* book in 1975.

Meetings in which social scientists were participating were held at Harvard and elsewhere, they were about the social danger of again introducing biological explanations about human behavior. Terrible accusations were made. Social scientists were not among the most militant as critics; the activists were mainly from the Sociobiology Study Group, which included Wilson's close biology colleagues Gould and Lewontin. Among the social scientists there was a deep sense of shock for all who had been seeing social science as fundamentally cultural and biology as a potential biological determinist danger. This was clearly manifest in the written and oral proclamations of the critics. Sociologists were completely untrained in biological reasoning at this point, and many felt revulsion, although others were curious. But for various reasons the critical initiative was completely left to the left wing defenders of the (scientific and moral/political) truth, and their arguments about dangers sounded true and convincing. In many respects, Pierre van den Berghe was right.

So what might a sociologist do about the situation? C Wright Mills would have said: use your sociological imagination! Do your best to try to find out what is actually going on! Use any method you think will work. Van den Berghe believed that sociologists don't like to deal directly with people, but rather with aggregated data. Still, interviewing is a legitimate tool of sociologists, it is part of their methods tool box. So in principle you should be able to interview anybody, simply as a sociologist doing research. So what about taking the bull by the horn—find out what those unfriendly seeming sociobiologists really know and think—and what is really going on? In a few years time, as the controversy grew, I ended up doing just this: interviewing protagonists on both sides of the debate, both in the US and Britain, and reading evolutionary biology to understand what they were talking about. To make sense of what people say, you have to use both empathetic understanding and analytical explanation. This is a strength of sociology. The result was first my dissertation and later a book *Defenders of the Truth*, which followed and analyzed the sociobiology controversy up to the year 2000.

As a specialist on this particular nature-nurture controversy (and the preceding IQ controversy which merged with the sociobiology one) I can say the following: The vehement and relentless polemic in the sociobiology controversy was not between different fields (such as sociology and biology), which is how one would maybe typically imagine such a controversy. Instead it was *within the very same field, biology*. The debate had to do with both scientific and moral/political beliefs on both sides, and these tended to be coupled together, so that a particular scientific position (say some sociobiological view, e.g. on adaptation) was automatically regarded as having a political basis and therefore classifiable as "bad science". The vicious nature-nurture controversy in psychology in the 1920s was also *within* a field, not between fields, and so was the early twenty-first century's controversy in anthropology around Napoleon Chagnon and his Yanomamo Indian research (see e.g. his own report, Chagnon, 2013).

In other words, different beliefs in nature vs. nurture may not be the only, or even main, reason for controversy in conflicts that look like nature vs. nurture debates. Each case needs separate investigation. I think I will end with one of my favorite quotes as to what may be going on in science in general—including social science. This insight comes from the French sociologist Pierre Bourdieu:

> The definition of what is at stake in the scientific struggle is thus one of the issues at stake in the scientific struggle, and the dominant are those who manage to impose the definition of science which says that the most accomplished realization of science consists in having, being, and doing what they have, are, or do (Bourdieu, 1975).

Appendix: The Sociobiology Debate, an "Atypical" Nature vs. Nurture Controversy

Are Nature-Nurture Controversies Actually Worldview Conflicts?

So-called "nature-nurture" or gene-environment controversies in science have been commonly viewed as politically motivated conflicts between conservative hereditarians and left-leaning "environmentalists" (later also called "culturists"). It may appear so, and the polemics in such controversies seemingly follow such a model. During the second part of the twentieth century the nature of human nature was debated in at last three such big controversies in an American or Anglo-Saxon setting, but with parallels in other countries as well.

The trigger in the first controversy was a number of popular ethology books on the innateness of traits such as aggression and territorial behavior (especially Ardrey, 1961, 1966), but it was especially Konrad Lorenz' *On Aggression* (1966) that became the focus of a broader debate. Lorenz declared that aggression was an inbuilt instinct also in human, and went on to discuss various ways to come to grips with this unfortunate fact. Then a few years later began the infamous IQ debate around the views of Arthur Jensen, an educational psychologist at Berkeley, who in an article in *Harvard Educational Review* had suggested that race, too, might to be included as factor when assessing educational performance (Jensen, 1969). And then in 1975 erupted the fierce sociobiology controversy around Harvard zoologist Edward O. Wilson's *Sociobiology: the New Synthesis*, an illustrated and popularly written "coffee table book" surveying recent research on social behavior of animals in the wild, and bringing in theories suggesting the behavior's genetic foundation.

There is continuity between these controversies, not only in a thematic but also in a real physical sense. Even though the object of criticism has varied (from ethology—Lorenz, to psychology—Jensen, to sociobiology—Wilson) it has often been the very same core critics involved, and they have wielded very similar arguments from one debate to the next. I would actually argue that we here have *two different total world views in conflict*; it is not a question of scientific evidence or criteria for a scientific theory, and it is not a question of mere paradigms in the Kuhnian sense (i.e. kept largely to an epistemological realm, however diffuse), but these are two conflicting separate whole systems of reasoning, each with their interconnected scientific and moral/political considerations, which is why I see "world view" as the best way to describe this.

While I do not think that this situation is the typical case for nature-nurture controversies on closer scrutiny it may turn out that something similar is going on in every one of them. In any case, here I am presenting a summary of my insights from the sociobiology controversy. It may perhaps be used as a model against which to compare other less overtly politicized or value-laden controversies. It is seldom, I believe, that opponents are so explicit in their criticism of each other.

The Promises and Dangers of Sociobiology

It is over half a century since the beginning of the famous sociobiology controversy began around Harvard zoologist E. O. Wilson. "Sociobiology" was defined as "the systematic study of the biological basis of all social behavior". The idea was that just like morphological features, animal behavior, too, could be analyzed as undergoing evolution. This in turn implied that it would now be possible to include *behavior* as well in the new Neodarwinian paradigm, which expressed evolution mathematically as a change in gene frequencies in populations. Through this effort Wilson hoped to continue the work of the architects of the Modern Synthesis (which took place between ca 1920 and 1950) (Mayr and Provine, 19 5, Segerstrale encyclopedia entry on Neodarwinism) in helping put all of evolutionary biology on a quantitative foundation and make it a real science.

However, it was *Sociobiology's* last chapter that gave rise to the infamous sociobiology debate, destined to last to the end of the century, when it largely merged with a conflict about evolutionary psychology, a new version of sociobiology oriented toward the workings of the human mind rather than invoking behavioral genes. The worst offender in this genre was probably Richard Dawkins who in *The Selfish Gene* (1976) vividly illustrated the idea of genes manipulating human behavior in order to replicate themselves using whatever images came to his mind (including Chicago gangsters). That book, just like Wilson's, was actually aimed at enlightening the general reader about exciting new research on behavior. The critics, however, saw danger lurking in the mere existence of these books.

Here a biological aside: It was in England that a major scientific breakthrough had taken place in theoretical biology in the 1960s. The long-standing mystery of animal altruism had been finally solved by British evolutionist William (Bill) Hamilton in 1964. His clever move was to shift the focus from the survival of the individual to the survival of the individual's *genes* instead, which meant this automatically involved other likely gene carriers, i.e. relatives. Interaction and helpfulness between related individuals could now dramatically explain the evolution of altruism, Darwin's unsolved puzzle. With the help of simple cost-benefit calculations and taking into account the relatedness between a helper and the recipient of a helpful act, Hamilton showed that from a gene-spreading point of view it made sense for a bird, say, to sacrifice itself by letting out an alarm call, if it in this way would be able to save a sufficiently large bunch of sufficiently close relatives. This caused a sharp shift from the prevailing research paradigm of "group selection" to the new paradigm of "kin selection" instead, and this would become the dominant paradigm for the next 40 years. This led to great flourishing in theory and research on social behavior (e.g., Trivers' "reciprocal altruism" or reciprocity (Trivers, 1971; and Hamilton and Axelrod's evolution of cooperation, and Tit-for-Tat strategy, 1981).

Growing Protests

What especially upset critics about *Sociobiology* was the last chapter which described and postulated all kinds of hypothetical genes also for human behaviors. Meanwhile Wilson's particular writing style did not always make clear what was fact and what speculation, which made the situations worse. Because for some critics, mere talk about genes would already trigger dangerous "biological determinist" thinking. A group of academic critics, "The Sociobiology Study Group", decided to take action by publishing a strongly worded letter in the popular *The New York Review of Books*, where they directly linked Wilson's *Sociobiol*ogy to conservative politics, racism, and Nazism (Allen et al., 1975). It surprised many that this group included several of Wilson's own Harvard colleagues, especially fellow-evolutionists Richard Lewontin and Stephen J. Gould. (Wilson later said he had not seen this coming at all and felt betrayed).

Now followed protest meetings and demonstrations at Harvard and elsewhere. The air was suddenly thick with anti-sociobiological sentiment. People were quite upset. A number of discussion and denouncement meetings were attended by people from all academic fields, particularly social scientists. The general sense was that Wilson had been violating an existing social taboo on dealing with humans in biological terms at all. The implicit understanding was that human behavior was to be legitimately discussed and explained in exclusively "environmentalist" terms—cultural ones, with Margaret Mead or behavior ones, with B.F. Skinner—but not biological, and certainly not genetic!

A more comprehensive critique of Wilson, named "Sociobiology – another biological determinism" was later produced in *BioScience* (Sociobiology Study Group of Science for the People, 1976), together with a response by Wilson, 1976). There Wilson was compared to earlier "determinists" such as Herbert Spencer and Konrad Lorenz. Wilson's emphasis on universal human traits was ridiculed. Wilson retorted by emphasizing the danger of complete environmentalism, which seemed to him the critics' position. "If culture is all that created human rights... then culture can equally well validate their removal, he warned, and went on to derive support from famous left wing academic Noam Chomsky, who had recently said he "considers extreme environmentalism to be a belief susceptible to dictatorships of both the left and the right" (Wilson, 1976).

When I said that the air at Harvard was thick with anti-sociobiological sentiment, where did I get that from? The answer is: I was there. I was a new graduate student in sociology, very much sharing in the general sentiment. It was overwhelming. I followed the action as I could as a "participant observer". It was only later that I would take a second look at this controversy as a detached sociological researcher.

The controversy about sociobiology soon scaled up to the national level in the form of a Sociobiology Symposium at the American Association for the Advancement of Science (AAAS) in Washington, D. C. in 1978. There a small group of activists from the International Committee against Racism (INCAR) suddenly rushed up on the podium displaying various slogans. One of them grabbed a conference pitcher of ice-water and poured it in Wilson's neck (he was sitting down), shouting: "Racist Wilson, you can't hide, we charge you with genocide!" Outside the symposium ballroom members of the Boston area Sociobiology Study Group, which had by then joined the national organization Science for the People offered various articles for sale, as well as issues of the new *Science for the People* magazine.

It would seem that most people at the time just bought into the critics' view of Wilson's political motives. Few actually read the book. In 1975 it was clearly too early to even mention the possibility of a biological basis for human behavior. There was still a strong prevailing "environmentalist" paradigm after the Second World War. Freud was very popular. And here now came Wilson suggesting in his last chapter that our most specific human characteristics could actually have a genetic basis: all the way from sex role divisions and aggressiveness to moral concerns and even our religious beliefs! Anathema!

Later the sociobiology debate became a major transatlantic scientific controversy over general principles of evolutionary biology as a science (particularly about the unit of selection and the role of adaptation as an evolutionary force). Now it was Dawkins and Steven J. Gould, rather than Wilson and Lewontin, who emerged as the chief duelists. The original debate about humans continued in a new guise as a conflict about "evolutionary psychology" a take-off from sociobiology. In the following I will be limiting myself to some core aspects of the complex sociobiology debate. Many more pieces of the puzzle are covered in my book *Defenders of the Truth: the Battle for Science in the Sociobiology Debate and Beyond* (Segerstrale, 2000a).

Bad Science Is Ideological and Leads to Bad Consequences

The members of Science for the People were genuinely convinced that sociobiology was, indeed, evil. The working logic of the critics involved a type of "cognitive coupling" between three things: bad science, ideological influences, and bad consequences. There was also a clear connection between the critics' criticism of sociobiology and their conception that "bad" science would be socially abused.

What then was "bad" science? That turned out to be the kind of science that the critics disliked: sociobiology, behavioral genetics, IQ research. Bad science was never the kind of science that the critics did themselves in their own labs. At the same time the following type of reasoning could be found in the critics' writings: IQ researchers and sociobiologists "could not" be interested in genuine science, because real science had to do with the molecular level. Therefore those researchers "must" be pursuing their research for ideological reasons—which could also explain why their science was bad. Quod erat demonstrandum!

In other words, the critics viewed it as a mora/political offense for their targets to publicly state their professional scientific views—because those views were erroneous from the critics' point of view!

Some examples: the idea of adaptation by natural selection was interpreted as a political conspiracy. Dawkins and others later tried to explain that the assumption that a trait is adaptive was a mere heuristic that made it possible do research at all (e.g., Dawkins 1982). For the group of critics, such a possibility was never even considered. Instead, major artillery was used already in the first letter of the Sociobiology Study Group, which charged that "for Wilson, what exists is adaptive, what is adaptive is good, therefore what exists is good." Gould and Lewontin later went on to criticize "the adaptationist program" in a series of scientific papers (e.g. Gould & Lewontin, 1979).

The Responsibility of the Scientist

Another issue dividing the two camps was the social responsibility of the scientist. The critics feared that any claims about genetic differences between individuals would be politically abused. Lewontin, the leading critic of Wilson, formulated the concern as follows:

At present our ignorance on this question is so enormous, our investigatory techniques so primitive and weak, our theoretical concepts so unformed, that it is unimaginable to me that lasting, serious truths about human nature are possible. On the other hand the need of the socially powerful to exonerate their institutions of responsibility for the problems they have created is extremely strong. Under these circumstances any investigations into the genetic control of human behaviors is bound to produce a pseudo-science that will inevitably be misused. (Lewontin, interview, the *Harvard Gazette*, November 3, 1975).

It was easy to see that Lewontin had a point. Theory-wise one had only to mention the "mismeasurement" of skulls resulting in racist and sexist theories (Gould, 1981), and various biologically based theories of racial supremacy. Policy-wise one could mention sterilization laws. And one might refer to the earlier nature-nurture controversy, which appears not to have been resolved by clear evidence.

Lewontin was not alone. This was actually the way the whole group of critics reasoned. In the sociobiology controversy could actually be found a rather unusual category of scientific practitioners, who in addition to their own scientific work regarded it as their duty to debunk what they saw as other scientists' "bad science" in fields often far away from their own. I have characterized this group as "weeders." In contrast, "planter" scientists typically did not feel that they needed to take action beyond their standard practice and left it to the democratic social process to decide about the use of scientific knowledge. This is why it came as a surprise when they were accused of "bad science".

The weeders turned weeding into a virtue. Steven J. Gould devoted himself to warning about the "mismeasure of man" and advocating "debunking" of false claims as a positive science (Gould, 1981) while Lewontin with two coauthors in *Not in Our Genes* (Lewontin et al., 1984), strove to demonstrate that the factual claims of the sociobiologists and IQ measurers were wrong.

Where did the weeders get their negativistic style from? The answer is: they deeply mistrusted the political system. Their political beliefs typically coincided with the New Left ideology of the 1960s and the political tenets of the American student and peace-movements. The American left-wing radicals had been influenced by the thinking of the European Critical School, which saw a power-elite manipulating the members of a mass society, which is why we should be alert and do Ideologiekritik! Ironically, from a European traditional Marxist point of view, most members of Science for the People would hardly be called "Marxists". Wilson, meanwhile, just chose the convenient label "Marxist" for all his opponents.

Biology as a Guide for Values

What was it that Wilson himself wanted with *Sociobiology?* It turns out that Wilson had a highly unusual agenda of his own. It was this agenda that made him write in a particular style, and sometimes move from "is" to "ought" in his writings—just as the critics said. But this program was not "political" in the narrow sense. Rather, it was a grandiose long-term plan for science and mankind. Underlying all this was a noble goal: to secure the future of mankind and life on Earth. With the help of biological insights into the truth of human nature, we would be able to make wiser choices and steer away from unfeasible cultural courses, perhaps even self-destruction. Wilson here continued an earlier theme from the 1960s of Lorenz and others.

My early interviews with Wilson made clear an additional personal reason. One of his larger goals was to establish "a genetically accurate and therefore completely fair code of ethics" as an alternative to the teachings of what he called "the theologians." He wanted to oppose the artificial teachings of religious leaders to what he regarded as a natural species-specific *human* ethical code. A former fervent Southern Baptist, he abandoned the Church as a teenager (Segerstrale, 1986).

"Politically Correct" Evolutionary Biology?

Just like Wilson, Gould and Lewontin, too, were pursuing a complex, combined moral and scientific agenda. Connected to their scientific objection to what they called "adaptationism" there was a moral/political concern. If everything is optimally adapted, there is no point in changing society. But if you instead of adaptation emphasize discontinuity, contingency, and chance, you indicate that in a radically new environment new types of individuals will flourish. It is not a question of the selection of the fittest, everybody gets his chance. Lewontin had even found an apt Biblical quote (from Ecclesiastes) to illustrate their position: "The race is not to the swift, nor yet bread to the wise ... but time and chance happeneth to them all."

On this view, Gould's continuous scientific search for theoretical alternatives to the adaptationist program can be seen as one long argument for social reform and social justice. Punctuated equilibria was his (and Niles Eldredge's) first dent in the adaptationist bulwark (Eldredge and Gould, 1972). Later Gould emphasized chance and contingency (historical accident) even more, arguing that in the Cambrian explosion many different phyla existed which later went extinct. The organisms that humans come from are traceable to only a single one of these simple organisms. In other words, humankind may never have come about. We came here entirely by chance (Gould, 1989).

If we look at Gould and Lewontin's overall strategy, then, it seems to have been to develop *the kinds of scientific arguments which, IF taken as social belief and IF acted upon, will conceivably have politically desirable social implications*! (Everybody getting their chance). They are simply taking the nasty social implications bull by the horns by providing "correct" science for the people.

Interestingly, Wilson, too, pursued this kind of agenda of "politically correct" science, at least part of the time. Already in Sociobiology he played down the significance of IQ and dismissed race as a useful biological concept. In his next book, *On Human Nature* (1978), he discussed "the cardinal value of the survival of the human gene pool" and the need for genetic diversity. He argued for universal human rights, "because we are mammals", and even observed that "the long-term consequence of inequity will always be visibly dangerous." He topped this off by advocating a more liberal sexual morality and defending homosexuality as socially useful. In other words, in blatant contrast to the critics' accusations of Wilson, the social implications he himself derived from his own theories were always liberal!

So we can see that both sides in the American sociobiology debate followed various types of normative strategies: they were both pursuing science and values together (deriving values from science, criticizing undesirable science, producing "politically correct" science, and so on). In contrast, the participants from the United Kingdom tended to adopt the exact opposite strategy: a strictly objectivist one, striving to keep science and values apart (e.g., Dawkins and Maynard Smith).

But there was a complication with Wilson. While he was naturally normative when writing in the spirit of his general moral-cum-scientific agenda, he shifted into an objectivist mode when he was writing as a scientist. Wilson had two hats, and it was not always clear which one he was putting on.

How Typical Is the Sociobiology Controversy as a Nature-Nurture Controversy?

So here we have covered various aspects of the sociobiology controversy, showing the quite heavy involvement of moral/political considerations on both sides. I think this may be an extreme case of such concerns, although they are probably an ingredient in a number of nature-nurture controversies. Indeed, in regular scientific disputes there might sometimes be additional moral/political reasons why a particular scientific view is preferred—or there may not, since scientists can have very strong views on scientific matters for epistemological or metaphysical strictly within the realm of science. So it is unusual to have scientists present their own moral/political beliefs, or attack what they claim to be those of their opponents—but in the case of the sociobiology and IQ controversies we do have clash of total world views, involving both scientific and moral/political convictions and going as far as to acknowledgement (or not) of particular facts (for details, see Chap. 13 in Segerstrale, 2000a, see also Segerstrale 1998).

One more assumption needs to be made, and that is that the mere statement of a fact implies a command for action. This was the constant concern of the opponents of sociobiology. In their view, the mere suggestion of a genetic basis for behavior would make people commit the naturalistic fallacy in their own beliefs and also affect social policy recommendations. That is why the critics were very negative even about plausibility arguments involving tentative facts, for instance in regard to IQ. And they may have been right. But what if there were to emerge irrefutable truths of some kind, say, on racial differences? I asked that question. One answer that I got from a leader of Science for the People was quite straightforward: "Then I would have to become a racist". (He immediately added that no such facts existed). That suggested to me that there may indeed be a sense of expectation to do something when you have a supposed fact, however tentative.

I then understood why there had been such alarm among the critics also when it came particularly to the issue of sex role differences (as dealt with briefly by Wilson in his *On Human Nature*, 1978). There Wilson explicitly stated his professional view, that there are probably some small differences, but added that he could also be wrong about this. He went on to point out that whatever the fact may be, "we" (society) can decide what to do with it. We can decide to enhance the differences, suppress them, or do nothing. This was an example of Wilson's own distinction between a scientific fact and its social implications, something that his opponent did not share, since they tended to focus on the potential abuse of scientific facts. So this is why they were in uproar about *On Human Nature*. I pointed out what Wilson had actually written about alternative actions. This had no effect. For my interviewee the important thing with Wilson was the simple fact that *"he said it*!" (something that he obviously shouldn't have done). Wilson's own qualifications of the case could be disregarded, because the cat was now out of the bag! Indeed, Wilson continued being cited and strongly attacked by Science for the People, always without his qualifications.

Obviously something would have to be done about this kind of perceived connection between biological facts and their negative social implications, particularly if this perception is widely shared in a society, and if there is a history of a particular interpretation of their meaning (which there is). However, the approach chosen by the critics of sociobiology and IQ unfortunately missed a good opportunity to do some social good. They could have been the "bearers of truth" about the long-standing gene-environment opposition having recently been recently overridden by the idea of gene expression—both are involved together. One could well guess that this was yet general knowledge among academics and even less the general population at the beginning of the 1970s, as the IQ controversy and the sociobiology controversy erupted. Instead they chose to in their rhetoric deliberately adopt exaggerated reasoning in terms of genetic determinism and gene-environment opposition, which they knew were by now completely false, for the hope of generating suitable political outrage. So through their participation *as scientists* as leaders of the opposition, and scoring moral/political points with many, they only helped reinforce existing mistaken beliefs and postpone the public's—and many academics'—understanding of recent biology, which could actually have been presented instead as conveying the positive message that the *environment* mattered!

Therefore, there was a strange situation at Harvard, where *biologists* were playing the crucial parts of both proponents and opponents, with only a few appearances of anthropologists (DeVore, Trivers), I do not recall any major impact of sociologists. This why I think that the sociobiology controversy (at least as it unfolded at Harvard and later during its first years) was not about nature-nurture as such, but rather about the projected social implications of biology, assuming the validity of "older" superseded truths. The local academics in general were not that ignorant, of course—many knew and felt that the wild accusations against Wilson had no merit, but nobody in his department or larger proximity came openly to his defense. In general, the sociobiology controversy was not handled well. I think people hoped it would just go away. When the sociobiology controversy became the project for my doctoral dissertation I had great difficulty finding anybody to be on my committee. In the end I found one brave political sociologist, one genuinely interested historian of science, and one kind former member of the Social Relations department. It turned out to be a friendly group who gave me just enough commentary to keep me on track.

References

Allen, E., et al. (1975, November 13). Against sociobiology. Letter. *The New York Review of Books, 182*, 184–186.

Ardrey, R. (1961). *African genesis*. Atheneum.

Ardrey, R. (1966). *The territorial imperative. A personal inquiry intio the animal origins of property and nations*. Atheneum.

Axelrod, R., & Hamilton, W. D. (1981). The evolution of cooperation. *Science, 211*, 379–403.

Barkow, J. (Ed.). (2006). *Missing the revolution: Darwinism for social scientists*. Oxford University Press.

Barkow, J. H., Cosmides, L., & Tooby, J. (1992). *The adapted mind*. Oxford University Press.

Bateson, P. P. G., & Hinde, R. (1976). *Growing points in ethology*. Cambridge University Press.

Bellah, R. N., Madsen, R., Sullivan, W. M., Swidler, A., & Tipton, S. M. (1995). *Habits of the heart: Individualism and commitment in American life*. University of California Press.

Benedict, R. (1934). *Patterns of culture*. Houghton and Mifflin.

Boas, F. (2011). *The mind of primitive man*. Macmillan.

Bourdieu, P. (1975). The specificity of the scientific field and the social conditions of the progress of reason. *International Social Science Information, 14*(5–6), 19–47.

Bowles, S., & Gintis, H. (2004). The evolution of strong reciprocity: Cooperation in heterogeneous populations. *Theoretical Population Biology, 69*, 17–28.

Boyd, R., & Richerson, P. (1985). *Culture and the evolutionary process*. University of Chicago Press.

Boyd, R., & Richerson, P. (2006). Solving the puzzle of human cooperation. In S. Levinson (Ed.), *Evolution and culture* (pp. 105–132). MIT Press.

Breland, K., & Breland, M. (1961). The misbehavior of orgnisms. *American Psychologist, 16*(November), 681–684.

Brigham, C. (1923). *A study of American Intelligence*.

Brigham, C. (1930). Intelligence tests of immigrant groups. *Psychological Review, 37*(March).

Buss, D. (2005). *Evolutionary psychology: The new science of mind* (2nd ed.). Allyn and Bacon.
Byrne, R. W., & Whiten, A. (1988). *Machiavellian intelligence: Social expertise and the evolution of intellect in monkeys, apes and humans*. Oxford University Press.
Cavalli-Sforza, L. L., & Feldman, M. W. (1981). *Cultural transmission and evolution*. Princeton University Press.
Chagnon, N. A. (2013). *Noble savages: My life among two dangerous tribes—The Yanomamo and the anthropologists*. Simon and Schuster.
Chagnon, N. A., & Irons, W. (Eds.). (1979). *Evolutionary biology and human social behavior: An anthropological perspective*. Duxbury.
Christakis, N. A. (2012). *Blueprint: The evolutionary origins of a good society*. Little, Brown Spark. Chronicle of Higher Education (January).
Clutton-Brock, T. (2009). Cooperation between non-kin in animal societies. *Nature, 42*(7269), 51–57.
Cole, S. (1994). Why sociology doesn't make progress like the natural sciences. *Sociological Forum, 9*, 133–154.
Collins, R. (1975). *Conflict sociology: Toward an explanatory science*. Academic.
Dahrendorf, R. (1959). *Class and class conflict in industrial society*. Stanford University Press.
Davenport, C. (1911). *Heredity in relation to eugenics*. Henry Holt.
Dawkins, R. (1976). *The selfish gene*. Oxford University Press.
Dawkins, R. (1982). *The extended phenotype. The gene as unit of selection*. Freeman.
Degler, C. (1991). *In search of human nature: The decline and revival of Darwinism in American social thought*. Oxford University Press.
Dunbar, R. (1996). *Grooming, gossip, and the evolution of language*. Harvard University Press.
Eibl-Eibesfeldt, I. (1975). *Ethology: The biology of behavior* (2nd ed.). Holt, Rinehart and Winston.
Eldredge, N., & Gould, S. J. (1972). Punctuated equilibria: An alternative to phyletic gradualism. In T. J. M. Schopf (Ed.), *Models in paleobiology*. Freeman Cooper.
Fehr, E., & Gachter, S. (2002). Altruistic punishment in humans. *Nature, 425*, 137–140.
Fox, R. (Ed.). (1975). *Biosocial anthropology*. John Wiley & Sons.
Fox, R. (1995). State of the art/science in anthropology. In P. Gross, N. Levitt, & M. Lewis (Eds.), *The flight from science and reason* (pp. 327–345). The New York Academy of Science, Johns Hopkins University Press.
Freedman, D. (1979). *Human sociobiology: A holistic approach*. Free Press.
Garfinkel, H. (1976). *Studies in ethnomethodology*. Prentice-Hall, Inc.
Gould, S. J. (1981). *The mismeasure of man*. W. W. Norton & Co.
Gould, S. J. (1989). *Wonderful life*. W. W. Norton & Co.
Gould, S. J., & Lewontin, R. C. (1979). The spandrels of San Marco and the Panglossian paradigm: A critique of the adaptationist programme. *Proceedings of the Royal Society of London B, 205*, 581–598.
Grant, M. (1916). *The passing of the great race*. C. Scribner's sons.
Hamilton, W. D. (1964). The genetical theory of social behavior. I and II. *Journal of Theoretical Biology, 7*, 1–16; 17–1–16; 32.
Hammerstein, P. (Ed.). (2003). *Genetic and Cultural Evolution of Cooperation*. Cambridge, MA/London: MIT Press.
Herrnstein, R. J., & Murray, C. (1994). *The bell curve: Intelligence and class structure in American life*. Free Press.
Homans, G. C. (1964, December). Bringing men back in. *American Sociological Review, 29*(6), 809–818.
Hopcroft, R. (Ed.). (2018). *Handbook of evolution, biology and society*. Oxford University Press.
Horowitz, I. L. (1993). *The decomposition of sociology*. Oxford University Press.
Humphrey, N. (1976). The social function of intellect. In P. P. G. Bateson & R. Hinde (Eds.), *Growing points in ethology*. Cambridge University Press.
Jensen, A. R. (1969). How much can we boost IQ and scholastic achievement? *Harvard Educational Review, 39*, 1–123.

Laland, K., Uller, T., Feldman, M., Sterelny, K., Muller, G. B., et al. (2014). Does evolutionary theory need a rethink? *Nature, 514*(October 8), 161–164.
Lewontin, R. C., Rose, S., & Kamin, L. (1984). *Not in our genes*. Pantheon Books.
Lipset, S. M. (1994). The state of American sociology. *Sociological Forum, 9*(2), 199–220.
Lieberson, S (1992) Einstein, Renoir, and Greeley: Some thoughts about evidence in sociology. American Sociological Review 57: 1–15.
Lopreato, J. (1984). *Human nature and biocultural evolution*. Allen and Unwin.
Lopreato, J., & Crippen, T. (1999). *Crisis in sociology: The need for Darwin*. Transaction.
Lorenz, K. (1966). *On aggression*. Methuen/Harcourt, Brace and World.
Lumsden, C. J., & Wilson, E. O. (1981). *Genes, mind and culture: The coevolutionary process*. Harvard University Press.
Maryanski, A. (1994). The pursuit of human nature in sociobiology and evolutionary sociology. *Sociological Perspectives, 37*, 375–389.
Maryanski, A. (1998). Evolutionary sociology. *Advances in Human Ecology, 7*, 1–56.
Maryanski, A., & Turner, J. H. (1993). *Human nature and the evolution of society*. Stanford University Press.
Masters, R. (1989). *The nature of politics*. Yale University Press.
Mazur, A. (1981). Biosociology. In J. Short Jr. (Ed.), *The state of sociology: Problems and prospects*. Beverly Hills.
Mazur, A. (2004). *Biosociology of dominance and deference*. Rowman & Littlefield.
Mazur, A., & Robinson, L. (1972). *Biology and social behavior*. Free Press.
Merton, R.K. (1957). Social Theory and Social Structure, rev.ed. Glencoe, Il: Free Press
Mesoudi, A. (2011). *Cultural evolution: How Darwinian theory can explain human culture and synthesize the social sciences*. University of Chicago Press.
Mills, C. W. (1959). *The sociological imagination*. Oxford University Press.
Mills, M. C., & Tropf, F. C. (2020). Sociology, genetics, and the coming age of sociogenomics. *Annual Review of Sociology 46*, 553–581.
Morris, D. (1967). *The naked ape: A zoologist's study of the human animal*. McGraw-Hill.
Myrdal, G. (1944). *An American dilemma: The Negro problem and modern democracy. Two vols*. New York.
Nelkin, D., & Lindee, M. S. (1995). *The DNA mystique*. Freeman.
Odum, H. (1951). *American sociology: The story of sociology in the United States through 1950*. Longman, Green.
Parsons, T. (1937). *The structure of social action*. McGraw-Hill.
Pigliucci, M., & Muller, G. (Eds.). (2010). *Evolution: The extended synthesis*. MIT Press.
Provine, W. (1973). Geneticists and the biology of race crossing. *Science 182*, 730–6.
Richerson, P., Baldini, R., Bell, A., Demps, K., Frost, K., Hills, V., Mathew, S., et al. (2016). Cultural group selection plays an essential role in explaining human cooperation: A sketch of the evidence. *Behavioral and Brain Science, 39*, e30. https://doi.org/10.1017/S0140525X1400106X
Richerson, P., & Boyd, R. (2005). *Not by genes alone: How culture transformed human evolution*. The University of Chicago Press.
Richerson, P., Boyd, R., & Henrich, J. (2003). Cultural evolution of human cooperation. In P. Hammerstein (Ed.), *Genetic and cultural evolution of cooperation*. MIT Press.
Richerson, P., Boyd, R., & Henrich, J. (2009). *Gene-culture evolution in a time of genomics*. PNAS early edition.
Riley, A. (2021). *Toward a biosocial science: Evolutiunary theory, human nature, and social life* (Part of Evolutionary Analysis in the Social Sciences, a 6 book series). Routledge.
Samelson, F. (1979). Putting psychology on the map: Ideology and intelligence testing. In A. Buss (Ed.), *Psychology in social context* (pp. 131–133). Irvington Publishers.
Sanderson, S. (2001). *The evolution of human sociality*. Rowman & Littlefield.
Sanderson, S. (2014). *Human nature and the evolution of society*. Westview Press.
Sapolsky, R. (2017). *Behave! The biology of humans at our best and worst*. Penguin Press.

Schmaus, W., Jesseph, D., & Segerstrale, U. (1992). The hard program in the sociology of scientific knowledge: A manifesto. *Social Epistemology, 6*(3), 243–265.

Segerstrale, U. (1986). Colleagues in conflict: An 'in vivo' analysis of the sociobiology controversy. *Biology and Philosophy, 1*, 1.

Segerstrale, U. (1993). Bringing the scientist back in: Toward an alternative sociology of science. In T. Brante & S. Fuller (Eds.), *Controversial science*. SUNY Press.

Segerstrale, U. (1998). Truth and consequences in the sociobiology debate and beyond. In V. Falger, P. Meyer, & J. M. G. van den Dennen (Eds.), Sociobiology and politics *Research in biopolitics* (Vol. 6, pp. 249–273). JAI Press.

Segerstrale, U. (2000a). *Defenders of the truth: The battle for science in the sociobiology debate and beyond*. Oxford University Press.

Segerstrale, U. (Ed.). (2000b). *Beyond the science wars*. SUNY Press.

Segerstrale, U. (2013). *Nature's oracle: The life and work of W. D. Hamilton*. Oxford University Press.

Segerstrale, U., & Molnar, P. (1997a). Introduction: Crossing the boundaries between nature and culture. In U. Segerstrale & P. Molnar (Eds.), *Nonverbal communication: Where nature meets culture* (pp. 1–26). Lawrence Erlbaum.

Segerstrale, U., & Molnar, P. (1997b). Biology with a human face: Nonverbal communication and social interaction. In P. Weingart, J. Richerson, S. D. Mitchell, & S. Maasen (Eds.), *Human by nature: Between biology and the social sciences* (pp. 181–193). Lawrence Erlbaum.

Segerstrale, U., & Molnar, P. (eds.) (2016/1997). *Nonverbal communication: Where nature meets culture*. Routledge.

Segerstrale, U. (2013). Nature's Oracle: The Life and Work of W.D. Hamilton. Oxford, UK: Oxford University Press.

Segerstrale, U. (Ed.) (forthcoming). *The evolution of cooperation and the role of Hamilton's rule*. Oxford University Press.

Simpson, I. H., & Simpson, R. L. (1994). The transformation of the American Sociological Association. *Sociological Forum, 9*, 259–278.

Skinner, B. F. (1953). *Science and human behavior*. MacMillan.

Sociobiology Study Group of Science for the People. (1976). Sociobiology - Another biology determinism. *Bioscience, 26*(3).

Stone, B. L. (2008). The evolution of culture and sociology. *The American Sociologist, 39*, 68–85.

Tiger, L. (1969). *Men in groups*. Random House, Thomas Nelson.

Tiger, L., & Fox, R. (1971). *The imperial animal*. Holr, Rinehart and Winston.

Tinbergen, N. (1951). *The study of instinct*. Oxford University Press.

Trivers, R. L. (1971). The evolution of reciprocal altruism. *Quarterly Review of Biology, 46*(4), 35–57.

Turner, J. H. (2020). *On human nature: The biology and sociology of what made us human* (Part of evolutionary analysis in the social sciences, a 6 book series). Routledge.

Turner, J., & Machalek, R. (2018). *The new evolutionary sociology: Recent and revitalized theoretical and methodological approaches* (Part of evolutionary analysis in the social sciences, a 6 book series). Routledge.

Turner, J., & Maryanski, A. (2008). *On the origin of society by natural selection*. Routledge.

Uller, T., & Laland, K. (Eds.). (2019). *Evolutionary causation: Biological and philosophical reflections*. MIT Press.

Van den Berghe, P. L. (1978). *Man in society: A biosocial view* (2nd ed.). Elsevier.

Van den Berghe, P. (1989). *Human family systems: An evolutionary view*. Elsevier.

Van den Berghe, P. (1990). Why most sociologists don't (and won't) think evolutionarily. *Sociological Forum, 5*, 173–185.

Velichkovsky, B. M., & Rumbaugh, D. M. (Eds.). (1996). *Communicating meaning: The evolution and development of language*. Lawrence Erlbaum Associates.

Wallace, W. L. (1983). *Principles of scientific sociology*. Aldine.

Weber, M. (1946, originally 1919). Science as a vocation. In H. H. Gerth, & C. W. Mills (eds). *From Max Weber: Essays in sociology* (pp. 129–156). Oxford University Press.

Weingart, P., Richerson, P. J., Mitchell, S. D., & Maasen, S. (Eds.). (1997). *Human by nature 1997: Between biology and the social sciences*. Lawrence Erlbaum.

West, S. A., Gardner, A., & Griffin, A. (2007a). Social semantics: Altruism, cooperation, mutualism, strong reciprocity and group selection. *Journal of Evolutionary Biology, 20*(2), 415–432.

West, S. A., Gardner, A., & Griffin, A. (2007b). Evolutionary explanations for cooperation. *Current Biology, 17*(16), R661–R672.

White, L. (1949). *The science of culture: A study of man and civilization*. Farrar, Strauss.

Wiegele, T. (Ed.). (1982). *Biology and the social sciences: An emerging revolution*. Westview Press.

Wilson, E. O. (1975). *Sociobiology: The new synthesis*. Harvard University Press.

Wilson, E. O. (1976). Academic vigilantism and the political significance of sociobiology. *Bioscience, 26*, 183–190.

Wilson, E. O. (1978). *On human nature*. Harvard University Press/Bantam Books.

Chapter 5
Cultural Sociology and the Relative Independence of Culture: An Ontological and Epistemological Analysis

Javier Pérez-Jara

Abstract This chapter examines ways to transcend simplistic and reductive perspectives on the nature vs. nurture debate by analyzing one of cultural sociology's central theses: the relative independence of culture. The degree to which culture is contingent upon social structures and natural processes distinguishes cultural sociology from various forms of reductionism and determinism. At the same time, its emphasis on culture's relative independence differentiates it from cultural idealism. This chapter focuses on the primary ontological and epistemological assumptions and implications underlying the thesis of culture's semi-autonomous power. To support this exploration, I propose a philosophical framework addressing key concepts such as upward and downward causation, ontological and epistemological discontinuities, asymmetries, and emergence. I argue that the relationships between biological and cultural factors form a subset of broader interactions within the asymmetrical and dynamic multidimensionality of human existence, which itself represents a specific instance of the universe's overarching asymmetrical multidimensionality.

Keywords Cultural sociology · Relative independence of culture · Vertical causation · Reductionism · Epiphenomenalism · Emergence

5.1 Introduction

Challenging monocausal and unidimensional approaches to human nature, this chapter argues for the existence of multiple ontological dimensions whose asymmetrical and dynamic interplay underlies the emergence of the human being. Biology—encompassing genes, epigenetics, and brain dimorphism—and culture—

J. Pérez-Jara (✉)
Department of Philosophy, University of Seville, Seville, Spain

Center for Cultural Sociology, Yale University, New Haven, CT, USA
e-mail: jperez16@us.es

J. Pérez-Jara, Í. Ongay (eds.), *Beyond Nature and Nurture*, Synthese Library 497,
https://doi.org/10.1007/978-3-031-87715-5_5

ranging from dietary customs and works of art to education and ideologies—are two of these dimensions.

Although other dimensions also contribute, they share a common characteristic: their interaction within a complex ontological framework marked by continuities, discontinuities, symmetries, and asymmetries. In this multidimensional context, vertical causality—both upward and downward—is central.

Following the cultural turn in social science, this paper explores culture as a dynamic system of meanings that are socially learned and shared. These meanings are linked to behaviors and to a vast array of natural and artificial phenomena, from the moon and trees to technological artifacts and biological processes like death. While it was once thought, as Cassirer and others claimed, that humans were the only cultural animals, recent scientific research has revealed remarkable developments in the study of animal cultures (Bonner, 1983; Bueno, 1996; Diéguez, 2011a, 2011b, 2014; Sapolsky, 2006; Viciana, 2021; Whiten, 2019).

Nevertheless, human culture exhibits an unparalleled level of complexity in the animal kingdom. Socially learned meanings are indispensable to the existential environment we inhabit (Dennett, 2017; Jablonka & Lamb, 2014; Mitchell, 2023; Tomasello, 2021). These meanings are intricately interconnected, forming complex patterns and structures that influence every aspect of daily life. From the rituals of greetings to the symbolic depth of technological artifacts, these cultural structures shape our actions and perceptions in profound ways. Far from being only mundane, they often evoke intense emotions, inspire hope and fear, and drive both heroic and reprehensible actions.

The relationship between cultural and non-cultural factors—such as biological processes, geographical and environmental features, and social structures—has been a central theme in social science, particularly since the advent of Marxist thought. Two well-known extremes dominate this debate. On one side, cultural idealism posits that cultural systems, such as religions, art forms, and languages, emerge, evolve, and decline independently of non-cultural factors. On the other, cultural epiphenomenalism asserts that culture is merely a passive reflection of non-cultural determinants.

Simplistic interpretations of Marxism exemplify this latter view, depicting culture as a superstructure entirely determined by the "material" infrastructure. Biologism provides another notable example, one with equally profound implications for political history. Both extremes oversimplify the complex interplay between cultural and non-cultural elements, which remains a fertile ground for theoretical exploration.

The thesis of the relative independence of culture is well established in fields such as cultural anthropology and sociology. In cultural anthropology, Malinowski exemplifies this perspective. While consistently emphasizing the functional primacy of basic human needs, he also defended the relative autonomy of culture and institutions from biological imperatives (Parsons, 1960, p. 64). In sociology, prominent Marxist thinkers, from Althusser to Bourdieu, similarly recognized the relative independence of culture from social structures.

However, Bourdieu's approach, like that of his mentor Althusser, was not entirely consistent. Ultimately, he argued that culture should be understood as an expression of class struggle (Alexander, 1995). This internal contradiction—despite Bourdieu's nuanced engagement with the "cultural turn"—echoes the theoretical inconsistencies in Marx's own work, which fluctuates between nuanced complexity, reductionism, and oversimplified binaries (Camprubí & Pérez-Jara, 2023; Pérez-Jara & Camprubí, 2022).

Positioned between the extremes of cultural idealism and cultural epiphenomenalism, cultural sociology asserts the relative independence of culture. This perspective emphasizes that different cultural systems can emerge under identical biological, geographical, and other non-cultural conditions. In other words, culture possesses a significant degree of contingency and autonomy relative to non-cultural factors.

Cultural sociologists have explored this phenomenon across various contexts, with cultural traumas serving as a prominent example. The inherent impact of natural disasters or war crimes does not, in itself, generate culturally traumatic narratives about those events (Alexander, 2001; Alexander & Pérez-Jara, 2021; Eyerman, 2001; Eyerman et al., 2013). Substantial cultural work is required to construct and establish these narratives. Non-cultural factors do not predetermine the processes of narration and signification, which explains why certain disasters are perpetually remembered in some societies, while others of comparable magnitude are overlooked or entirely ignored.

5.2 The Thesis of Culture's Relative Independence Involves Essential Ontological and Epistemological Assumptions

According to Jeffrey Alexander and Philip Smith's "strong program", cultural sociology must embrace a meaning-centered approach, rejecting reductionist frameworks that attempt, unsuccessfully, to explain cultural structures and processes as mere "projections" of human biology, social structures, or "material factors." Unlike the sociology of culture, which treats culture as a dependent variable shaped by these factors, the "strong program" advocates interpreting culture as an independent variable in sociological analysis (Alexander, 2003; Alexander & Smith, 2002).

However, the concept of the relative independence of culture carries a degree of semantic ambiguity that warrants careful consideration. While cultural sociology positions culture as an epistemologically independent variable, it simultaneously recognizes, in contrast to idealism, culture's ontological reliance on non-cultural factors. But to what extent? Is this relative dependence fixed, or does it fluctuate across different and evolving contexts?

Crucially, whether discussing biological processes or social structures, recognizing non-cultural factors implies rejecting two extremes: the idea that culture is all-encompassing (what might be called the *hubris of culture*) and the notion of culture as a self-contained metaphysical domain. Instead, both human and animal cultures, with all their diversity and complexity, emerge as ontological dimensions shaped by a dynamic interplay of continuities and discontinuities with other dimensions of the universe—namely, the physical, chemical, biological, and social.

Therefore, the most effective ontological framework for supporting the thesis of culture's relative independence from non-cultural factors is one that avoids both hypostatizing culture and reducing it to its constituent elements. Denying cultural idealism is arguably easier than refuting downward reductionism (Alexander, 2005). For a reasonable audience, it is not difficult to show that cultural systems cannot exist without social structures or human beings equipped with functioning brains. But what happens with reductionism? Despite minor and partial reductions, the ontological irreducibility between different dimensions of the universe necessitates a rejection of ontological aggregationism—the notion that something is merely the sum of its components. This view underpins commonsensical claims such as "liquid water is H_2O." However, as John Stuart Mill (1952) famously pointed out, ice and steam are also composed of H_2O but are not liquid water. The structure of a whole, along with its mechanisms and interactions with its environment, is as significant as its components in determining its nature.[1]

For instance, a functional family may consist of the same individuals as a dysfunctional family or even a group of unrelated people. Similarly, a recently deceased animal retains the same chemical and biochemical components as a living body. Yet life cannot be reduced to an aggregation of chemical elements alone. Biological structures, mechanisms, and interactions with the external ecosystem play a crucial role. It is this specific interplay that enables the emergence of properties absent in the isolated components of a living organism.

These qualitatively irreducible properties are commonly referred to in ontology as *emergent properties*—characteristics absent in the components of a whole (Broad, 1925; Bunge, 1977, 2010; Humphreys, 1997; Romero, 2018; Roos, 2024). The universe offers countless examples of emergence: the brightness of stars, which does not exist in their individual components; the capacity of cells to replicate, self-repair and maintain homeostasis, absent from their chemical constituents; human agency, perception, and complex emotional capacities, which cannot be found in bones, organs, cells, or their chemical building blocks (let alone in atoms, quarks, or leptons); and numerous features of human societies that are absent in their individual members.

Entities that exhibit emergent properties can themselves be considered *emergent entities*. Molecules and cells, for instance, are not present at the subatomic level of quarks and gluons, just as galaxies do not exist at the level of individual stellar objects (Romero, 2022). In contrast, *metaphysical neophobia*—the claim that nothing genuinely new can arise in the universe—maintains that the emergent

[1] For a general but detailed analysis of these dimensions, see Bunge (1977, 2003) and Romero (2018). For an in-depth exploration of the essential role of mechanisms in the phenomenon of emergence, see Glennan (2024).

properties of a system are latent or potential within its components. However, this *potentialist* view fails to adequately account for the phenomenon of emergence (Santos, 2015b).

Ontological emergence represents the primary philosophical challenge to downward reductionism and the aggregationism it entails. While ontological aggregationism is not entirely wrong, it fails to account for emergent properties. To avoid hypostatizing metaphysics, we must acknowledge that an entity cannot exist without its components: there are no animals without cells, no stars without atoms, and no societies without individuals. However, ontological aggregationism is incomplete because it overlooks the existence of emergent properties. Against various forms of reductionism—ranging from physio-reductionism and chemo-reductionism to bio-reductionism, socio-reductionism, and culture-reductionism—each dimension of reality exhibits a degree of relative ontological autonomy. Against the claims of ontological aggregationism, *being made of x is not the same as being x* (Pérez-Jara, 2024b; Harman & Pérez-Jara, 2022). In the case of human culture, for example, although it cannot exist without the chemical, physical, biological, and social processes that support it, this does not mean culture is merely an epiphenomenal reflection of those processes.

The famous Parmenidean and Lucretian axiom *ex nihilo nihil fit* has often led philosophers and scientists to adopt ontological neophobia—the belief that the universe excludes the possibility of genuine ontological novelty—under the mistaken assumption that the emergence of any new entity implies *creatio ex nihilo* (Pérez-Jara et al., 2022). This ontological neophobia has profound epistemological implications: it fosters the view that anything seemingly "new" must ultimately be ontologically reducible to its constituent elements. When applied to the realm of culture, this perspective implies that cultural systems can be fully reduced to social structures, which, in turn, can be reduced to individual psychology, neurobiology, chemistry, and ultimately, the laws of physics. The notion that physics could explain the fall of the Roman Empire or marriage patterns in Singapore may seem absurd, but it is the logical conclusion of radical downward reductionism. Consequently, many scholars resist taking downward reductionism to its extreme. Instead, they adopt a more moderate form of micro-reductionism, such as reducing culture to social structure but not further to genetic or biochemical levels. In other words, downward reductionism does not necessarily extend all the way down, which we could describe as *total micro-reductionism.*[2]

[2] Since millennia, the demons of reductionism are persistent. And tempting. They are difficult to expel from both philosophy and science. And the demons of reductionism, like the evil spirits that haunted Saint Anthony in the desert, are tempting because they (let's admit it) make a point, since, again, nothing can exist or be explained without its components. Moreover, some reductions are legit, which complicates the issue even more. Nevertheless, when there is qualitative novelty (i.e. when a whole has properties whose components lack), then a total ontological reduction is not possible (Ongay, 2018, 2019; Pérez-Jara, 2022, 2024a, 2024b; Romero, 2018). In these contexts, an epistemological reduction should only be conducted when it knows its limitations and often narrow scope (for instance, when partially explaining psychological depression by some imbalances of neurotransmitters).

What are social structures made of? They emerge from complex interactions between individuals. And what are individuals made of? Many social theorists would hesitate to answer this question—either under the (often misguided) belief that it is irrelevant to their research or because of a lack of philosophical expertise. In contrast to total micro-reductionism, which reduces culture to physics, partial micro-reductionism maintains that ontological and epistemological reductions are relatively independent. In other words, at certain levels of scientific explanation, we can understand mental processes through neurobiological mechanisms without committing to ontological reductionism or eliminative materialism (Bunge, 2003; Ongay, 2022).

Philosophers often paradoxically present reductionism in reductionistic terms, ignoring upward reductions. Upward reductions, or macro-reductions, occur when the reality of something is understood in terms of the larger whole to which it belongs—that is, when we reduce from the micro to the macro level. Upward or macro-reductionism is essentially synonymous with radical holism and can manifest in various forms, ranging from metaphysical claims ("we are mere, illusory drops in the divine ocean!") to political ideologies ("you are insignificant; only the nation matters!").[3]

The discussion of upward or macro-reductionism is crucial for understanding the relationship between cultural and non-cultural factors. Is there something ontologically "above" culture to which it could be reduced? This is a common perspective among theologians, for example, who reduce certain cultural systems—such as moral codes or mythological constructs—to the category of divine or supernatural revelation. Conversely, atheists and agnostics might apply downward reductionism by reducing these same revelations to human-made cultural systems. However, this does not imply that micro-reductionism is more prevalent or persistent than macro-reductionism in the study of culture. For instance, many contemporary radical constructivists reduce physical, chemical, and biological knowledge to socially constructed and historically contingent constructions. From this radical standpoint, natural sciences do not describe an objective, mind-independent reality; instead, every "natural" phenomenon—from quantum fields and black holes to sexual dimorphism—becomes a contingent cultural construct.

5.3 Vertical Causal Processes and Their Ontological Muzzles

The ontological analysis of cultural sociology's thesis on the relative independence of culture requires a serious engagement with a fundamental ontological concept: vertical causality. As I have previously argued, cultures are more than mere epiphenomenal reflections of social structures, class struggles, or biological processes.

[3] In this approach, organizational or contextual dependency is seen as a necessary but insufficient condition for ontological emergence (Santos, 2015b).

Cultural processes possess the power to influence certain non-cultural phenomena. They shape not only our minds and social relations but also our biology and, to a significant extent, even the Earth itself, as emphasized in the core thesis of the Anthropocene. Rejecting flat ontologies, which ignore the essential asymmetrical structure of reality, requires us to seriously consider the thesis of culture's semi-autonomous causal power. This, in turn, necessitates a rigorous exploration of vertical causation—that is, causal relationships between processes operating at different levels of emergence. Everywhere we look, wholes are ontologically constrained by their parts. Or, more precisely: the emergent properties of a system are constrained by the system's components, structure, processes, and environment. Physical processes govern chemical reactions, chemical processes drive biological activities, and so on. This phenomenon is known as upward causation.

Upward causality is central to understanding what I refer to as "reality muzzle(s)" elsewhere. Such ontological muzzles act as constraints against the excesses of radical constructivism and relativism (Pérez-Jara, 2022, 2024a, 2024b; Pérez-Jara & Camprubí, 2022). Human operations and social and cultural artifacts—including scientific theories and models—are necessarily constrained, at least, by physical, chemical, and biological realities. It is true (and this is the kernel of truth in radical constructivism and relativism) that our knowledge of the external world is far from immediate; it is constructed, filtered, and subject to various cognitive and ideological distortions. However, no construction is a creation *ex nihilo*; rather, it draws from pre-existing materials that impose their ontological nature, like a muzzle, on the process. Thanks to our ontological and epistemological dependence on external material reality, we can discern that some constructions are better and more accurate than others in partially grasping that reality and our place within it. Without this grounding, we risk slipping into self-contradictory epistemologies (Camprubí, 2022; Pérez-Jara, 2014, 2022; Romero et al., 2022; Teixidó, 2024; Teixidó & Carcacía, 2024). While the ancient dream of ontological transparency in human knowledge is little more than an epistemological fairy tale, no less erroneous is the opposite myth of total ontological opacity.[4]

Given the plurality of ontological muzzles, we can identify at least physio-muzzles, chemo-muzzles, bio-muzzles, and socio-muzzles (Pérez-Jara, 2024a, 2024b). Regardless of how flexible or powerful cultural narratives may be in shaping our cognition, emotions, and practices, any ideology that denies death will ultimately collide with the harsh reality of its inevitability. No matter how deeply ingrained the psycho-social mechanisms of suggestion and brainwashing may be, death remains an insurmountable bio-muzzle, much like entropy functions as a physio-muzzle.

[4] What we could call an anthropic matrix or anthropic illusion has fascinated thinkers from the radical Greek Sophists, Buddhist phenomenalists, and classical skeptics, all the way to Nietzsche, the radical pragmatists, and most postmodern thinkers. A classical critique of these approaches is that, according to their assumptions, they cannot justify themselves without violating their most basic epistemological principles.

Of course, the metaphor of ontological muzzles serves, as I have already stated, to counter the excesses of radical constructivism. Reality is not infinitely malleable. However, physical, chemical, biological, social, and cultural-artificial realities also function as *ontological enablers* of many things. In short: depending on the context, reality *enables* certain states while *muzzling* others.

Given this, the crucial question concerning culture's semi-autonomous power of influence is: Does vertical causality operate only in one direction, from the bottom up—i.e., upward causation? Physicalism, or vulgar materialism, denies downward causation—the idea that the behavior of parts can be at least partially determined by the behavior of the whole (Campbell, 1974, 1990). At most, a physicalist can only acknowledge vertical causation within the physical level, since that level is divided into ontological sublevels. However, we have seen how the ontological phenomenon of emergence denies physicalism. Radical reductionism is flawed, even within the realm of physics, in which there are also qualitatively irreducible novelties (Drossel, 2021; Romero et al., 2022). Physicalism, and the epiphenomenalism it entails, distorts both reality and our understanding of it by denying the ontological qualitative novelty of supra-physical realities, from chemical substances and biological organisms to cultural systems.

Physicalism entails a flat ontology, as it posits that all phenomena ultimately reduce to the properties and interactions of physical entities. However, not every flat ontology aligns with physicalism. In Bruno Latour's Actor-Network Theory (ANT), for instance, relational assemblages constitute the primary ontological structure of the world (Latour, 2005). ANT rejects the hierarchical privileging of certain entities—such as physical matter—over others. Instead, it frames the world as composed of networks of relations in which physical, social, and conceptual entities interact on an equal footing.

From this perspective, physical assemblages are merely one type among a diverse array of relational realities. There are no ontological asymmetries between physical and non-physical assemblages: everything—from ideas, empires, and living beings to black holes and quasars—exists at the same ontological level. Latour's ontology thus challenges the physicalist assumption that material realities dominate or underpin non-material ones (Latour, 1993). By granting equal ontological status to all entities, ANT emphasizes the dynamic interplay of actors in networks, whether they are tangible, conceptual, or social. But the prize to pay is too high: to deny the asymmetrical structure of reality enabled by different levels of emergence.[5] While this flexible framework allows Latour and similar thinkers to sidestep both simplistic materialism and idealism, symmetrical relationism can become a precarious theoretical quicksand (Camprubí, 2022).

Of course, this does not mean that symmetry should be downplayed in our understanding of reality. Symmetry plays a crucial role in numerous physical and chemical laws and processes, as well as in biotic systems. However, these

[5] And this is not to mention that we also find asymmetries within each level of emergence—asymmetries that we could refer to as horizontal (Ongay & Pérez-Jara, 2022).

symmetries coexist with the structural asymmetries enabled by emergent levels of organization. The complexity of these debates has given rise to a wide array of perspectives on the relationship between emergence and vertical causality, highlighting profound divisions in metaphysical and ontological frameworks. Key topics of discussion include the distinction between weak and strong emergence and their relationship with downward causation (Bedau, 1997, 2002; Bedau & Humphreys, 2007; Gozzano, 2017); the rejection of layered reality, advocated by Heil (2003); the possibility of emergence without layers, defended by Voosholz (2021); ontological emergence without vertical causation, proposed by Wong (2020); and the relational-horizontal ontological perspective developed by Santos (2015a). These diverse approaches reflect the ongoing challenge of reconciling the concept of emergence with broader metaphysical theories.

These diverse viewpoints reflect the ongoing challenges in reconciling the concepts of emergence, causality, and ontology. They highlight fundamental disagreements over whether emergence requires hierarchical layers, how causation operates across or within levels, and whether emergent phenomena can possess causal powers independent of their base components.

A detailed philosophical discussion on the complexities of emergence and non-flat ontologies is better suited for another context. For the purposes of this approach, it suffices to note that emergence necessitates rejecting both symmetrism and relationism. While relations are indeed fundamental—*contra substantialism*—in the ontological phenomenon of emergence, multiple factors are critical to explaining qualitative novelty. These include the components of a system, its structure, mechanisms, and interactions with the environment (Bunge, 1977, 2003, 2006a, 2006b, 2010; Romero, 2018). Crucially, such novelty entails asymmetrical relations, highlighting a hierarchy of influence and interdependence (Pérez-Jara, 2022, 2024a, 2024b).

Critiques of the "layered" view of reality and the notion of vertical causality are valid only when aimed at reified interpretations of the universe's "levels." However, it is entirely feasible to conceptualize the universe's asymmetrical structure and emergent dimensions without falling victim to what I term *Medusa's Effect*—the tendency to petrify any dimension of reality into a completely autonomous thing (Pérez-Jara, 2022; Pérez-Jara & Camprubí, 2022).[6]

To consider downward causation as problematic or even impossible, one must first conceive of physical, chemical, biological, social, and artificial levels as distinct, overlapping strata where causal closure occurs. However, the emergent properties of biotic realities, for instance, lack absolute autonomy from chemical and physical processes. Metaphors like strata, levels, or layered structures—when taken too literally—can be misleading. In a mountain, a building, or a pie, the layers exhibit a degree of ontological independence that chemical compounds, for instance,

[6] In this last approach, organizational or context dependency would be a necessary but insufficient condition for ontological emergence (Santos, 2015b).

do not have from physical processes. Some of John Heil's critiques of levels in emergentist ontology stem from this misunderstanding (Heil, 2003, 2012).

It is entirely possible to conceptualize the asymmetrical nature of physical, chemical, biological, social, and cultural matter without falling into semantic confusion. Abandoning the concept of "level" does not necessarily entail the collapse of the emergence framework. The notion of "ontological level" is not univocal; it can be understood without resorting to hypostatized metaphysics.

Finding a suitable metaphor to illustrate the complex interplay of continuities, discontinuities, symmetries, and asymmetries that structure reality is no easy task. However, alongside other philosophers, we can treat an ontological level neither as an independent entity nor as something more or less real than adjacent levels, as in Neoplatonic metaphysics. If we take the ontological irreducibility of many processes in the universe seriously, we can define an ontological level as a collection of entities sharing certain properties. Thus, levels of organization can be characterized in relation to a set of common properties (Romero, 2022).

In the ontological phenomenon of emergence, there exists a dynamic interplay of continuity and discontinuity. Discontinuity manifests in the qualitative "jumps" from physics to society and culture, while continuity underpins the connections across these domains. If emergence were solely defined by discontinuity, it would veer into creationism—a misconception that drives many thinkers toward neophobic, aggregationist ontologies. Conversely, if emergence implied only continuity, neophobic aggregationism would prevail.

The falsehood of reductionism underscores the necessity of discontinuity, while the falsehood of creationism highlights the importance of continuity (Pérez-Jara, 2022). Thus, there is no genuine dilemma between continuity and discontinuity. Radical continuism would erase the universe's plurality and qualitative novelty. Even physicalism concedes the existence of horizontal discontinuities. Conversely, radical discontinuism would make the universe impossible, as it would eliminate causality and all interactions between entities (Bueno, 1972; Ongay, 2024; Pérez-Jara, 2022).

The complex, dynamic interplay of continuities and discontinuities prevents us from hypostatizing any dimension of the universe, preserving both its diversity and coherence.

5.4 Time to Break the Epiphenomenal Mirror!

Another potential objection to downward causation deserves careful consideration. If we take upward causation for granted and recognize how physical processes constrain and influence chemical processes—which, in turn, shape biological processes and so on—should we not view supra-physical causal relations as merely epiphenomenal? For instance, if physical processes determine the emergent properties of chemical processes, should chemical causality not be considered epiphenomenal, a kind of fatalistic mirror of physical causality?

This reasoning has led some thinkers to deny the existence of free will. They argue that if freedom does not exist in the realm of physics—where quantum stochasticity does not equate to "freedom"—then it cannot exist in human psychology, as physical processes would fully determine mental states. However, this view overlooks a critical aspect of organizational mereological complexity: certain properties are preserved while others are lost as we move across emergent levels of organization. For example, an electron has a spin and a quantum number, but a cell does not. Similarly, while living beings are composed of physical and chemical processes that are not alive, a society composed of living beings is not itself alive. Certainly, a cell is still composed of elementary particles with properties like spin and quantum numbers, just as a society is ultimately composed of living individuals. The emergence of biotic or social systems does not fundamentally alter the intrinsic properties of their constituent parts. However, some of these properties do not manifest at the new emergent levels. For instance, stochastic processes are crucial at the quantum level, enabling non-causal (though ontologically legal) phenomena such as muon decay. In contrast, quantum stochasticity is absent at higher ontological levels, such as the biological and social domains, where all observed phenomena are causally determined. This interplay highlights how emergent properties cannot be reduced to their constituent processes, and how causation operates differently across levels. Just as societies influence their individual members, individuals' actions and thoughts reciprocally affect the chemical and physical properties of their bodies.[7]

I will refer to the notion that a top organizational level cannot influence its components as *upward or down-top fatalism*. This perspective is not confined to physics; it can be observed across various disciplines. For example, consider radical biologicism in the social sciences, which posits that society and culture are merely epiphenomena of biology. This view remains pervasive today, ranging from its most extreme proponents (Foster & Tennant, 2022; Goldberg, 1973, 1993) to its more moderate advocates (Waal, 2022).[8]

The ontological reality of downward causation must be acknowledged alongside the structural asymmetry between upward and downward causation. In other words, the causal loop connecting these two directions is not reciprocal. The overwhelming

[7] Other authors who deny free will, such as Sapolsky (2023), acknowledge the existence of emergence and downward causality. However, they impose an impossibly high standard for discussing agency and free will: either we possess absolute free will and agency (an obviously unattainable condition), or we are entirely devoid of them. In contrast to these approaches, I advocate for a perspective that recognizes agency as a quality inherent to all living beings. The agential dimension of life can be examined across various contexts, ranging from basic processes like bacterial chemotaxis and plant tropisms to complex decision-making mechanisms (see Dennett, 2017; Mitchell, 2023; Nuño de la Rosa, 2023).

[8] This upward fatalism implies what, metaphorically speaking, we could also call the "hubris" of different dimensions of matter: the hubris of physical matter in physicalism or the hubris of biological matter in biologicism. Which epistemologically is easily translated to the hubris of the physicalist or the hubris of the biologist in her biased and reductionistic account of reality. Thus, we could talk, at least, about physiohubris, chemohubris, biohubris, and sociohubris in both their ontological and epistemological sides.

dominance of upward causation is undeniable. A drastic physical or chemical change can obliterate all life on Earth in an instant. For example, if the temperature of the room where I am writing this paper were to rise by thousands of degrees Fahrenheit, everything would be reduced to plasma, devoid of any social or biological properties. Life exists only within extremely narrow thermal, density, and gravitational parameters.

In contrast, living beings—even with advanced technologies like nuclear power—lack the capacity to fundamentally alter stars, galaxies, or the general nature of the universe. Material forces constrain us far more than we constrain them. Yet, despite this structural asymmetry, the reality of vertical two-way causation refutes epiphenomenalism in its various forms: whether in the philosophy of nature (biotic matter as an epiphenomenon of physical matter), the philosophy of mind (mental processes as epiphenomena of electrochemical brain activity), or the philosophy of culture (culture as an epiphenomenon of social structure or class struggle).

Some authors do, indeed, exaggerate the power of downward causation by hypostatizing a particular ontological level. Idealism—both in its subjective and objective forms—is a notable example of this. In objective idealism (or Platonism), mathematical and logical structures are said to "shape" the various dimensions of matter, while in subjective idealism or spiritualism, autonomous mental processes are believed to shape the brain, or even matter in general. However, the notion that mathematical structures can directly enter into causal relations with matter is far from clear (Bueno, 2021). Moreover, the idea that mathematical entities can influence material systems independently of human beings (or at least artificial intelligence) is utterly absurd (Pérez-Jara, 2022; Romero et al., 2022). Conceptual artifacts cannot *bypass* the cognitive systems that think of them.

Rather, abstract thinking shaped by mathematical constructs can exert a top-down influence through physical manipulations by humans in their environments—actions that are, of course, only possible through the neurobiological activity of humans. Nonetheless, the excesses of idealism should not lead us to dismiss downward causation along with the bathwater of Platonism and spiritualism. The deniers of downward causation not only adopt a flawed ontology, but also embrace a self-contradictory epistemology. Knowledge and science require intentional action, which is only possible if top-down causation is real (Ellis, 2021; Ellis & Gabriel, 2021).

Animals' biotic systems are undoubtedly a necessary condition for the emergence of social meanings. Even if AI systems were capable of generating social meanings, they would still be dependent on human beings, as the creators of these systems. However, culture possesses the power to influence social interactions and even human biology, as evidenced by phenomena ranging from the Baldwin effect to epigenetic changes and various other bio-social interactions (Dennett, 2017; Loverdo & Viciana, 2018). Many biological traits of humans are, or could be, shaped by cultural systems. These include certain features of the brain connectome wiring, the modulation of neurotransmitter and hormone production essential to mental processes, body composition, changes in bone structure, reproductive traits, immune system functioning, and the activation of specific genes, among others (Mitchell, 2018; Ongay, 2008, 2011).

There is no universal or homogeneous pattern in vertical causation. In the context of the nature versus nurture debate, this means that downward causation can sometimes be more influential, as seen in the power of culture to shape individual behaviors and preferences. In other instances, upward causation takes precedence, such as in cases of severe genetic disorders, complications during pregnancy or birth that lead to intellectual disabilities, or significant brain injuries. For example, the inability of two-year-old children to become great philosophers or politicians cannot be explained by cultural factors alone; it is rooted in biological processes related to brain development and the cognitive capacities that emerge from it. Conversely, factors such as crime rates, divorce rates, and obesity are, on average, more influenced by sociocultural factors than by biological ones.

To summarize: sociocultural systems influence neurobiological processes, biological processes can significantly alter chemical compositions, and chemical reactions can generate notable physical effects, such as increases in temperature. These downward processes are as real as the irreducible ontological novelties that support them. Moreover, downward causation operates not only across ontological levels but also within them, particularly within sub-levels of organization. For instance, macro-social systems shape micro-social systems, such as families; families influence individuals; organ systems regulate cellular activity; and cells, in turn, determine the sequences, combinations, and arrangements of their DNA. This dynamic interplay highlights the intricate, multi-layered nature of causation both across and within the levels of reality.

5.5 Macro-Asymmetries and Micro-Asymmetries Between Human Biological and Cultural Traits

The ontological multidimensionality discussed in this chapter epistemologically aligns with the necessity of scientific interdisciplinarity, countering the aforementioned epistemological reductionisms (physicalism, sociologism, biologism, psychologism, economicism). I have examined how the ontological dimensions that constitute the universe—and, therefore, the human world—are structurally asymmetrical, challenging the symmetry principle advocated by certain flat ontologies. This asymmetry is both general and specific. Macro-asymmetries exist between physical, chemical, biological, social, and cultural-artificial levels. While social systems depend on biological, chemical, and physical processes, physical processes themselves lack social properties.

On the other hand, micro-asymmetries are more variable and depend on specific, dynamic contexts. For instance, in the case of depression, chemical factors may sometimes play a more significant role (e.g., depression triggered by toxins or pollution). At other times, biological factors take precedence, such as a strong genetic predisposition, brain tumors, insufficient sleep, or disrupted circadian rhythms. In other cases, social and cultural influences may be more dominant, such as pessimistic religious or ideological narratives, bullying, or personal losses. In most cases, depression emerges as a multidimensional phenomenon, driven by a combination of factors that only rarely contribute in equal measure. This ontological perspective has an epistemological counterpart: at times, social psychology offers greater explanatory power than neurobiology or genetics in understanding the origins of depression; in other instances, cultural sociology may serve as the primary framework for explaining the phenomenon, and so forth. To summarize: in macro-asymmetries, the lower levels (such as physical matter) always holds greater causal power than the upper levels (such as social matter). In contrast, in micro-asymmetries, the causal power can, and often does, flow in the opposite direction. It is at this point that the causal power of culture can reveal its full magnitude.

The interplay between macro-asymmetries and micro-symmetries is central to the nature versus nurture debate. Since the emergence of genetics, the nature versus nurture debate has often been framed as a contest between genes and the environment. This genetically-updated version of the ancient physis/nomos debate presented human psychology as a combination of a priori (genetic) and a posteriori (environmental/experiential) factors. However, this simplistic view resulted from hypostatizing genes—treating them as independent agents, detached from their contexts. Current scientific research clearly demonstrates that this is not the case: the influence of genes depends on the contexts in which they operate. Biology has long moved beyond genetic determinism. Since the work of Conrad Waddington and Jean Piaget, human development, including brain development, has been understood as both genetic and epigenetic (Carey, 2013; Francis, 2012; Jablonka & Lamb, 1995; Moore, 2015; Sapolsky, 2017). Epigenetics itself is shaped by environmental factors, including society and culture. For instance, biographical experiences, nutrition, and exposure to toxins can activate or suppress certain genetic predispositions. In other words, in humans, as in other complex organisms, biology provides a set of conditions for potential development rather than a predetermined fate. These potentials are actualized through intricate interactions with non-biological factors, such as cultural systems. The relationship between genes and psychological traits—like the genotype-phenotype connection—is not direct but mediated by a complex interplay of supra-biological factors (social interactions, cultural structures) and infra-biological factors (such as environmental toxins and solar radiation). The situation becomes even more complex when considering that epigenetic changes can be inherited (Moore, 2015). However, this does not imply unlimited ontological plasticity, as some overly optimistic authors suggest. Human plasticity is constrained by both supra- and infra-biological limits, which will be explored further in other chapters of this book.

Contrary to what seems to be the argument put forth by authors like David S. Moore, there is no inherent symmetry between genes and environments in their interaction. Some proponents of ontological symmetrism claim that, in every case, genes and environments are equally influential in shaping the final form of a trait. However, once this symmetry is rejected, it becomes clear that the relative importance of genes and environments can vary depending on the circumstances. Neither genetics nor epigenetics operates in a metaphysical vacuum. There is always some kind of asymmetry in the causal interactions between different dimensions of reality, including the development of a human being. Furthermore, it is contradictory to argue that epigenetics in humans is still poorly understood while simultaneously claiming that no human traits are more genetically determined than others (Moore, 2015).

Symmetrism, though, does contain an element of truth: (1) In addition to the symmetry observed in many physical and chemical laws, as well as in many aspects of biological systems, it is undeniable that all human traits have a genetic foundation, as genes contribute to the development of every trait, and (2) These traits are acquired over time, as organisms initially lack them and later develop them (Bateson & Mameli, 2006; see also Ongay's Chapter in this book). However, the fact that genetic, epigenetic, and environmental factors all interact simultaneously to facilitate the emergence of mental processes and behaviors does not imply that these factors carry equal weight in every situation. For example, while human genes allow for the possibility of becoming philosophers under certain environmental conditions, cats' genes—despite what might seem initially obvious—do not. Within the human species, certain individuals have a stronger genetic predisposition to mental illnesses or impairments, even within similar environments. If an alien were to visit Earth and observe a person speaking Korean, it would be misleading to claim that the person's hair color and language are equally shaped by genetic, epigenetic, and environmental factors. The language we speak (and our specific ideological beliefs) seems more strongly influenced by environmental factors than traits like our nose shape or hair color, though even these physical traits emerge from complex interactions between genes and the environment. Severe neurodevelopmental disorders and brain injuries serve as compelling counterexamples to radical constructivism and symmetrism, demonstrating that biological factors can often exert greater causal influence than cultural ones in shaping human behavior and development.

Nevertheless, even in these extreme cases, it is important to recognize that many neurodevelopmental disorders are influenced by sociocultural factors. While the specific causes of these disorders are biological, society and culture—operating as supra-biological realities—can shape biological processes, including brain development, through downward causation. Nervous systems provide the foundation for social interactions and cultural meanings, while these very social interactions, mediated by cultural contexts, are essential for the ontogenetic development of the human nervous system.

5.6 Conclusion: Overcoming the Nature vs Nurture Debate

Empirical scientific research consistently challenges both biological and structural reductionism by demonstrating that cultural systems exert a significant degree of contingency and causal influence on non-cultural realities, from the functioning of human brains to entire ecosystems. Empirical findings also contradict the claims of strong social constructivism and cultural idealism by showing that non-cultural processes—ranging from environmental factors like toxins and genetic and epigenetic influences, to broader social structures—play a critical role in shaping cultural phenomena. To fully comprehend this complex and dynamic interplay between culture and non-cultural realities, philosophical theorizing is essential. It enables us to navigate the nuances of these interactions and formulate a coherent understanding of the interconnectedness of biological, social, and cultural dimensions.

This chapter has proposed a comprehensive ontological and epistemological framework for interpreting the central thesis of cultural sociology: the semi-autonomous causal power of culture. While significant and valuable work has been done on the philosophical underpinnings of cultural sociology, much of it has overlooked two crucial ontological features that I consider essential: the asymmetrical and dynamic multidimensionality of human beings, and the vertical causal processes that operate within this multidimensionality. In addressing these features, I have explored key ontological phenomena connected to them, particularly emergence, in contrast to flat ontologies and neophobic metaphysics, as well as the complex interplay of continuities and discontinuities that structures the world, challenging both radical monism and pluralism.

Pluralism has become a prominent concept in contemporary ontology and epistemology, with a broad range of philosophers endorsing it, including Bruno Latour, John Dupré, Markus Gabriel, Nancy Cartwright, Gustavo Bueno, Graham Harman, and many others. However, like any significant philosophical idea, pluralism is not a monolithic term—it encompasses various interpretations. While many thinkers advocate for the multidimensionality of human beings, there is no consensus on how this multidimensionality should be understood. In my view of reality, ontological continuities (the driving force behind monism) are as crucial as ontological discontinuities (the foundation of pluralism). Moreover, the universe contains many different types of continuities and discontinuities. As I have argued elsewhere (Pérez-Jara, 2022), we must move beyond the simplistic binary of monism versus pluralism.

My philosophical framework offers a fresh perspective on the nature vs. nurture debate by sidestepping theoretical dead ends such as dualisms, reductionisms, hypostatizations, and fatalisms of various types. Social interactions and cultural meanings are emergent phenomena that are deeply intertwined with physical, chemical, and biological processes. As I have argued, these phenomena are not mere epiphenomena; they possess genuine downward causal power. However, this power is neither symmetrical in general (the physical world exerts an influence on human culture that culture does not reciprocate) nor in specific cases—each instance involves unique asymmetries between bottom-up and top-down causal interactions. The conclusion is clear: Reality is not only very complex and dynamic; it is also profoundly asymmetrical in many contexts.

The general principle of contextualized asymmetry leads to important epistemological conclusions. For example, difficulties in speech may, in some cases, be primarily explained by neurobiological factors—such as drug use, a brain tumor, or brain injury—while in other cases, psychological or sociological factors may provide a more fitting explanation, particularly when the brain functions "normally" according to certain average standards.

Contrary to the simplistic and sterile view of reality promoted by metaphysical neophobia, human beings and their cultural systems represent profound qualitative novelties in the universe, arising from asymmetrical, multi-level, semi-discontinuous, and both horizontal and vertical causal processes. Ignoring this dynamic ontological interweaving—and its epistemological implications—can severely hinder the debate on culture within human multidimensionality. While it may seem tempting to reduce the complex questions surrounding humanity's ontological status to simplistic binaries like nature vs. nurture, such an approach not only reflects theoretical laziness but also carries practical risks.

References

Alexander, J. (1995). The reality of reduction: The failed synthesis of Pierre Bourdieu. In J. C. Alexander (Ed.), *Fin de Siècle social theory*. Verso.

Alexander, J. C. (2001). Toward a theory of cultural trauma. In J. Alexander et al. (Eds.), *Cultural trauma and collective identity* (pp. 1–30). University of California Press.

Alexander, J. C. (2003). *The meanings of social life: A cultural sociology*. Oxford University Press.

Alexander, J. (2005). Why cultural sociology is not 'idealist': A reply to McClennan. *Theory, Culture & Society, 22*(6), 19–29.

Alexander, J., & Pérez-Jara, J. (2021). Mito y rito en el corona-drama internacional. Jeffrey Alexander en diálogo con Javier Pérez Jara. In J. del Llano & L. Camprubí (Eds.), *Sociedad Entre Pandemias* (pp. 257–258). Gaspar Casal Foundation.

Alexander, J., & Smith, P. (2002). The strong program in cultural theory: Elements of a structural hermeneutics. In J. Turner (Ed.), *Handbook of sociological theory*. Kluwer Academic.

Bateson, P., & Mameli, M. (2006). Innateness and the sciences. *Biology and Philosophy, 21*(2), 155–188.

Bedau, M. (1997). Weak emergence. *Philosophical Perspectives, 11*, 375–399.

Bedau, M. A. (2002). Downward causation and the autonomy of weak emergence. *Principia, 6*(1), 5–50.

Bedau, M. A., & Humphreys, P. (Eds.). (2007). *Emergence: Contemporary readings in philosophy and science*. MIT Press.

Bonner, J. T. (1983). *The evolution of culture in animals*. Princeton University Press.

Broad, C. D. (1925). *The mind and its place in nature*. Kegan Paul.

Bueno, G. (1972). *Ensayos materialistas*. Taurus.

Bueno, G. (1996). *El animal divino. Ensayo de una filosofía materialista de la religión*. Pentalfa.

Bueno, O. (2021). Mathematics and measurement: Causation and the mind. In J. Voosholz & M. Gabriel (Eds.), *Top-down causation and emergence* (pp. 65–75). Springer Verlag.

Bunge, M. (1977). Treatise on basic philosophy. In *The furniture of the world* (Vol. 3). Reidel Publishing Company.

Bunge, M. (2003). *Emergence and convergence: Qualitative novelty and the unity of knowledge*. University of Toronto Press.

Bunge, M. (2006a). *Chasing reality: Strife over realism*. Toronto University Press.

Bunge, M. (2006b). *Chasing reality. Strife over realism*. University of Toronto Press.

Bunge, M. (2010). *Matter and mind: A philosophical inquiry*. Springer.

Campbell, D. T. (1974). 'Downward causation' in hierarchically organized biological systems. In F. J. Ayala & T. Dobzhansky (Eds.), *Studies in the philosophy of biology* (pp. 179–186). Macmillan Press.

Campbell, D. T. (1990). Levels of organization, downward causation, and the selection-theory approach to evolutionary epistemology. In E. Tobach & G. Greenberg (Eds.), *Scientific methodology in the study of mind: Evolutionary epistemology* (pp. 1–17). Erlbaum.

Camprubí, L. (2022). Materialism and the history of science. In G. E. Romero, J. Pérez-Jara, & L. Camprubí (Eds.), *Contemporary materialism: Its ontology and epistemology* (Synthese library) (Vol. 447). Springer. https://doi.org/10.1007/978-3-030-89488-7_8

Camprubí, L., & Pérez-Jara, J. (2023). La arqueología como arqueología y la relativa independencia de la cultura. *Revista De Occidente, 510*, 36–53.

Carey, N. (2013). *The epigenetics revolution: How modern biology is rewriting our understanding of genetics, disease, and inheritance*. Columbia University Press.

Dennett, D. (2017). *From bacteria to bach and back: The evolution of minds*. W. W. Norton & Company.

Diéguez, A. (2011a). *La evolución del conocimiento: de la mente animal a la mente humana*. Biblioteca Nueva.

Diéguez, A. (2011b). La filosofía de la biología como campo interdisciplinar. *Ludus Vitalis, XIX*(36), 261–265.

Diéguez, A. (2014). Pensamiento conceptual en animales. In A. Diéguez & J. M. Atencia (Eds.), *Naturaleza animal y humana* (pp. 83–114). Biblioteca Nueva.

Drossel, B. (2021). Strong emergence in condensed matter physics. In J. Voosholz & M. Gabriel (Eds.), *Top-down causation and emergence*. Springer.

Ellis, G. F. R. (2021). Physics, determinism, and the brain. In J. Voosholz & M. Gabriel (Eds.), *Top-down causation and emergence* (pp. 157–214). Springer Verlag.

Ellis, G. F. R., & Gabriel, M. (2021). Physical, logical, and mental top-down effects. In J. Voosholz & M. Gabriel (Eds.), *Top-down causation and emergence* (Synthese library) (Vol. 439).

Eyerman, R. (2001). *Cultural trauma, slavery and the formation of African American identity*. Cambridge University Press.

Eyerman, R., Alexander, J. C., & Breese, E. B. (2013). *Narrating trauma: On the impact of collective suffering*. Routledge.

Foster, M., & Tennant, D. B. (2022). *It's good to be a man: A handbook for godly masculinity*. Canon Press.

Francis, R. C. (2012). *Epigenetics: How environment shapes our genes*. W. W. Norton & Company.

Glennan, S. (2024). The mechanisms of emergence. In J. L. Cordovil, G. Santos, & D. Vecchi (Eds.), *New mechanism. History, philosophy and theory of the life sciences* (Vol. 35). Springer. https://doi.org/10.1007/978-3-031-46917-6_11

Goldberg, S. (1973). *The inevitability of patriarchy: Why the biological difference between men and women always produces male domination*. William Morrow and Company.

Goldberg, S. (1993). *Why men rule: A theory of male dominance*. Open Court.

Gozzano, S. (2017). The compatibility of downward causation and emergence. In F. Orilia & M. P. Paoletti (Eds.), *Philosophical and scientific perspectives on downward causation* (pp. 296–312). Routledge.

Harman, G., & Pérez-Jara, J. (2022). Materialism is false/materialism is not false. In G. E. Romero, J. Pérez-Jara, & L. Camprubí (Eds.), *Contemporary materialism: Its ontology and epistemology*. Springer.

Heil, J. (2003). *From an ontological point of view*. Oxford University Press.

Heil, J. (2012). *The universe as we find it*. Oxford University Press.

Humphreys, P. (1997). How properties emerge. *Philosophy of Science, 64*(1), 1–17. https://doi.org/10.1086/392533

Jablonka, E., & Lamb, M. (1995). *Epigenetic inheritance and evolution. The Lamarckian dimension*. Oxford University Press.

Jablonka, E., & Lamb, M. (2014). *Evolution in four dimensions* (revised edition: Genetic, epigenetic, behavioral, and symbolic variation in the history of life). Bradford Books.

Latour, B. (1993). *We have never been modern*. Cambridge: Harvard University Press.
Latour, B. (2005). *Reassembling the social: An introduction to actor-network theory*. Oxford University Press.
Loverdo, C., & Viciana, H. (2018). Cultural transmission and biological markets. *Biology & Philosophy, 33*(5–6), 40.
Mill, J. S. (1952). *A system of logic*. Longmans, Green & Co.
Mitchell, K. (2018). *Innate: How the wiring of our brains shapes who we are*. Princeton University Press.
Mitchell, K. (2023). *Free agents: How evolution gave us free will*. Princeton University Press.
Moore, D. S. (2015). *The developing genome. An introduction to behavioral epigenetics*. Oxford University Press.
Nuño de la Rosa, L. (2023). Agency in reproduction. *Evolution & Development, 25*(6), 418–429.
Ongay, ´i. (2008). Entre el hábito y el instinto. Cuestiones gnoseológicas y ontológicas concernientes a las ideas de "conducta" y "evolución". *El Basilisco, 39*, 3–36.
Ongay, ´i. (2011). Ni innato ni aprendido. Capas básicas y capas metodológicas de las ciencias de la conducta. *El Catoblepas, 110*, 11.
Ongay, ´i. (2018). Making a case in favor of not yet applied science. *Science and Education: Academic Journal of Ushynsky University, 27*(7–8), 821–824.
Ongay, ´i. (2019). Is simplicity a myth? Mach and bunge on the principle of parsimony. In M. R. Matthews (Ed.), *Mario Bunge: A centenary festschrift* (pp. 163–180). Springer.
Ongay, ´i. (2022). Mind and matter. In G. E. Romero, J. Pérez-Jara, & L. Camprubí (Eds.), *Contemporary materialism: Its ontology and epistemology* (Synthese library) (Vol. 447). Springer. https://doi.org/10.1007/978-3-030-89488-7_7
Ongay, ´i. (2024). Cause and effect in biology, culture, and the (extended) mind: A coevolutionary approach. *Review of Evolutionary Political Economy, 5*, 445–471. https://doi.org/10.1007/s43253-024-00127-z
Ongay, ´i., & Pérez-Jara, J. (2022). Emergent materialism implies continuism/emergent materialism does not imply continuism. In G. E. Romero, J. Pérez-Jara, & L. Camprubí (Eds.), *Contemporary materialism: Its ontology and epistemology* (Synthese library). Springer. https:/ /doi.org/10.1007/978-3-030-89488-7_12
Parsons, T. (1960). Theory of social systems. In R. Firth (Ed.), *Man and culture: An evaluation of the work of Bronislaw Malinowski*. Routledge.
Pérez-Jara, J. (2014). *La filosofía de Bertrand Russell*. Pentalfa.
Pérez-Jara, J. (2022). Discontinuous materialism. In G. E. Romero, J. Pérez-Jara, & L. Camprubí (Eds.), *Contemporary materialism: Its ontology and epistemology* (Synthese library) (Vol. 447). Springer. https://doi.org/10.1007/978-3-030-89488-7_3
Pérez-Jara, J. (2024a). The ontology of coevolution beyond economic systems. *Review of Evolutionary Political Economy, 5*, 425–444. https://doi.org/10.1007/s43253-024-00124-2
Pérez-Jara, J. (2024b). Quantum mechanics and inclusive materialism. *Philosophies, 9*(5), 140. https://doi.org/10.1007/s43253-024-00124-2
Pérez-Jara, J. And Camprubí, L. 2022. Science and Apocalypse in Bertrand Russell: A cultural sociology. : Lexington Books.
Pérez-Jara, J., Romero, G. E., & Camprubí, L. (2022). What is materialism? History and concepts. In G. E. Romero, J. Pérez-Jara, & L. Camprubí (Eds.), *Contemporary materialism: Its ontology and epistemology* (Synthese library) (Vol. 447). Springer. https://doi.org/10.1007/978-3-030-89488-7_1
Romero, G. E. (2018). *Scientific philosophy*. Springer. https://doi.org/10.1007/978-3-319-97631-0
Romero, G. (2022). Systemic materialism. In G. Romero, J. Pérez-Jara, & B. Camprubí (Eds.), *Contemporary materialism: Its ontology and epistemology*. Springer.
Romero, G., Pérez-Jara, J., & Camprubí, B. (2022). *Contemporary materialism: Its ontology and epistemology*. Springer.
Roos, M. (2024). Emergence, levels, and hierarchy. In *Principles of complexity economics. Classroom companion: Economics*. Springer. https://doi.org/10.1007/978-3-031-51436-4_9

Santos, G. C. (2015a). Upward and downward causation from a relational-horizontal ontological perspective. *Axiomathes, 25*, 23–40.

Santos, G. (2015b). Ontological emergence: How is that possible? Towards a new relational ontology. *Foundations of Science, 20*(4).

Sapolsky, R. (2006). Social cultures among nonhuman primates. *Current Anthropology, 47*(4), 641–656.

Sapolsky, R. (2017). *Behave: The biology of humans at our best and worst*. Penguin Press.

Sapolsky, R. (2023). *Determined: A science of life without free will*. Penguin Press.

Teixidó, Ó. (2024). *Introducción a la filosofía de la ciencia sistemática en psicología*. Psara Ediciones.

Teixidó, O., & Carcacía, I. (2024). La filosofía sintética: ¿qué es y por qué todo filósofo dedicado a la investigación debería adoptarla? *Logos: Anales del Seminario de Metafísica, 57*(2), 1–20.

Tomasello, M. (2021). *Becoming human: A theory of ontogeny*. Belknap Press.

Viciana, H. (2021). Animal culture: But of which kind? *Studies in History and Philosophy of Science, 90*, 208–218.

Voosholz, J. (2021). Top-down causation without levels. In J. Voosholz & M. Gabriel (Eds.), *Top-down causation and emergence* (pp. 269–296). Springer Verlag.

Waal, F. d. (2022). *Different: Gender through the eyes of a primatologist*. Norton & Company.

Whiten, A. (2019). Cultural evolution in animals. *Annual Review of Ecology, Evolution, and Systematics, 50*(1), 27–48.

Wong, S. L. (2020). Ontological emergence without vertical causation. *Axiomathes, 30*(5), 501–514.

Chapter 6
Psychogenesis in Evolution and History

José-Carlos Sánchez-González

Abstract This chapter proposes a criterion for critically addressing the nature-culture duality, replacing it with a non-dual conception in which evolution, which includes organisms capable of experience and learning, is the matrix of culture, of artifact production and of the start of history. The thread that makes it possible to understand this continuity is psychogenesis, that is, adaptive, innovative behavioral processes. Psychogenetic processes are mediated by biological and social conditions and in turn participate in evolution (by organic selection) and in the production of artifacts, action systems and cultural meanings. The proposal is supported by recent developments in epigenetics, comparative psychology's critique of instinct, recent findings about animal cultures and the long tradition of genetic (developmental) psychology. It includes a critical review of the usual conception of organic selection as a mere transformation of habit into instinct. It also includes a critique of materialistic-monistic conceptions of biology and the causal-deterministic explanations of behavioral processes. Psychogenetic processes are not determined unidirectionally by the genome, the brain, the environment, cultural habits and norms, language, power, etc., nor by the "sum" of them all, but contains—mediated by all of them—its own empirical dynamics of determination. As an alternative to deterministic explanations, the chapter offers a distinction between and an integration of probabilistic and (psycho)genetic explanations. After the epigenetic revolution, the articulation of evolutionary biology with the behavioral and social sciences by incorporating a genetic point of view seems a little more feasible.

J.-C. Sánchez-González (✉)
Department of Psychology, University of Oviedo, Oviedo, Spain
e-mail: jocasan@uniovi.es

J. Pérez-Jara, Í. Ongay (eds.), *Beyond Nature and Nurture*, Synthese Library 497,
https://doi.org/10.1007/978-3-031-87715-5_6

6.1 Overcoming the Nature-Culture Duality, an Unfinished Task

Recent contributions in evolutionary biology have significantly expanded the possibilities for rethinking the problematic nature-culture duality. Many of these contributions go hand in hand with the decline of neo-Darwinism and call for the incorporation of new elements in evolution. Undoubtedly, animal cultures are one of these new elements (Avital & Jablonka, 2000; Mesoudi et al., 2006; Richerson & Boyd, 2008; Whiten, 2021). Behavior, as a complex developmental system capable of generating novel adaptation and transmitting it from generation to generation, seems to be another decisive element to be incorporated.

The range of behavioral adaptations is extraordinarily broad. It includes the generation of adaptive habits as well as habit transmission by multiple routes of social influence from one individual to another, such as attentional guidance, facilitation of learning by observation, emulative copying, imitative copying and even teaching (Whiten, 2021). In some species it also includes habits related to the production and use of physical structures that enable better exploitation of environmental resources, such as niches (Laland et al., 2019; Olding-Smee, 1988) and artifacts in general, which are the rudiments of material culture. All this opens up a novel scenario in which we find processes and relationships between artifacts and the behavior of new generations born into a world already transformed by the presence of those same artifacts, whose use must be learned.

The complexity of the processes of experience and learning that emerge with this new scenario goes far beyond anything the Modern Synthesis could have imagined. Although learning was never denied by the Modern Synthesis and was even recognized in foundational texts (Huxley, 1942), it played a secondary role there. Given the priority of genomic and selectionist explanations, the inevitable theoretical tendency was to approach animal behavior as just another selected trait, an "instinct" that, if anything, would be completed by simple associative, Pavlovian or instrumental learning processes.

In order for behavior to be credited with producing adaptive novelty, including what Ginsburg and Jablonka (2019) call the "origin of consciousness," behavior seems to be called upon to play a role in evolution that is much too big for this narrow framework. In her proposal for a theory of evolution based on the adaptive plasticity of phenotypes, West-Eberhard (2003, 2005) emphasizes the evolutionary role of behavior as a developmental system capable of producing learning. She conceives of learning as a peculiar process of self-selection of which emotions or attention are part. And she says that learning is the most plastic and powerful system for generating Darwinian adaptation, since learned behavioral adaptations (e.g., new migratory routes, new food sources, new foraging strategies) can sustain the survival of a population for generations, regardless of whether genomic changes occur to facilitate the development of the adapted behavior in future generations. In the event that heritable changes do in fact occur, it would be a case of not only Darwinian adaptation, but also Darwinian evolution.

The Extended Synthesis, as presented to us in the work of Pigliucci (2001) and Pigliucci and Müller (2010), is an impressive attempt, in tune with the new evolutionary developmental biology, EvoDevo, and spectacular new developments in epigenetics, to overcome the limitations of the Modern Synthesis. Pigliucci and Müller highlight West-Eberhard's contribution but do not share West-Eberhard's special emphasis on behavior as the main engine of Darwinian adaptation. In fact, behavior still plays no decisive role in the Extended Synthesis and EvoDevo in general but is presented as just another manifestation of phenotypic plasticity. EvoDevo focuses instead on the processes of tissue and organ development, which contain their own dynamics and their own constraints: thus, the genome, although a decisive part of the process, does not contain the pre-formed or encoded result, the trait. The result is epigenetic, the product of incessant processes of adjustment and self-regulation that weave different developmental paths within a range or norm of reaction.

Couldn't exactly the same be said of behavioral development, as a continuous process of reorganization of a system of habits through experiential regulations? This is exactly what Jean Piaget and James Mark Baldwin would have said, as would the psychologists and biologists who have taken an epigenetic approach to the study of behavior since the end of the nineteenth century. Regardless of the recent spectacular achievements of epigenetics and EvoDevo, it must be recognized that the idea of epigenesis, that is, the idea that development is not an unfolding of something pre-formed but the dynamic process of trait formation, is much older, and that there is a long controversy, dating back to Aristotle, between preformation and epigenesis (Robert, 2004; Young, 2013). The critical reaction to neo-Darwinism throughout the twentieth and twenty-first centuries has undoubtedly offered new insights into phenotypic plasticity and epigenetic processes (Jablonka & Lamb, 2005; Pigliucci, 2001; Waddington, 1961; West-Eberhard, 2003). In fact the controversy is still raging on, and the present work is just another episode in its development (Fernández, 1978, 1988, 2005).

Developmental psychology, or genetic psychology, took an epigenetic perspective precisely in order to explain the supposed "faculties of the human soul" as the result of an empirical process of development, supported in turn by an evolutionary, Darwinian perspective. Thus, the adjective "genetic" is to be understood here as referring to the empirical process of generation, not the unfolding of something given or pre-formed, whether deep in the genome or in the soul. Developmental psychology is an epigenetic approach to behavior in a evolutionary framework which incorporates the complexities of human cognition into the order of natural processes, in accordance with Darwin's requirement that the human mind must be explained in continuity with other forms of animal cognition. Today developmental psychology is a central field of human psychology and is sometimes labeled Developmental Science. It studies, for example, the development of human attention, perception, emotions, socialization, symbolic thought, language, reasoning, judgment and moral sense (Lerner, 2012; Overton et al., 2015).

The genetic psychologist does not see behavior as a linear effect of external causes (brain input, environmental stimulation, mechanical association determined by the mere presence and repetition of environmental stimuli, etc.), just as the Evo-Devo biologist does not see tissue or organ as a linear effect of genetic programming or environmental perturbation but as the resultant of adjustment processes relying on the whole organism and environmental variables and having a regulatory logic of their own, by which behavioral outcomes are determined. Both the biological development of a tissue and the psychological development of a habit (or system of actions) are processes in which the organization of the tissue or habit takes place, undoubtedly taking into account all the antecedent factors. It is not therefore a question of denying determination, but of seeing it as a process of epigenesis that has its own dynamics.

Now, thus presented, it might seem that a bridge could be built to enable evolutionary developmental biology to complete the Extended Synthesis and incorporate behavior, as a developmental system, according to the contributions of genetic and developmental psychology. But the fact is that no such thing has happened. The bridge dangles between the two shores, not quite complete. And this has an important effect when it comes to dealing with the nature-culture problem. Without an adequate solution for the behavioral problem, the question of culture in nature remains unclear. A reductionist "solution" to animal behavior and culture, setting aside all human culture and history (including language, symbolic and purposive activity, values, knowledge, etc.), would still leave us trapped in the nature-culture duality. Human culture would still belong to another 'kingdom'.

The difficulties in completing the bridge are structural. They are systematic, correlated, endlessly discussed problems of biology, behavioral sciences and social sciences, often concerning the scope and limits of mechanistic conceptions of life (Henning & Scarfe, 2013; Plotkin, 2004). When West-Eberhard draws content from psychology to expose the complexity and flexibility of behavior, she reflects the current state of the discipline, where mechanistic traditions (associative, connectionist and computational traditions, or even neural reductionism) have considerably more adherents than do epigenetic approaches like developmental psychology.

On the side of biology, the reason for the problem is—as will be argued in this chapter—that the basic explanatory scheme on which EvoDevo and the new Extended Synthesis (including West-Eberhard) rely is in essence an extension of the Modern Synthesis's explanatory scheme: a deterministic scheme, now significantly enriched with the idea of multilevel determination, quite unlike the Modern Synthesis, which tends to prioritize unidirectional gene-trait determination.

This explanatory approach, inspired by Developmental Systems Theory (Oyama, 1985; Oyama et al., 2001), has been extraordinarily fruitful in that it has made it possible to contemplate and integrate new phenomena. The product to be explained is then presented as the result of the complex interactions of all the levels that play a role in the process. Behavior would also be a resultant of these interactions, for example, as a complex result of genomic, neurological and environmental factors. The problem is that such a scheme may conceal the specific genetic logic of the interactions that occur in behavioral development. This kind of explanation certainly takes into account all the antecedent factors, as we have said, but it explains the behavioral result through its own gestation, by means of the sequence of experiential regulations involving "trial and error" and past experience.

Current comparative psychology, represented by authors such as Gottlieb and Lickliter, shares the developmental systems approach and the tendency to offer the explanation of behavioral change as the result of the sum of multilevel interactions, often induced by an environmental disturbance. Comparative psychology has had a long and intricate history since its origins in Darwin's work. Throughout that history it has been an enemy of behaviorist environmentalism and neo-Darwinian geneticism alike. It has developed the critique of instinct, i.e., the critique of the idea that a behavioral pattern appears, complete and fully functional, without the participation of experience (Lehrman, 1953). And it has done so precisely from an epigenetic perspective, emphasizing the pre- and postnatal experience needed for the pattern to emerge with full functionality. Today its rejection of innatism has become a radical critique of Evolutionary Psychology, which Lickliter considers a holdover from selectionism and an obstacle to a necessary future "evolutionary developmental psychology" (Lickliter, 2008; Lickliter & Honeycutt, 2013). Nevertheless, despite this radical epigenetic outlook, for Lickliter explanations of behavioral development in comparative psychology are, in line with Gottlieb, the result of multilevel interactions, as in Developmental Systems Theory.

6.2 The Bridge Offered by Organic Selection, and the Subsequent Endless Misunderstanding

The theory of organic selection, or "Baldwin effect," was the first post-Darwin systematic outline of a bridge between evolutionary theory and general theory about adaptive behavior, i.e., between life sciences and behavioral and social sciences. It was formulated in 1896 by James Mark Baldwin and his contemporaries Henry Fairfield Osborn and Conwy Lloyd Morgan, but it was Baldwin who further elaborated on the idea, to which he contributed a new general theory of psychological development.

If we can speak of an unfinished work or, to use the same metaphor as before, a bridge that fails to span the shores, it is because organic selection contains the complete blueprint for the bridge. It also lies at the base of current attempts to complete the structure, assigning behavior a decisive role in adaptation and evolution, as sought, for example, by West-Eberhard. Since its formulation, organic selection has not ceased to be used, in one way or another, by authors of the most diverse tendencies, from neo-Darwinism to zoosemiotics, passing through artificial life and Evolutionary Psychology. K. Popper, C.H. Waddington, R. Cairns, J. Maynard-Smith, D. Dennett, J. Hoffmeyer, K. Kull, S. Pinker, R. Dawkins, H.C. Plotkin, B. Hall, G.E. Hinton and S.J. Nowland, T. Deacon and M.J. West-Eberhard are just some of those who have used it in one way or another.

There is, however, a special problem with organic selection. It is a problem that pervades the controversies of the twentieth and twenty-first centuries and threatens to persist and continue to sow confusion until the end of time. There are essentially two possible interpretations of organic selection, which we can call the *restricted* version and the *broad* version (Sánchez & Loredo, 2007). Which version one favors depends on how much Baldwin one has read, but it depends even more on one's philosophy of biology and psychology. Those who favor more of a mechanistic view of evolution and the reduction of animal behavior to instinct tend to use the restricted version, especially if they have only read "A New Factor in Evolution" (Baldwin, 1896). Those who, on the contrary, are more sensitive to the processes of genesis and even social factors use the broad version, if they have also read enough Baldwin to defend themselves from the pressure of the majority who believe that the restricted version is the one true possible view.

The *restricted version* says that learning processes can produce novel habits that, if maintained generation after generation, will over time enable selection for random genomic variations that will eventually support, reinforce or even replace habit with instinct. This is the version offered by Baldwin in his famous article "A New Factor in Evolution" (Baldwin, 1896), where the priority objective was to offer a non-Lamarckian explanation of the impact of acquired traits on evolution and particularly on instinct formation.

Use of the restricted interpretation is associated with mechanistic, environmentalist and selectionist conceptions of learning, where new habits are in essence responses determined by the environment, for example, as a result of the environmental selection of randomly emitted responses. Skinner, in fact, saw a deep analogy between the two worlds, between phylogenetic contingencies (random genetic variation and environmental selection) and the ontogenetic contingencies that could explain learning (random emission of responses by the organism and subsequent selective reinforcement by the environment). Pinker and Bloom give a bold, radical, modern formulation of the restricted version: "The process whereby environmentally-induced responses set up selection pressures for such responses to become innate, triggering conventional Darwinian evolution that superficially mimics a Lamarckian sequence, is sometimes known as the Baldwin Effect" (Pinker & Bloom, 1990, p. 722).

Development and genesis disappear here. The new behaviors that emerge have the mission of laying the groundwork for natural selection to make them innate. Essentially, they arise under the old "blind variation and selective retention" scheme. The scheme fits it with both Modern Synthesis and Skinnerian environmentalism. Significantly, however, Baldwin used the expression "organic selection" for the first time to refer precisely to the fact that it is the organism, through its conditions of sensitivity and experience, that selects, organizes and integrates responses. That is why selection is "organic."

The scheme also fits in with Evolutionary Psychology's adaptationism and the reduction of human culture to memetics, where the dynamics of the genesis and continuous transformation of artifacts, techniques and ideas is reduced to a matter of variation and selection of memes (Costall, 2014). It is a complete program for the reduction of nature and culture to the god of chance, as a "solution" to genesis.

The *broad version* rejects the rigidity of the restricted view. It says that the processes of learning and experience constitute a plastic system whose destiny is not to disappear into a fossilized myriad of instincts, but to continue to produce adaptation in the face of new circumstances. It is worth bearing in mind that Baldwin laid the foundations for his genetic psychology before writing "A New Factor in Evolution." He later brought all the threads of the theory together and published it in his book *Mental Development in the Child and the Race* (Baldwin, 1906).

Baldwin tries to explain the evolution of plastic behavioral systems, which enable us to account for the peculiar plasticity of human behavior. He does not endeavor to create a theory of the random fabrication of rigid instincts with no plastic components. That there are strong specializations or channelings for learning is something that has been put beyond all doubt today, but that there is no instinct that is completely rigid or unrelated to experience is something that we can see today much more clearly than could Baldwin and his contemporaries. Indeed, experimental work on habituation and reflex sensitization processes as well as theoretical advances such as the work Daniel Lehrman reports in his famous article criticizing Lorenz's theory of instinct (Lehrman, 1953) both show that the absolute rigidity of instinctual patterns, in which ethology once believed, does not exist. Ethology, then, is not generally a discipline essentially distinct from comparative psychology, insofar as it studies behavioral processes of adaptation of Darwinian organisms, including learning, which takes place on the basis of the aforementioned species-specific specializations or channelings.

Throughout his work Baldwin emphasizes this broad vision of organic selection, where behavioral plasticity generates novel habits that are maintained generation after generation (either by repeated individual learning or by social influence) and thus contribute to the selection of heritable variations that directly or indirectly support such habits, without an ensuing loss of plasticity and without the complete, *obligatory and invariable* substitution of habit for instinct. This is the core idea of organic selection. Baldwin articulates it by distinguishing between coincident variations and correlated variations.

Coincident variations are inheritable changes that directly support the adaptive habit. For example, an inheritable change in food preferences may help later generations learn to make better or faster use of a food source already familiar to previous generations. Thus, coincident variations generate these specialized dispositions or instinctive channelings.

Correlated variations are changes that support the "lifestyle" or resource exploitation system to which habits have led, for example by strengthening the tail, claws and beak and lengthening the tongue of the woodpecker. Correlated variation helps to understand the co-evolution of various traits, not all of which are directly related with the habit, but all of which are useful for consolidating an already-existing exploitation system.

The restricted version must then be seen as an extreme case of the process: the complete replacement of the habit by an equivalent instinct. And it must be seen within the conceptual framework of the late nineteenth century, which was far from the sophistication of EvoDevo and contemporary epigenetics. Organic selection is, then, a process in which behavioral plasticity does not disappear, which is precisely what makes it possible to generate habits, either to repeat the learning of the habit of the past or to generate, once again, some variation or novelty. Adaptive behavioral innovations can in turn succeed without the need for an immediate incorporation of heritable variations (as in fact happens even in the restricted version), thanks to processes of social influence, which are an intrinsic part of individual development in social animals, especially humans.

In the "slightly revised" version of "A New Factor in Evolution" that Baldwin incorporates into the collection of papers appearing in 1902 in his book *Development and Evolution*, he says:

> This [learning and social transmission of habits], then, is a form of ontogenetic accommodation; it aids to keep these creatures alive, and so to produce definite change in the way explained above. It is, therefore, a special, and from its wide range an extremely important, instance of the operation of the general principle of organic selection. But it has further value: it keeps alive a series of functions which either are not yet, or never do become, congenital at all. (Baldwin, 1917, p. 104).

Organic selection thus incorporates social inheritance, which contributes to the fixation of heritable variations, be they many or few, including the possibility that for the time being there are none. Social inheritance is not specifically human, but it is especially applicable to the human case, where a great many behavioral variations do not seem to massively carry over coincident heritable variations (although some do, e.g., adult tolerance to lactose in livestock cultures).

Human language quite possibly evolved this way. Firstly, on the cerebral side, an increasingly complex system of neural dispositions has evolved, selected to ensure rapid language learning and effective language use generation after generation, instead of a special module's having been produced by "lucky random variation." Secondly, language includes a system of rules of pronunciation and syntactic arrangement and a system of semantic assignment that have to be respected and maintained generation after generation. And thirdly, in terms of behavioral plasticity, language implies an uninterrupted chain of actions of relearning, use and social transmission. The three realities are irreducible and necessary in order to understand the phenomenon. Wherever there has been a break in social transmission and use, language has disappeared or has been reduced to unintelligible inscriptions on stone, the "ruins" of a written language.

Terrence Deacon (1997, 2003) uses these ideas at first, acknowledging organic selection as an inspiration. West-Eberhard's (2003) theory of evolution also employs the broad version of organic selection, where behavioral processes can generate Darwinian adaptation over generations, relying on the available plasticity without waiting for lucky mutations, and can ultimately lead to Darwinian evolution if somehow the development leading to the adaptations remains heritable (according to processes now defined by epigenetics).

Waddington and Schmalhausen, forerunners of EvoDevo, single out processes (stabilizing selection, genetic assimilation) that do not explicitly rely on behavior but do point to an idea that is analogous to organic selection, although not "essentially similar," as claimed by Gottlieb (1992). The idea is that there may be selection of heritable factors that facilitate the appearance in offspring of relatively novel courses of development (e.g., physiological courses of development) that have already emerged as successful responses to environmental perturbations.

Thus, to summarize, the difference between the restricted and the broad version is that the broad version (1) is not reduced to a process of replacing a habit with an instinct, (2) emphasizes animal behavioral plasticity and brings in a psychological theory to explain it, developmental or genetic psychology; (3) incorporates processes of social influence, or "social inheritance", to explain the maintenance and transmission of habits over generations and (4) explains the incorporation of heritable genomic variations that often complement or support the habit or even, in certain situations, replace it completely, as suggested by the restricted version (which therefore is a particular case of the broad version).

In general, the broad version of organic selection can be said to refer specifically to the adaptive behavioral processes of genesis and social transmission and to the role of these processes as "selectors" of eventual heritable, coincident or correlated variation. Organic selection does not specify which mechanisms produce heritable variation; it simply adopts the scientific results offered by biology. In Baldwin's time and during neo-Darwinism, these mechanisms were thought of in terms of random variation of discrete genes for discrete traits, but developmental biology has substantially transformed that view with epigenetic processes and even epigenetic inheritance phenomena (Jablonka & Lamb, 2005), which enrich our understanding of variation and inheritance but do not replace or override the logic of the processes of behavioral genesis. Organic selection is perfectly compatible with modern epigenetics and with eventual phenomena of epigenetic inheritance, stabilizing selection or genetic assimilation.

The authors who have most clearly used the broad version of organic selection come from fields closer to Baldwin's psychological and social work, such as Moore (2003) and Weber (2003), and even biosemiotics (Hoffmeyer & Kull, 2003). More recently, authors such as Le (2019) and Calandrini and Marsico (2020) also point to the broad version.

What Baldwin brought especially to organic selection, as we have highlighted, was a new systematic psychology, a psychology of development that laid the groundwork for the modern tradition of genetic psychology and decisively influenced Jean Piaget's genetic child psychology and Vygotski's sociohistorical psychology. But Baldwin, a fervent Darwinian, was not only aiming at a genetic psychology of the infant mind; he sought at the same time to create a genetic comparative psychology, that is, a theory of habit development "simple" enough to be applied to elementary processes of experience in the animal world, including even unicellular organisms, as H.S. Jennings did under the influence of Lloyd Morgan and Baldwin (Jennings, 1904).

Let us pause for a moment to consider the current landscape of learning capabilities beyond human beings. The experimental procedures of classical, or Pavlovian, conditioning have reached an extraordinary degree of sophistication after a century of research by the associative learning theory tradition. It is now possible to operationalize and distinguish various processes, such as first-order conditioning, blocking, shadowing, sensory preconditioning, conditioned inhibition, conditioned taste aversion and second-order conditioning. So, a whole variety of conditioning phenomena have been demonstrated in all vertebrate species experimented with and in a considerable number of species belonging to various invertebrate *phyla*. In protists, despite the long tradition of research begun by Jennings, there is very scanty evidence of conditioning; what evidence there is still needs replication. In plants the evidence is practically nil. This does not mean that elementary processes of behavioral plasticity do not exist, but that Pavlovian associative procedures have found their operational limit, just as that experiments on imitation in echinoderms would probably find theirs. Non-associative processes of habituation and sensitization, involving intensification or attenuation of a reflex as a result of sensory experience, have been demonstrated in paramecia and amoebae. Phenomena of enhancement in the execution of a pattern, e.g., escape, are also found as an effect of successive repetitions. Such phenomena could be interpreted as the most elementary forms of development (Loy et al., 2021).

The research strategy in comparative psychology or ethology consists in *reapplying* to a new species, for example an annelid, an experimental procedure that is already sufficiently demonstrative of a certain process in another species, for example, a mollusk or a mammal. The procedure operationally defines the particular process (blocking, imprinting, sensory memory, taste aversion learning, attachment, object permanence, emulation, etc.), and, when re-applied to the new species, with the necessary adjustments, enables us to *affirm* any similarities and differences we find. What is not part of the comparative strategy (and what "Lloyd Morgan's canon" suggested avoiding) is both the gratuitous *attribution* of "superior" processes to new species (anthropomorphism) and the *reduction* of a lavish variety and complexity of processes to a single, simple process (reductionism). To Morgan's disappointment, reductionism was the choice of many, starting with the behaviorists (Costall, 1993). But the pendulum has recently swung in the opposite direction. Today, attributions of psychism, consciousness and cognition are reappearing in biology. Although the phenomena they describe are often astonishing, their operationalization is in some cases deficient. Sometimes they rely on computational interpretations of complex physiological processes, or the whole of life, as a system that "captures" and "uses" "information," as sometimes happens in the field of Minimal Cognition (Loy et al., 2021). In philosophy panpsychism has taken on new vigor through the work of Goff (2019). In the attempt to solve the problem of dualism posed in metaphysical terms (how consciousness can arise from matter), Goff attributes consciousness or psychism to the totality of the cosmos, since time immemorial, from quarks on up. This is certainly not comparative psychology, neither from the methodological point of view aforementioned, nor from the ontological point of view. Atoms can do nothing to "survive" or "adapt". Psychology has to do primarily with action, and where there is no action there is no reason for there to be consciousness.

6.3 Probabilistic and Genetic Explanations

We have already indicated that, when explaining behavior in the framework of Developmental Systems Theory, Evo-Devo biologists and comparative psychologists tend to offer a picture of behavior as the effect or result of multilevel interactions. However, we have been arguing that genetic explanations are indispensable and essential in behavioral sciences. Of course, physical or physiological factors must be taken into account, but the explanation of behavioral change does not stem from the mere presence of these factors. It stems from the assimilation and functional use that the organism can make of these factors, according to its behavioral resources.

A set of spots in the distance is a sign of the presence of prey only to the experienced eye of a hunter. The hunter can therefore use the spots to estimate distance and direction and initiate his approach. In the process of approaching, the hunter will correct his position relative to his prey over and over again, keeping in mind wind direction and terrain features that he can use as concealment. The process contains continuous adjustments. Hunting successes or failures will modify the hunter's habits or strategies. The physical and physiological factors (wind, hunger, exhaustion, etc.) are part of the game but do not determine the result beforehand. The result must be arrived at, through these incessant self-regulations, each of which hinges on the perceived state of the world at the time, at the ever-present risk of failure.

In fact, a process of behavioral genesis, unlike a "spiritual creation," is very modest. It consists in repeating an old habit. In behavioral genesis, difficulties that arise as the habit is repeated necessitate variations or readjustments, insofar as possible; if those variations or readjustments are successful, they can become fixed by structurally transforming the previous habit. It is not a sum of loose associations or responses, nor a computational task of a specific domain, but structured growth that weaves a system of operations, as Piaget would say (go forward-wait; go straight-round; stay leeward and hidden or be seen and detected, etc.).

Now, we believe that it is possible to clarify the apparent insurmountable contradiction between genetic and deterministic explanations in behavioral sciences. To do so, we must distinguish between *genetic explanations* and *probabilistic explanations*. The explanations that are usually presented in psychology as cause and effect are not, strictly speaking, a matter of cause and effect; they are probabilistic explanations, as emphasized by comparative psychologists themselves, starting with Gottlieb. There is no causal scheme whatsoever, if by strict causality we mean something analogous to the physical sciences' explanation, i.e., a complete chain of transformations leading from an antecedent state to the final state whose explanation we seek.

Getting back to our hunter, the outcome of the process is not written, nor can it be determined in advance by a behavioral scientist. On the contrary, it has to be performed by our hunter with contextual adjustments, in exactly the same sense that scientists cannot predict the outcome of new research or fresh reflections, but laboriously arrive at their outcome by performing the process.

Similarly, in human psychology, we would say that it is inaccurate to claim that we can predict with causal accuracy that a given person will drop out of school because he or she belongs to a dysfunctional, low-income family with an alcoholic father. Undoubtedly, factors of this type induce us to affirm that the person's developmental process is probably heading towards a greater or lesser degree of academic failure. But, as in the case of our hunter or our scientist, the process must take place. It is not pre-determined. Even knowing in advance that each hunt, for example in leopards, has a 60% probability of failure.

Probabilistic explanations are therefore not causal. They are, however, perfectly rational, legitimate and extraordinarily useful, because when similar conditions are given for organisms that share similarities (of species, development, resources, etc.), the results are statistically similar. But they cannot hide the underlying genesis process. If the result finally appears, it is because the developmental process that generated it has taken place. Often, even when there are very similar starting conditions, the process does not occur identically, according to a "standard," because even to arrive at the standard result, relatively different, idiosyncratic developmental paths have had to be followed.

Genetic explanation is a notion of major importance in developmental psychology from its beginnings, for example in W. Wundt or F. Sander, and notably in H. Werner (1940, 1956), to the present, for example in the in the extensive theoretical work of J. Valsiner (2013). In the behavioral sciences, therefore, we have two kinds of explanations, probabilistic and genetic. Both are legitimate. They involve knowledge and the capacity for prediction and intervention. They work on both understanding regularities and understanding the specific genesis that the organism has to carry out, which enables us to understand regularities as well as variation and divergence. What we do not find in behavioral sciences are deterministic explanations in the classical sense of natural sciences like Newtonian mechanics and chemistry. Perhaps this way, through the coordination and justification of the need and relevance of both types of explanation, the possibility of completing the bridge that organic selection outlined more than a century ago becomes a little clearer.

6.4 Evolution and History

Organic selection, in the broad sense, with the comparative genetic and social psychology that goes with it, is the best available outline of a bridge that could integrate biology and behavioral sciences, and that could therefore provide Extended Synthesis with the connection its needs to complete its process of "extension." Furthermore, it could help the behavioral and social sciences reach a better definition of how they both fit in with evolution, in order to avoid the misstep that goes along with reductionism, which is the mistake of divorcing the social sciences entirely from evolution.

In the twentieth century, there have been various episodes of this divorce. For example, there is the rupture of Boas's anthropology with biology, in view of the conceptual poverty of biological or psychological theories that multiplied instincts to explain any human behavior. Another more recent and particularly significant episode of the divorce is expressed in the proliferation of *social constructivisms*, often linked to relativistic positions on human action and knowledge. These positions are often based on irrationalist views of human behavior. For example, the sociology of knowledge, with its Strong Program, takes for granted that human behavior in general, including scientists' behavior, is governed by interests but not by reasons. The requirement of behavioral consistency and rationality seems, however, intrinsic to action in evolution and therefore in history. In general, social constructivisms lack a theory of action supported by an evolutionary perspective, and they revolve around cultural categories such as "discourse," "power" and "negotiation." They often explicitly reject the need or desirability of evolutionary support, under the assumption that any connection with biology would imply reductionism, and they ignore the old non-reductionist "discourse" of genetic psychology (Loredo & Sánchez, 2012; Sánchez & Loredo, 2009).

What is important for the nature-culture problem is that, with a broad reading of organic selection, we pave the way for a non-reductionist explanation of animal cultures, which allows us to see the continuity with hominid cultures and finally with human culture.

Moreover, cultural products are not only stable habits, but eventually artifacts, "industries," which in certain species, particularly in the genus *Homo*, have gradually become an essential adaptive strategy. Nature, in the biological realm, already incorporates components of culture, provided that we articulate them through a non-reductionist theory of behavior. Nature and culture are no longer two opposing kingdoms. The opposition of realms was based precisely on a view of nature or organic life as an exclusively "molecular" reality, as Dobzhansky and his colleagues posited when defining evolution as a change in the distribution of gene frequencies in populations (Dobzhansky et al., 1977). Correlatively, then, culture should be defined as the realm of the spirit, of things like intelligence, symbols, language and values.

As opposed to the nature-culture duality, the evolution-history series, in the sense we have been expounding, expresses much more clearly the continuity of "substance" while recognizing qualitative changes everywhere. These changes would not be understandable without taking seriously the behavioral processes of genesis of new habits or systems of action. In evolution, as in history, there are molecular components and processes (genomic ones, physiological ones, etc.), there are systems of action, and there are artifacts. The critique of the substantial nature-culture duality implies a pluralistic ontology, also for a sufficiently "extended" evolutionary biology. The genetic, constructivist tradition, including pragmatism, has been prone to more or less systematic pluralist approaches. Today there is a plurality of pluralisms, among which we can highlight the philosophy of biology of

J. Dupré (2012). In the Spanish literature, the philosophical materialism of G. Bueno develops a sophisticated pluralistic ontology with constructivist basis, although Bueno's ontology is essentially disconnected from the psychogenetic approach and the question of organic selection in the sense in which we are presenting it here (Bueno, 1972, 2016; Ongay, 2022; Pérez-Jara, 2022).

The criteria for defining what history is or when it begins vary widely. The point here is not to analyze them, but just to emphasize a well-known behavioral idea: history arises in continuity with previous cultural processes, when systems of habit or action appear that are reapplied on top of already-existing systems of habit or action. For example, when linguistic practices appear (first oral practices and then written practices), they begin to define and fix the deeds, works, customs, norms and knowledge of a community. Thus, in general, new systems of action, which come to codify and manage old systems of action and artifacts, generate new levels of productivity, social organization and shared consciousness.

We might therefore sort behavioral systems and culture very roughly into three levels. The first level is that in which culture is practically restricted to *stable habits that are repeated and propagated by social influence*. We have already commented that current research on animal cultures shows that the processes of social influence are extraordinarily varied, from the most elementary, such as preference biases, to processes of emulation or imitation or even instruction, where there is a deliberate intent to lead another's behavior to certain results, as seems to occur in the case of the teaching of hunting techniques in killer whales (Whiten, 2021). This is the first level of what we can call culture, perhaps the most common throughout evolution.

At the second level we find the *production of simple artifacts and processes of social influence that enable the stabilization of the system of habits by which these artifacts are used and reproduced*. Classifications could be set up according to very diverse criteria: operation complexity, artifact structure, forms of social transmission, etc. In any case, generally speaking, starting from the level-1 processes, certain populations develop habits in manufacturing and using artifacts. These artifacts, portable or otherwise, are repeated consistently over generations, thanks to processes of imitation and instruction from parents to children, and are transformed, usually slowly. The results that anthropologists, paleontologists and historians have found are sequences of artifact forms, for example, the series of thousands of years old stone tools belonging to capuchin monkeys (Falótico et al., 2019), or axes and flint points that we know from various *Homo* species.

At the third level, the history level, we find new systems of action that begin to operate on the previous level's own cultural production: on the physical production of artifacts, resources and food; on memories and scattered narratives; on procedures and skills that must be preserved. All this opens up a new level of organization of group consciousness about the group's origin, its "past," but also about the use of techniques and knowledge with which collective action is projected, which begin to define what is possible or expected, i.e., "the future."

6.5 Habit and Artifact: A Complex Dynamic

Let us get a little more specific about the genetic dynamics that lie at the base of the production and transmission of cultural artifacts in non-human animals, in hominids or in humans. We will do so through a set of related concepts: mediation, diagramming, scaffolding, openness and functional pulse of artifacts.

Mediation When there are artifacts that are part of the culture of a population or group (including trails, burrows and other analogous physical structures), the group's overall behavior is mediated by those artifacts, in the sense that the group could not achieve its current levels of adaptation without them, but not in the sense that it is determined by them. In general the same can be said whatever the complexity of the artifact: we wingless vertebrates cannot fly, be it simple gliding or sustained propelled flight, without making use of some enabling artifact; and the character and quality of the artifact dictate decisive possibilities of and limits on flight, but they do not completely determine the functionality of flight: the destination, the trajectory, the load, etc.

Diagramming By "diagramming" we mean the basic system of operations for using an artifact, the operations that must be performed or else the artifact completely loses its functional reality, i.e., disappears as a functional object even if it remains as a physical entity. For example, a termite fishing pole has to be inserted lengthwise, and its twigs must be removed first so that it does not get stuck. This is how expert adults do it. An inexperienced ape using a stick of any old size and shape is not yet in possession of a "termite fishing pole." Each individual must learn the essential use of the artifact, its diagramming, regardless of whether he learns (as he probably will) additional elements, idiosyncratic variants that sometimes extend the object's diagramming, for example, how to jiggle the stick in the termite mound so that the termites attack it more furiously and more can be caught on each cast.

Scaffolding The influence of other, more expert individuals is often decisive, and such influence varies widely, as research on animal cultures shows (Whiten, 2021). But the same happens with the psychology of human development, which emphasizes the social and cultural constitution of our mind (Valsiner & Rosa, 2007; Valsiner & Van der Veer, 2000). In the human case, and perhaps in other mammalian species, there are deliberate processes of instruction of learners by expert adults (Thornton & Raihani, 2008). The experts' operations are aimed at guiding the learner's developmental process toward the predictable next steps, preventing the learner from straying into errors or complications. Adults know from their own experience how to anticipate possible errors or complications. In developmental psychology this relationship is called scaffolding. Scaffolding seeks, among other things, to help the learner reach diagramming. Relationships of social influence are thus intrinsically oriented to the maintenance of "objectivity," that is, the reconstruction and safeguarding of the knowledge acquired by the group.

Openness But the artifact, objective culture, does not determine action or reduce behavioral plasticity, because without behavioral plasticity diagramming cannot be reconstructed and artifacts cannot be used contextually. Action and artifact belong to two different levels, in a complex relationship. The artifact is born of the action (habit existed without artifacts) and now exercises a mediating influence that undoubtedly opens up new possibilities for the action. But action is not reduced to mere fulfillment or execution of diagramming; plasticity is not governed or determined by diagramming, just as we said that habit is not crystallized in instinct. The plasticity of the action remains necessarily present in the use of the artifact and can always overcome and transform learned use. We can call this dimension "openness." Openness is nothing more than the acknowledgement of genetic potential for reorganization or adaptive readjustment of the system. Thus, once artifacts exist, action depends heavily on them (through mediation), but an artifact does not explain or determine an action, nor does an artifact exist independently of all action.

Functional Pulse of Artifacts Thus, artifacts, or cultural realities in general, do not lead an autonomous, purely "objective" existence. An artifact exists insofar as subjects reconstruct their reality by learning its basic diagramming. Operations, including systematic scaffolding operations (teaching), are aimed at maintaining the stability and consistency of the artifact's use. Furthermore, through use (which involves contextual testing of the artifact), operations can transform the artifact. This is an example of openness.

Through successful differential uses and adjustments, the various genera and species of hammers have "evolved" (the carpenter's hammer, the formworker's hammer, the sheet metal worker's hammer, the sculptor's mallet, the engraving hammer). These transformations involve both changes in habit and changes in the artifact's physical structure. Metaphorically speaking, artifacts live and evolve through action, which, as a classic historian of technique says, is never pure speculative or fanciful creativity (Basalla, 1988). Action realizes artifacts by using them, and eventually it transforms them. We might call this the functional pulse of artifacts, to emphasize that artifacts are fully real only through actions, that they are not self-subsistent entities, and that without actions artifacts "die" and decay into mere physical remains: ruins, debris, rotten wood, marks on stone, scrap metal or ink strokes.

The morphology of flint points evolved under regulated carving operations, through generations of masters who knew the phases, the sequence and the angle of the blows to be applied to the stone. Without these regulated operations (diagramming), scaffolded in instructional processes yet capable of openness, there would be no history of flint knapping. The operations, which some would say are ethereal (behavior is said to leave no trace) are as real as the stones themselves. Modern experimental archaeology, in a kind of reverse behavioral engineering, reconstructs the diagramming of the Paleolithic master: thus and thus the edges of the stone are cleaned and the flint core is prepared, thus the flake is obtained, thus the edges are struck to obtain the sharp denticulation.

Artifacts therefore have no entity or existence that is entirely independent of operations, and of course they are not operators, because they do not perform actions. They are not, for example, Latourian "actants." Latour is another example on how to "divorce" the social sciences from a genetic theory of operations; in his case the divorce begins with a disregard for Piaget's psychology (Latour, 1996). But perhaps the most controversial artifact is the computer. Computers are not beings capable of performing operations in the sense we are explaining, either, because they are not organisms. This reason, though obvious, is not trivial. Goff's panpsychism notwithstanding, most probably action and thought are attributes of life, a developmental system. Computers do not need, do not desire, do not feel, do not suffer. They compute, but they do not perform meaningful actions. Searle would say that they do not think. Computation is not a model capable of accounting for operations; even computational psychology seems to accept the limitations of computation in explaining the central processes (Fodor, 2000) that (as we would say) are not controlled by an instructional system and must break out of the diagram and generate solutions even at the risk of error.

6.6 A Very Brief Conclusion Considering How the Bridge Would Change the Shores

Profound and largely revolutionary transformations in biology have opened the door to the recovery of epigenetics and development, to the integration of "animal cultures" in the evolutionary process and to the incorporation of a theory of behavior as an essential factor of Darwinian adaptation. The deterministic conception of behavior, linked to a reductionist view of life and evolution (molecular materiality), is the main obstacle to the completion of this revolution.

To help put this conception behind us, we have emphasized the radical anti-metaphysical significance of genetic explanations in psychology, and we have stressed their initial linkage to a plan of comparative psychology capable of showing minimal or elementary forms of development in the animal kingdom. We have also developed the distinction between genetic and probabilistic explanations in the study of behavioral processes, which enables us to accept the contribution of traditional explanations that have been presented to us as "causal" or "deterministic," although we have reinterpreted them as resulting from empirical processes of development involving self-regulation and genesis.

The theory of organic selection lies at the base of attempts to integrate evolution and behavior from the late nineteenth century on. The theory's restricted interpretation still leads to a reductionist view of life and evolution, and it can be used to continue to maintain a substantial wall between nature and culture. The broad interpretation allows behavior to be introduced into evolution as a developmental, genetic process that accounts for behavioral innovation, the sustaining of habits through generations, the variation and functional transformation of habits, and the complex processes of social influence and artifact mediation. The Extended Synthesis would be better extended by integrating these other realms of reality in evolution: behavior, as a process of genesis, and artifacts, as realities that mediate behavior and at the same time stem from and depend on behavior.

The perspective we have presented can contribute to the critique of the substantial nature-culture duality, replacing it with the evolution-history sequence, where there are incessant qualitative changes but no substantial leaps. The evolution-history continuum also has implications for the social sciences, which are often divorced from the life sciences, because it allows—and requires—the social sciences to incorporate a reductionism-free evolutionary perspective.

References

Avital, E., & Jablonka, E. (2000). *Animal traditions. Behavioral inheritance in evolution.* Cambridge University Press.

Baldwin, J. M. (1896). A new factor in evolution. *American Naturalist, 30*, 441–451, 536–553.

Baldwin, J. M. (1906). *Mental development in the child and the race. Methods and processes.* MacMillan.

Baldwin, J. M. (1917). *Development and evolution.* Macmillan.

Basalla, G. (1988). *The evolution of technology.* Cambridge University Press.

Bueno, G. (1972). *Ensayos materialistas.* Taurus.

Bueno, G. (2016). *El Ego trascendental.* Grupo Helicón.

Calandrini, A. R., & Marsico, G. (2020). Rediscovering James Mark Baldwin. *Psychology Hub, 37*(1), 5–15. https://doi.org/10.13133/2724-2943/16896

Costall, A. (1993). How Lloyd Morgan's canon backfired. *Journal of the History of the Behavioral Sciences, 29*(2), 113–122.

Costall, A. (2014). The 'meme' meme revisited. In T. Breyer (Ed.), *Epistemological dimensions of evolutionary psychology* (pp. 89–101). Springer. https://doi.org/10.1007/978-1-4939-1387-9_4

Deacon, T. W. (1997). *The symbolic species. The co-evolution of language and the brain.* Penguin.

Deacon, T. W. (2003). Multilevel selection in a complex adaptive system: The problem of language origins. In B. H. Weber & D. J. Depew (Eds.), *Evolution and learning. The Baldwin effect reconsidered* (pp. 81–106). MIT Press.

Dobzhansky, T., Ayala, F. J., Stebbins, G. L., & Valentine, J. W. (1977). *Evolution.* Freeman.

Dupré, J. (2012). *Processes of life. Essays in the philosophy of biology.* Oxford University Press.

Falótico, T., Proffitt, T., Ottoni, E. B., Staff, R. A., & Haslam, M. (2019). Three thousand years of wild capuchin stone tool use. *Nature Ecology & Evolution, 3*(7), 1034–1038.

Fernández, T. R. (1978). Culturas animales. *El Basilisco: Revista de materialismo filosófico, 1*, 17–31.

Fernández, T. R. (1988). Conducta y evolución : historia y marco de un problema. *Anuario de Psicología, 39*, 99–136.

Fernández, T. R. (2005). Sobre la historia natural del sujeto y su lugar en una Historia de la Ciencia. A propósito de Robert J. Richards y el romanticismo de Darwin. *Estudios de Psicología, 26*(1), 67–104.

Fodor, J. (2000). *The mind doesn't work that way: The scope and limits of computational psychology.* MIT Press.

Ginsburg, S., & Jablonka, E. (2019). *The evolution of the sensitive soul: learning and the origins of consciousness.* MIT Press.

Goff, P. (2019). *Galileo's error: Foundations for a new science of consciousness.* Vintage.

Gottlieb, G. (1992). *Individual development and evolution: The genesis of novel behavior*. Oxford University Press.

Henning, B. G., & Scarfe, A. C. C. (2013). *Beyond mechanism. Putting life back into biology*. Lexington Books.

Hoffmeyer, J., & Kull, K. (2003). Baldwin and biosemiotics: What intelligence is for. In B. H. Weber & D. J. Depew (Eds.), *Evolution and learning. The Baldwin effect reconsidered*. MIT Press.

Huxley, J. S. (1942). *Evolution. The modern synthesis*. Harper.

Jablonka, E., & Lamb, M. (2005). *Evolution in four dimensions. Genetic, epigenetic, behavioral, and symbolic variation in the history of life*. MIT Press.

Jennings, H. S. (1904). *Contributions to the study of the behavior of lower organisms*. Carnegie Institution.

Laland, K. N., Odling-Smee, J., & Feldman, M. W. (2019). Understanding niche construction as an evolutionary process. In T. Uller & K. N. Laland (Eds.), *Evolutionary causation: Biological and philosophical reflections* (pp. 127–152). MIT Press.

Latour, B. (1996). Trains of thought—Piaget, formalism and the fifth dimension. *Common Knowledge, 6*(3), 170–191.

Le, N. (2019). Organic selection and social heredity: The original Baldwin effect revisited. In *Proceedings of the ALIFE 2019: The 2019 conference on artificial life* (pp. 515–522). ASME. Online. https://doi.org/10.1162/isal_a_00214

Lehrman, D. (1953). A critique of Konrad Lorenz's theory of instinctive behavior. *The Quarterly Review of Biology, 28*(4), 337–363.

Lerner, R. M. (2012). Developmental science: Past, present, and future. *International Journal of Developmental Science, 6*(1-2), 29–36.

Lickliter, R. (2008). The growth of developmental thought: Implications for a new evolutionary psychology. *New Ideas in Psychology, 26*, 353–369.

Lickliter, R., & Honeycutt, H. (2013). A developmental evolutionary framework for psychology. *Review of General Psychology, 17*(2), 184–189. https://doi.org/10.1037/a0032932

Loredo, J. C., & Sánchez, J. C. (2012). Neither dichotomies nor dualism: simply genesis. *Integrative Psychological and Behavioral Science, 46*(3), 410–423. https://doi.org/10.1007/s12124-012-9193-z

Loy, I., Carnero-Sierra, S., Acebes, F., Muñiz-Moreno, J., Muñiz-Diez, C., & Sánchez-González, J. C. (2021). Where association ends: A review of associative learning in invertebrates, plants and protista, and a reflection on its limits. *Journal of Experimental Psychology. Animal Learning and Cognition, 47*(3), 234–251.

Mesoudi, A., Whiten, A., & Laland, K. N. (2006). Towards a unified science of cultural evolution. *Behavioral and Brain Sciences, 29*(4), 329–347.

Moore, C. L. (2003). Evolution, development, and the individual acquisition of traits: What we've learned since Baldwin. In B. H. Weber & D. J. Depew (Eds.), *Evolution and learning: The Baldwin effect reconsidered* (pp. 115–139). MIT Press.

Olding-Smee, F. J. (1988). Niche-constructing phenotypes. In H. C. Plotkin (Ed.), *The role of behavior in evolution*. MIT Press.

Ongay, ´i. (2022). Mind and matter. In G. E. Romero, J. Pérez-Jara, & L. Camprubí (Eds.), *Contemporary materialism: Its ontology and epistemology* (Synthese library) (Vol. 447). Springer. https://doi.org/10.1007/978-3-030-89488-7_7

Overton, W. F., Molenaar, P., & Lerner, R. M. (2015). *Handbook of child psychology and developmental science: Theory and method* (Vol. 1). Wiley.

Oyama, S. (1985). *The ontogeny of information: Developmental systems and evolution*. Duke University Press.

Oyama, S., Griffiths, P. E., & Gray, R. D. (2001). *Cycles of contingency: Developmental systems and evolution*. MIT Press.

Pérez-Jara, J. (2022). Discontinuous materialism. In G. E. Romero, J. Pérez-Jara, & L. Camprubí (Eds.), *Contemporary materialism: Its ontology and epistemology* (Synthese library) (Vol. 447). Springer. https://doi.org/10.1007/978-3-030-89488-7_3

Pigliucci, M. (2001). *Phenotypic plasticity. Beyond nature and nurture*. Johns Hopkins University Press.
Pigliucci, M., & Müller, G. B. (2010). *Evolution, the extended synthesis*. MIT Press.
Pinker, S., & Bloom, P. (1990). Natural language and natural selection. *Behavioral and Brain Sciences, 13*(4), 707–727.
Plotkin, H. (2004). *Evolutionary though in psychology: A brief history*. Blackwell.
Richerson, P. J., & Boyd, R. (2008). *Not by genes alone: How culture transformed human evolution*. University of Chicago Press.
Robert, J. S. (2004). *Embryology, epigenesis & evolution: Taking development seriously*. Cambridge University Press.
Sánchez, J. C., & Loredo, J. C. (2007). In circles we go: Baldwin's theory of organic selection and its current uses. A constructivist view. *Theory and Psychology, 17*(1), 33–58.
Sánchez, J. C., & Loredo, J. C. (2009). Constructivisms from a genetic point of view: a critical classification of current tendencies. *Integrative Psychological and Behavioral Science, 43(4)*, 332–349. http://www.springerlink.com/content/17v200168510x4n7/?p=b0c1bb9ab68e4f68af0b3ded92bf4c0f&pi=2
Thornton, A., & Raihani, N. J. (2008). The evolution of teaching. *Animal Behaviour, 75*(6), 1823–1836.
Valsiner, J. (2013). Breaking the arrows of causality: The idea of catalysis in its making. In K. R. Cabell & J. Valsiner (Eds.), *The catalyzing mind: Beyond models of causality* (pp. 17–32). New York, NY, Springer.
Valsiner, J., & Rosa, A. (2007). *The Cambridge handbook of sociocultural psychology*. Cambridge University Press.
Valsiner, J., & Van der Veer, R. (2000). *The social mind: Construction of the idea*. Cambridge University Press.
Waddington, C. H. (1961). Genetic Assimilation. Adv. Gen., *10*, 257–293.
Weber, B. (2003). Emergence of mind and the Baldwin effect. In B. H. Weber & D. J. Depew (Eds.), *Evolution and learning. The Baldwin effect reconsidered*. MIT Press.
Werner, H. (1940). *Comparative psychology of mental development*. Harper and Brothers.
Werner, H. (1956). Microgenesis and aphasia. *The Journal of Abnormal and Social Psychology, 52*(3), 347–353. https://doi.org/10.1037/h0048896
West-Eberhard, M. J. (2003). *Developmental plasticity and evolution*. Oxford University Press.
West-Eberhard, M. J. (2005). Phenotypic accommodation: Adaptive innovation due to developmental plasticity. *Journal of Experimental Zoology Part B-Molecular and Developmental Evolution, 304B*(6), 610–618.
Whiten, A. (2021). The burgeoning reach of animal culture. *Science, 372*(6537), eabe6514.
Young, J. L. (2013). The Baldwin effect and the persistent problem of preformation versus epigenesis. *New Ideas in Psychology, 31*(3), 355–362. https://doi.org/10.1016/j.newideapsych.2012.07.002

Chapter 7
Neither Nature Nor Nurture: On the Inadequateness of Innatism and Environmentalism in Biology and Psychology

Íñigo Ongay

Abstract This chapter constitutes a philosophical examination of the long contest between nature and nurture in the areas of concern of Behavioral Biology, Ethology and Psychology. The paper considers the roles of the notions of innateness and acquisition in the causal production of behavior. While there seems to be broad agreement to the effect that the dichotomy between what is innate and what is learned is misguided and obsolete, this essay concerns the epistemological and ontological reasons why neither nativism nor environmentalism constitute heuristically fruitful views in accounting for the origination of behavior. The chapter starts off by providing a historical presentation of some of the cornerstones of both innatism and environmentalism in the development of classical Ethology and Behaviorist Psychology. It then proceeds to examine how the debate between innatism and environmentalism has subsequently been dismantled by more recent approaches in the Behavioral Sciences. Finally, the author ends up with some philosophical comments concerning the empirical and conceptual pitfalls of both traditions and the inadequateness of the dichotomous framework of the dispute when addressing the causal channels by which the development of behavioral patterns takes place.

Keywords Innateness · Acquisition · Ethology · Behavioral psychology · Philosophy of science

7.1 Innatism and Environmentalism

The dispute between environmentalism and innatism, as it has been discussed in the realms of Ethology, Psychology and Behavioral Biology, constitutes a chestnut of old vintage. There is wide agreement that the battle has long ended although the

Í. Ongay (✉)
Faculty of Social and Human Sciences, University of Deusto, Bilbao, Spain
e-mail: Inigo.ongay@deusto.es

J. Pérez-Jara, Í. Ongay (eds.), *Beyond Nature and Nurture*, Synthese Library 497,
https://doi.org/10.1007/978-3-031-87715-5_7

consensus becomes less than universal when it comes to determining which of the warhorses won the war. If the debate is to be construed as pertaining to what causes what in the origination of behavioral patterns in human and non-human animals, there is the appearance of a universal concurrence pointing to the effect that both genes and the environment need to be given their proper roles within the context of a complex interplay between biological, psychological and socio-cultural factors acting conjointly to bring about behavior. What is more: given the complexity of the feedback loops that obtain in these sorts of causal processes it is to be expected that such factors will not be always (if even) easy to disentangle.

This representation of the query may seem alright as long as it goes but it is well to notice a further complication. So conceived, the dispute is a nice example of a dismal battle fought over goths. On the one hand, the idea that all the patterns of behavior deployed by human and non-human animals are single-handedly determined by genes seems silly enough in light of all that Evolutionary Developmental Biology has taught us about how real biological systems work (Aubret et al., 2004; Soha, 2020; Stotz & Griffiths, 2018; Sun et al., 2022). The environmentalist smiles disdainfully at the notion that the behavior of organisms should be accounted by genetic factors in a vacuum of the ecological environments where the organisms bearing those genes operate. It is high time, she musters, we abandoned the innatist hypothesis to start paying attention to the plasticity of development when it comes to explaining the behavioral phenotypes of humans and other animals. After all, one of the many take-home messages from current cutting-edge research in the arenas of theoretical biology and philosophy of biology (Dupré, 2018; Laland & Brown, 2018) suggests that things are definitely not so simple as innatism depicts them.

Now, does innatism really favor the claim that the influence of the environment is negligible for the development of behavioral traits? Ethologists and Evolutionary Psychologists inclined to defend the innatist thesis will presumably deny that this is so. It is true that genes don't produce any effect on an environmental emptiness but that trivial and rather uncontroversial observation is not what is at stake here. Instead, the type of sophisticated innatism that is worth-discussing claims that all things considered, inherited evolutionary tendencies should not be ignored when considering why individual organisms act the way they do. If the psychologist, the sociologist or the cultural anthropologist ignores the causal contributions of genetic factors when confronted with human behavioral systems, they do so at their own peril.

If all that there is to the debate were the admittedly sketchy presentation that I have provided in the previous paragraph, it would seem difficult to escape the rather depressing conclusion that both sides are tilting at windmills here. After all, saying that both environmental and innate factors are to be taken into account in elucidating what gives rise to complex behavioral patterns has the appearance of an undeniable truism of little empirical significance. Doubtlessly, in point of historical fact there might have been biologists and social scientists in the past who have embraced one or the other of the unnuanced alternatives, but when one rejects both of them and endorses a more refined view of how behavior is brought about there seems to be no issue left. It is clear then that there is a sense in which innate factors come crucially into play but that does not deny that in another sense environmental triggers are equally critical.

Is there anything more substantial that remains to be discussed after these obvious truisms are put to one side.? In what follows I will identify a way in which the very framework of the dispute is seriously astray both conceptually and empirically. That may seem as a foregone conclusion after the manner the contest has been characterized so far but the reader would do well to pay attention at a little plot-twist here. My contention will not be that the debate is flawed because both parties are partially correct—On the contrary, I will argue that both positions are irremediably wrong no matter the degree and subtility of nuance with which the dichotomy between nature and nurture is enriched.

7.2 Ethology and Innatism: Some Historical Background

I have just presented the nature-nurture debate as involving a dichotomous partition of the causal factors driving animal and human behavior. In this regard, whereas innatism argues that behavioral patterns depend primarily on the contribution of inherited dispositions, environmentalists by large forge an opposite line to claim that it is the ecological context where organisms live that is in charge of directing the development of the ways those individuals act. It may seem from such a sketchy presentation of the issue that the controversy concerns the sciences of Biology and Psychology. Yet, it is well to remember that some of the ideas being discussed here have a history that long predates the establishment of these scientific fields.

Perhaps one of the most relevant precedents in the long history of innatism lies in the divide between rationalists and empiricists within the context of modern philosophical thinking. In this scenario, whereas most rationalists in the wake of Descartes or Leibniz favored a form of nativism which countenanced innate ideas, empiricists like John Locke or David Hume among others insisted that all the contents of the human mind be understood as coming from experience alone. It has become something of a harmless cliché of the history of philosophy to assert that Immanuel Kant's *Critique of Pure Reason* will later on come to offer a theoretical synthesis between both sides of the quarrel explaining the degrees to which the transcendental apparatus of the mind and the empirical data coming from outside are equally pivotal in the origination of knowledge.

Relevant as it is for the history of philosophy, this debate was epistemic in character in that the main motives at stake had to do with the scope, limits and origins of knowledge claims. It is worth considering however that contenders in the more recent incarnation of the nature-nurture controversy didn't have this abstract epistemic point as their main motivation for thinking about the roles that inherited and environmental factors may bear to accounting for patterns of behavior. Instead of such comparatively more theoretical epistemic interests, if we think of pioneers of Ethology like Douglas Spalding, Charles Otis-Whitman, Wallace-Craig, Erich von Holst or Oskar Heinroth, their main concern when they turned their attention to the notion of inherited behavior was explaining a range of phenomena related to what animal organisms do in the wild. In that regard the naturalist program of innatism was supposed to stand or fall depending on some relevant empirical details of animal behavior which allegedly could not be explained solely in terms of the experience that the animals in question may have had during their ontogeny.

Austrian ethologist Konrad Lorenz features prominently in the historical construction of this naturalistic program of innatism. Although there is evidence (Callebaut, 2003; Celentano, 2018; Lorenz, 1941) that certain *Kantian ideas* had influence over Lorenz's theoretical development, the chief motivational force driving his theory of innate behavior was to treat behavioral phenotypes as though they were morphological traits bearing evidence on the phylogenetic history of the animal species deploying them. This inclination to explain behavior in physiological terms combined with Lorenz's comparative and taxonomic approach to the study of organismal activity (Brigandt, 2005) led to a strongly mechanistic framework in which patterns of behavior are divided into three components, namely: the accumulation of internal *readiness* for a specific action, the activation of an Innate Releasing Mechanism (IRM) and the occurrence of a consummatory act which is the purpose of the action. Understood in this way, inherited behavioral patterns are not just innate in that they seem independent of the past learning experience of the individual in question as shown by the typical deprivation experiments that Lorenz conducted with geese, ducks and fishes during the 1930s, they are also fairly rigid in the sense that they don't appear to be significantly modellable by the environment. Think, for example, of the proverbial case of the imprinting process in precocial birds. Think further of the stereotypic and spontaneous egg-rolling behavior of the goose. Think lastly of the agonistic behavior in robins and other avian species which are often released in the absence of a sing stimulus. Here is how Konrad Lorenz puts succinctly this point about the *spontaneity and invariance* of the goal-directed *fixed action patterns*:

> What particularly seemed to demand an explanation was a very unexpected correlation which evidently existed between the spontaneity and the invariability of the innate behaviour patterns in question. According to the vitalists' opinion, spontaneous activity must automatically be considered as purposive, and must, therefore, show adaptive variability in the pursuit of its end or goal. According to the mechanists' opinion, any innate and individually invariable behaviour pattern must be considered to be a chain of unconditioned reflexes and must, therefore, be totally devoid of spontaneity. In defiance of both these dogmas, just those highly differentiated innate sequences of movements that not only were absolutely invariable individually, but even in phylogeny were rather conservative and slowly changeable characters, proved to possess a very distinct and striking kind of spontaneity. At the level of superficial observation, the innate behaviour patterns in question appeared reflex-like in that they were set off by a sort of 'trigger-action' in a very specific stimulus situation. But on closer inspection it became apparent that these activities are, at bottom, to a very high degree independent from external stimulation. Captive animals, deprived of the normal object or releasing situation of an innate behaviour pattern, will persist in discharging the same sequences of movements at a very inadequate substitute object or situation. The longer the normal stimulation is withheld, the less necessary it becomes, in order to set off the reaction, to supply all of the stimuli pertaining to it. The

> longer the reaction does not go off, the finer the trigger that releases it seems to become set. In other words, the threshold of the stimulation necessary to release this type of innate reaction, is not a constant, but is undergoing a continuous process of lowering, going on throughout the time during which the reaction is not released. This gradual lowering of threshold does, in a good many cases, actually reach the theoretically possible limit of zero, that is, the activity in question will finally go off in vacuo, with an effect somewhat suggestive of the explosion of a boiler whose safety valve fails to function. This occurrence has been termed 'Leerlaufreaktion' in German, vacuum reaction and explosion reaction. (Lorenz, 1950, 247)

It is sometimes maintained that a natural way to interpret Lorentz's view of the invariability of the spontaneous activity of organisms is to associate it with the more traditional concept of *instinct* (Brigandt, 2005; Griffiths, 2004; Richards, 1974). While it is true that Lorenz sometimes employs that notion (Lorenz, 1956, 1963), a word of caution may come handy here. To be strictly fair to his account one needs to recognize that Lorenz's conception of instinctual behavior is defined in a largely negative fashion: instincts are species-specific purposive patterns of action that don't depend in their *maturation* on the learning history of the individual raising them. It becomes clear then in light of this morphological view of behavior why these inherited unlearnt actions are to be given evidential priority in casting light on the evolutionary history of the species over other more environmental sensitive forms of conduct which depend on the individual experience of the organism.

When it came to flesh out positively such a negative conception of instinctual actions, Lorenz provided his famous psycho-hydraulical model of motivation to describe the inner functioning of the mechanism releasing innate action patterns. Despite certain attempts to empirically vindicate some of the ideas behind it (Kudryavtseva, 2004), there seems to be broad consensus that the model itself remains overtly speculative (Hinde, 1966; Lehrman, 1953). This by itself needn't be taken as a signal that Lorenz's theory of instinctively innate behavior is flawed beyond the point of repair. After all many otherwise sensible scientific hypotheses have historically kept speculative until they finally got empirically backed up by the evidence at hand. Unfortunately, however, there is a second line of concern regarding the conceptual viability of Lorenz's account. One way of singling out this second problem is by turning to Daniel S. Lehrman's famous critique of Lorenz's theory in his 1953 paper (Lehrman, 1953)

Lehrman's seminal paper presents a twofold line of attack on Lorenz's theory of innate behavior. First, he contends that isolating a pattern of action from any source of developmental experience the organism may receive is a tall order. The typical deprivation experiments can never achieve quite this. Take for example the apparently "innate" pecking behavior in chicks raised in isolation in light of Kuo's proposal about the pre-natal experience that chicks are typically exposed to from the first stages of their embryonic development. Take also the nest-building behaviors in rats which could seem "innate" except for the experience of those rodents with licking their own genitalia. What this observation entails is that the researcher should do well to stop sort of assuming that a given deprivation experiment implies that the animal has been deprived from all experience whatever. Instead of jumping to such a premature conclusion, it is best to ascertain more modestly *which exact sources of developmental experience* have in fact been controlled for.

So far Lehrman's objection to Lorenz's idea of innateness is epistemic in nature as it suggests that when appraising the results of a given experiment you'd better avoid reading too much into what it implies and what it doesn't. In this sense, philosophers of science may feel reminded of some familiar points regarding the inherent contrastive character of hypotheses evaluation (Sober, 1990) and the threat of the only-game in town fallacy when employing abductive reasoning (Achinstein, 1990). Although those are all nice epistemic points to bear in mind, it is worth noticing that there is nonetheless an additional more ontological remark that Lehrman's paper has in offer. This second angle has to do with the conceptual intelligibility or lack thereof of the dichotomy between inheritance and learning when it comes to determining causes of behavioral sequences in the life of an organism. The idea that Lehrman advances in this regard is that it is unclear what it would take for an action to be inherited. Presumably, so Lehrman argues, Lorenz does not mean to say that a miniature of the "instinctive behavior" is physically present in the zygote or that alternatively "instincts" are to be construed as a vitalist entelechy guiding the individual organism in a certain direction. However, when those prima facie bizarre preformist assumptions are eschewed, it becomes hard to see what heuristic benefits the innatist position may contribute to the analysis of behavior. Instead of falling prey to the allure of such idle dichotomous partitions between what is learnt and what is innate, scientists studying animal behavior should be advised to focus their attention on the *development processes* by which the actions themselves come about. Lehrman offers a good description of the situation in what follows:

> The "instinct" is obviously not present in the zygote. Just as obviously, it is present in the behavior of the animal after the appropriate age. The problem for the investigator who wishes to make a causal analysis of behavior is: How did this behavior comes about? The use of "explanatory" categories such as "innate" or "genetically fixed" obscures the necessity of investigating *developmental processes* in order to gain insight into the actual mechanisms of behavior and their interrelation. The problem of development is the problem of the development of new structures and activity patterns from the resolution of the interaction of *existing* structures and patterns, within the organism and its internal environment, and between the organism and its outer environment. At any stage of development, the new features emerge from the interactions between the *current stage* and the environment. The interaction out of which the organism develops *is* not one, as is often said, between heredity and environment. it is between *organism* and environment! And the organism is different at each different stage of development. (Lehrman, 1953, 345)

The reason I find this characterization apposite is that Lehrman is not signaling here at the empirical unsoundness of Lorenz's scientific project. Rather than pinpointing the empirical inadequacy of innatism, the claim Lehrman advances is more fundamental as it would, if true, show that the very notion of hereditary behavior on which the *Lorentzian project* is based remains conceptually misguided. Even though Lorenz never felt compelled to leave aside the dichotomy between innate and learned actions by these and other criticisms of the divide (Lehrman, 1970; Schneirla, 1966), historically, there is little doubt that the field of Ethology was never the same after them (Burkhardt, 2005; Griffiths, 2004; Ongay, 2010, 2021; Vicedo, 2009, 2013). In the 1950s and 1960s a new generation of Ethologists coming prominently from the English-speaking world took to heart the discredit of the notion of innate hereditary behavior replacing it by a brand-new wealth of concepts regarding how organismal activity originates and is maintained in their environments.

7.3 The Ups and Downs of Psychological Environmentalism

Much like innatism, environmentalist views have had a similarly long history which starts off with such classic formulations of the blank slate thesis as those of Aristotle, John Locke and David Hume to name but a few. However, it is perhaps with the advent of modern scientific psychology that the view itself saw a more systematic and fine-tunned articulation. Innatists wanted to explain the origination of certain individual behaviors in a way that avoided making reference to the experience of the organism with the environmental clues it faces. Environmentalist psychology conveys a parallel albeit inverse declaration of independence as it postulates that animal and human behavior needs to be accounted for without reference to the innate propensities of the individual in question insofar as these inherited dimensions of behavior go well beyond the scientific scope of the science of psychology. The driving force motivating this neat demarcation of what counts as heuristically respectable subject matter in the realm of psychology lies in the historical success of such areas of endeavor as those of psycho-reflexology and classical conditioning (Boakes, 2003; Coleman, 2002; Loucks, 1937; Madsen, 1988) or learning psychology after the discovery of the law of effect by Edward L. Thorndike (1927). In the wake of those pivotal theoretical constructions, the advent of classical behaviorism (Watson, 1913, 1930) came with the development of a view of psychology in which the plasticity of the behavioral patterns any organism could acquire by means of associative learning was conceived as unlimitedly open-ended. Not only that: aside from innate mechanisms of the sort that *Lorenzian* classical Ethology had contemplated, the behaviorist tradition also opposed the kind of inner mental drives that had featured prominently in previous introspectionist schools of psychological thinking (but see Costall, 2013 for a historical contrast). The following excerpt from Wanton's 1930 programmatic book *Behaviorism* gives us a concise and vivid characterization of the reasons why this is so:

> Psychology as the behaviorist views it is a purely objective experimental branch of natural science. Its theoretical goal is the prediction and control of behavior. Introspection forms no essential part of its methods, nor is the scientific value of its data dependent upon the readiness with which they lend themselves to interpretation in terms of consciousness. The behaviorist, in his effort to get a unitary scheme of animal response, recognizes no dividing line between man and brute. The behavior of man, with all its refinement and complexity, forms only a part of the behaviorist's total scheme of investigation. (Watson, 1913, 158).

This paragraph may seem to offer a suitable portrayal of the methodological purview of objective psychology, but it is important to keep straight what classical behaviorism of such a Watsonian coinage implies and what it does not. First, different from other versions of behaviorism like G. Ryle's logical behaviorism (Ryle, 1949), Watson does not intend to make a semantic contention about the behavioral meaning of mentalistic terms in use in psychology. Furthermore, he does not mean to commit behavioral psychology to the metaphysical claim that inner forces of motivation just don't exist. Instead of trading in these more substantive theses, Watson seems to be interested in isolating a methodological piece of advice on what scientific psychology should be concerned about. The methodological policy at stake insists that the psychologist concentrate her attention on the goal of predicting and controlling what the individual does by modifying the relevant external stimuli and learning history responsible for the elicitation of the actions. According to this methodological line of thinking, psychology as the science of behavior should keep silent about mentalistic contents and innate factors: not that they need to be declared inexistent once and for all, but in any event, their purported private and unobservable character puts them outside of the scope of what is amenable to scientific treatment unless the external and public behavior of the individual gives out a source of indirect evidential basis to access them.

Viewed in this way, the kind of methodological behaviorism that Watson did favor may seem somewhat modest and therefore disappointing. In fact, it *was disappointing* to some later figures in the history of objective psychology. One of such figures, B.F Skinner (1938, 1956), instead of treating mental inner contents as epistemic black boxes not to trespass introspectively for methodological reasons as Watson had done, proceeded more resolutely to get rid of them wholesale by turning his attention to the behaviors of the organisms "*in their own right*". Skinner's opening gambit suggests that the combination of classical and operant conditioning should not be interpreted merely as a methodological canon of how psychology should be understood. In turn, according to Skinner's framework learning behavior figures in the center of the psychological stage as the entirety of what the subject matter itself is all about. Notice that according to this paradigm of radical behaviorism there is nothing (scientifically interesting) to the origination and maintenance of behavior except for what can be captured in virtue of the principles of associative and operant learning. To be sure, the reason this radical depuration of learned behavior as a dimension to be studied by purely behavioristic means came to dominate the field of psychology for a while was the technological success that Skinner (1948, 1959) and others achieved in conditioning animals exposed to the proper reinforcement protocols within the experimental milieu of the operant conditioning chamber to elicit actions sometimes fairly surprisingly far from what these organisms would do in the wild. However, it became clear in due time that even under rigidly controlled experimental conditions there were certain definite limits to what organisms could learn to do. It further became apparent that some of those limits were related to species-specific constraints on learning (Bowers, 2018; Domjan, 2012; Shettleworth, 1972), biological preparedness (McNally, 1987, 2016; Seligman, 1970, 1971), selective association effects in taste-aversion studies (Garcia

& Koelling, 1966) and other similar examples of *misbehavior* of organisms (Breland & Breland, 1961) running afoul of the standards of what radical behaviorism deemed possible (Herrnstein, 1977a, see also Skinner, 1977 and Herrnstein, 1977b for replies). This dismaying situation alongside the development of the so-called cognitive revolution after the famous debate between B.F. Skinner and N. Chomsky (Chomsky, 1959; Skinner, 1957) ended up hammering the last nail into the coffin of the project of radical behaviorism and paving the way to a state of affairs where the traditions of Ethology and Psychology of Learning will seemingly tend to merge with each other (Bateson & Hinde, 1976; Burkhardt, 2005; Fernández Rodríguez, 1980, 1988; Hinde, 1966).

7.4 Some Philosophical Comments

So far, I have presented the contest between innatism and environmentalism in a somewhat historical manner. The take-home message to be extracted from what I have said in the previous sections is that interpreted in their most restricted forms, both positions are clear dead ends in the behavioral sciences. This doesn't mean that they are not amendable in a manner that makes them sound more sensible. One way to present a more plausible and perhaps modest version of nativism and environmentalism is resorting to the square of opposition in traditional logic. Notice that in light of this familiar artifact, although innatism and environmentalism may be construed as conveying two mutually *contrary statements* which cannot be true at the same time (namely: "all behavior is innate" or alternatively "all behavior is learnt and none is innate") they don't need to be interpreted that way. A more temptingly sensitive alternative is to understand them as implying *subcontrary statements*, in which case they both could possibly be true at the same time so long as they are made to refer to distinct patterns of behavior: some behaviors are innate and some others acquired by a learning process. There is an analogy to be drawn here from Kant's two last antinomies in his *Critique of Pure Reason* (Kant, 1998) which could similarly be both true when circumscribed to two different realms in Kant's ontological map of the universe. In this vein, it is worth noticing that the possibility remains open to assert that there are innate and learnt behavioral patterns and that each of these sorts of behavior will conceivably respond to a different variety of contingencies. For example, the late B.F Skinner's proposal that contingencies of reinforcement and contingencies of survival can coexist peacefully doing their respective explanatory work in regard to the ontogeny and the phylogeny of behavior (Skinner, 1977) points out to a similar division of labor between both positions.

This proposal may sound more reasonable in comparison to the unnuanced views about innatism and environmentalism discussed earlier but it also raises a question which should give the reader pause. The difficulty here concerns whether it is so easy a task to compartmentalize animal behavior neatly into entirely innate and completely learned actions. Taken at a face value, the obvious answer to this question is *no*. This should bring the idea home that yes, there are some causal factors affecting *any* behavioral pattern we can think of that are genetic in nature and thus independent of the experience of the individual, while some others depend the environment in which the organisms live. One of the attractions of such a compromise is that it seems to nicely accommodate an aspect of the agreement depicted at the beginning of the paper: everybody appears to concur that the debate between innatism and environmentalism does not represent an *either-or* discussion but a *both-and* type of problem.

This much may sound obvious and therefore disappointingly trivial. Now it is time to turn our attention to a somehow less trivial issue: once the proverbial truisms about biological propensities and learned environmental factors being both important causal contributors to behavior are put aside, are there any other empirically substantive issues left to discuss? One general consideration to bring about here from the customary conceptual toolkit of the philosopher of science is that saying that two kinds of causal factors are necessary to deliver a given effect leaves open the question of whether they are both equally important (Pérez-Jara, 2024a, 2024b). Assuming that there are two factors coming into play, equality of causal relevance may obtain among them but doesn't need to do so. Bearing that in mind, one substantive question to be addressed even in the light of the truistic consensus just depicted is if the respective contributions of genes and the environment to the development of behavioral traits are both on the same footing when it comes to their causal relevance.

It is worth seeing that when looking piecemeal at individual behavioral patterns of particular organisms the suggestion that I have just described has the alluring resemblance of an empirically tractable task. After all, countless animal and plant breeders as well as experts in quantitative genetics have long been effectively disentangling the contributions of genes and environmental factors when it comes to many phenotypic traits of biological individuals and so it might be tempting to think that doing the same with behavioral traits represents an issue identically amenable to the methods of science.

There are nonetheless two perplexities that this task raises. To isolate the first problem, it is well to return to a point raised by Lehrman in his critique of Lorenz's notion of instinct. In his cutting edge 1953 paper Lehrman claimed that the concept of innate behavior was conceptually unclear and then asserted, rather ironically, that Lorenz presumably wouldn't have it that "instincts" are physically present in the zygote. Here, it is natural to expect more sophisticated innatists to note that this argument is surely a straw-man. Rather than this overtly outlandish preformist claim, what biologically oriented scientists of behavior have in mind when they talk about instincts is genetic predispositions to act in certain ways. One natural way to interpret this response resorts to the familiar philosophical concept of dispositional properties (Choi, 2006). Someone might argue that the behavioral and social sciences (from studies on addictive behaviors to criminology) are fraught with examples of genetically controlled *dispositions* to act in certain ways perhaps in the form of *hereditary pre-programations* (Eibl-Eibesfeldt, 1970), instincts to learn (Marler, 1999) or epigenetic *rules* (Wilson, 1978).

However, even if Lehrman's irony goes too far, that does not undermine the strength of his criticism of the idea of innate behavior. I take the kernel of truth of that line of critique to imply that there is something seriously odd about the very notion of a hereditary behavioral pattern. To see why, consider that the notion of behavior does not make psychological sense except in connection to the organism's experience in its environment (Fuentes Ortega, 2004; Ginsburg & Jablonka, 2019; Ongay, 2022). If this conceptual dependence obtains, it is hard to see how behavior can be hereditary since, quite straightforwardly, no experience can be inherited in a literal sense. This does not mean that there are not inherited conditions that make behavior possible: after all, the entire phenotypic structure of any organism (from its morphology to its physiology) is inherited in a certain sense but that does not show that behavior itself is hereditary. By means of example consider the typically stereotypical filial imprinting behavior in precocial birds (McCabe, 2019) or take alternatively mating behavior patterns in sticklebacks (Tinbergen, 1953): while it is clear that the visual system of the individuals in question is controlled by means of a complex interplay of genetic and epigenetic systems of inheritance (Jablonka & Lamb, 2020) alongside multiple other mechanisms of non-genetic inheritance (Mameli, 2004, 2005; Laland et al., 2016) the moves displayed by the individuals are not themselves morphological or physiological structures to be passed off to their offspring.

A second problem concerns the notion of environment. If we are to construe the notion in a broad enough fashion conveying a negative term for *everything* that is *not genetic*, then this concept entails that there is nothing about an organism that does not depend on a multifarious array of environmental features from the amount of solar irradiance to the force of gravitation to the chemical composition of the air, soil or water in which the individual lives. The same, too, applies to the concept of acquisition; understood in a broad sense, there is no single trait that is not acquired in a way or another. In this regard, as argued by P. Bateson and M. Mameli all traits are genetically influenced because genes participate in the development of them all. Yet at the same time, all traits are acquired since at first the organism does not have the trait and then at a subsequent point of time the organism comes to have the trait (Bateson & Mameli, 2006). Such a ubiquity of acquisition does not mean to say that inheritance is any less all-encompassing. It means that the distinction itself, whatever its remanent attractiveness, gets dismantled when we consider the development of what organisms are like and how they act (Lamm & Jablonka, 2008).

The upshot of this discussion is not that there is something principally wrong to the project of trying to make headway on ascertaining the genetic and environmental conditions for the development of particular behavioral patterns. Quite on the contrary, it is all right to attempt to do that as long as those conditions don't get hypostatized as two independent and distinct causal forces driving, however additively, the way behavioral systems develop. Borrowing the jargon of philosopher Gustavo Bueno (1978) the concepts of innateness and acquisition are *conjoined* which means that they need to be made to refer to each other to remain heuristically helpful. As Bueno analyzes them, typical cojoined concepts in the history of various scientific fields present an apparent dual structure that may easily lead to the misguiding impression that they oppose each other. However, upon closer inspection, it turns out that they can only be rightly understood inter-definitionally and not by reducing them to one another. By mean of example, *points and lines* in Euclidean geometry or *rest and motion in* classical Galilean mechanics represent instances of such a set of cojoined concepts to be defined by reference to each other. I think that much the same holds true about inherited innate components and acquisition: an organism acquires environmental experience and thus elicits behavior in virtue of pre-existing structures which are themselves both inherited and acquired at the same time and remain prone to be modeled by selective regimes in part constructed by its own organismal activity.

In contrast, when the concepts of innateness and acquisition are viewed in a substantialist way, the dichotomy itself becomes a trivial and idle example of a divide that belongs in the realm of folk Biology and Psychology (Bateson & Mameli, 2006, 2007, 2011; Ongay, 2011, 2021) because in point of fact *everything is both of those things at the exact same time*. It is true that some scientists studying animal behaviors on the ground may well have paid considerable lip service to innatism or to environmentalism in discussing the nature of their scientific disciplines, but when we turn to these scientists' accounts of animal individuals' actions what one observes is the distinction between innateness and acquisition falling apart. There is an analogy from the history of physics worth considering here. Granted that I. Newton's metaphysical ideas about absolute space and time ended up being dismantled, but such a refutation leaves fundamentally unaltered whatever kernel of truth that Newtonian mechanics may have retained. One more general lesson to be extracted both from the controversies between innatism and environmentalism in the behavioral sciences and the analogy from Newtonian physics is that one would always do well to distinguish the practice of science from what scientists *say* when they *philosophize* about their fields.

7.5 Concluding Remarks

Konrad Lorenz's form of innatism and B.F. Skinner's behaviorist version of environmentalism were not irremediably doomed to failure out of purely a-priori reasons. Quite on the contrary, there was a real substantive question to be asked about animal behavior being genetically or environmentally independent and it turned out to be neither. When these views of the origination of behavioral patterns got refuted, it happened on a combination of empirical and conceptual grounds showing that animal activities couldn't be easily categorized in virtue of any binary compartmentation of inherited and acquired properties. Something that is worth seeing in that respect is that such more sophisticated view of behavior aligns rather nicely with a pluralistic ontology that avoids the risk of over-simplification (Camprubí & Pérez-Jara, 2023; Harman & Pérez-Jara, 2022; Pérez-Jara, 2022)

In a well-known passage of his *Aesthetics*, German philosopher G.W.F. Hegel wrote that "(...) art, considered in its highest vocation, is and remains for us a thing of the past" (Hegel, 1975). In a similar vein, I think that the debate nature-nurture, *viewed as two independent and different causal forces modeling behavior*, represents something equally obsolete. The reason why it is important to qualify the previous sentence is that this paper has criticized the distinction in its substantialist binary manifestation but that does not entail that other non-binary ways of unraveling the multiplicity of components in the development of behavioral traits could not look perfectly respectable. In the 1950s British blue-tits and robins developed the habit of peaking the foil tops of milk bottles on the doorsteps of English houses to drink the layer of cream beneath. This behavior then became rampant among individuals of the same species all over the UK thus stabilizing a cultural tradition which subsequently affected the selective regimes of the birds in an evolutionary feedback loop with genetic and non-genetic factors coming into play (Avital & Jablonka, 2000; Viciana, 2021). It is well not note that this feedback loop takes place by means of what James Mark Baldwin classically called *organic selection* (Baldwin, 1857, 1896; Ongay, 2024; Pérez-Jara, 2024a). Saying that such a tradition is innate in the sense of being developmentally independent from the experiences of the birds during their extended ontogeny would be obviously mistaken but it would also be flawed to deny that the acquisition of the opening technique relies on many features of avian physiology and morphology. In this respect, it is a quasi a-priori truth that this and other behavioral patterns are both genetically and environmentally controlled in a multitude of ways but saying that does not even start to tackle the real casual processes that bring behavior about.

References

Achinstein, P. (1990). The only game in town. *Philosophical Studies. An International Journal for Philosophy in the Analytic Tradition, 28*(3), 179–201.

Aubret, F., Shine, R., & Bonnet, X. (2004). Adaptive developmental plasticity in snakes. *Nature, 431*(7006), 261–262.

Avital, E., & Jablonka, E. (2000). *Animal traditions. Behavioral inheritance in evolution.* Cambridge University Press.

Baldwin, J. M. (1857). Organic selection. *Science, 5(121)*, 634–636.

Baldwin, J. M. (1896). A new factor in evolution. *American Naturalist, 30*(354), 441–451.

Bateson, P. P. G., & Hinde, R. A. (Eds.). (1976). *Growing points in ethology*. Cambridge University Press.

Bateson, P., & Mameli, M. (2006). Innateness and the sciences. *Biology and Philosophy, 21*(2), 155–188.

Bateson, P., & Mameli, M. (2007). The innate and the acquired: useful clusters or a residual distinction from folk biology? *Developmental Psychobiology, 49*(8), 818–831.

Bateson, P., & Mameli, M. (2011). An evaluation of the concept of innateness. *Philosophical Transactions of the Royal Society B, 366*, 436–443.

Boakes, R. A. (2003). The impact of Pavlov on the psychology of learning in English-speaking countries. *Spanish Journal of Psychology, 6*(2), 93–98.

Bowers, R. I. (2018). Constraints on Learning. In J. Vonk & T. Shackelford (Eds.), *Encyclopedia of animal cognition and behavior*. Springer.
Breland, K., & Breland, M. (1961). The misbehavior of organisms. *American Psychologist, 16*(11), 681–684.
Brigandt, I. (2005). The instinct concept of the early Konrad Lorenz. *Journal of the History of Biology, 38*(3), 571–608.
Bueno, G. (1978). Conceptos conjugados. *El Basilisco, 1*, 88–92.
Burkhardt, R. W. (2005). *Patterns of behavior: Konrad Lorenz, Niko Tinbergen and the founding of ethology*. Chicago University Press.
Camprubí, L., & Pérez-Jara, J. (2023). La arqueología como arqueología y la relativa independencia de la cultura. *Revista de Occidente, 3*.
Choi, S (2006). Dispositions. In *The Stanford encyclopedia of philosophy.*
Chomsky, N. (1959). Review of verbal behavior. *Language, 1*, 26–58.
Coleman, S. R. (2002). Circumstances and themes in the history of classical conditioning. In J. W. Moore (Ed.), *A neuroscientist's guide to classical conditioning*. Springer.
Callebaut, W. (2003). Lorenz's philosophical naturalism in the mirror of contemporary science studies. *Ludus Vitalis, 9*(20):26–55.
Costall, A. (2013). Introspection and the myth of methodological behaviorism. In J. W. Clegg (Ed.), *Self-observation in the social sciences*. Transaction.
Celentano, M. (2018). From Konrad Lorenz's "Phylogenetic Apriorism" to the birth of evolutionary epistemology. In: O. Andreica, A. Olteanu, (Eds.), *Readings in Numanities. Numanities - Arts and Humanities in Progress,* (Vol. 3). Cham: Springer.
Domjan, M. (2012). Biological or evolutionary constraints on learning. In N. M. Steel (Ed.), *Encyclopedia of the sciences of learning*. Springer.
Dupré, J. (2018). Human nature: A process perspective. In E. Hanon & T. Lewens (Eds.), *Why we disagree about human nature*. Oxford University Press.
Eibl-Eibesfeldt, I. (1970). *Ethology: The biology of behavior*. Holt, Rinehart, & Winston.
Fernández Rodríguez, T. R. (1980). Conductismo y etología. Una introducción a la polémica entre Hernstein y Skinner. *Estudios de Psicología, 1*, 40–45.
Fernández Rodríguez, T. R. (1988). Conducta y evolución. Historia y marco de un problema. *Anuario de Psicología, 39*(2), 99–136.
Fuentes Ortega, J. B. (2004). Con ocasión del I centenario del nacimiento de B. F. Skinner: crítica de la relación entre el análisis funcional de la conducta y la filosofía del conductismo radical. *Cuaderno de Materiales, 21*.
Garcia, J., & Koelling, R. A. (1966). Relation of cue to consequence in avoidance learning. *Psychonomic Science, 4*(3), 123–124.
Ginsburg, S., & Jablonka, E. (2019). *The evolution of the sensitive soul. Learning and the origin of consciousness*. The MIT Press.
Griffiths, P. E. (2004). Instinct in the 1950s. The British reception of Konrad Lorenz's theory of instinctive behavior. *Biology and Philosophy, 19(4)*, 609–631.
Harman, G., & Pérez-Jara, J. (2022). Materialism is false/materialism is not false. In G. E. Romero, J. Pérez-Jara, & L. Camprubí (Eds.), *Contemporary materialism: Its ontology and epistemology. synthese library* (Vol. 447). Springer. https://doi.org/10.1007/978-3-030-89488-7_13
Hegel, G. W. F. (1975). *Aesthetics. Lectures on fine arts*. Clarendon Press.
Herrnstein, R. J. (1977a). The evolution of behaviorism. *American Psychologist, 32*(8), 593–603.
Herrnstein, R. J. (1977b). Doing what comes naturally: A reply to Professor Skinner. *American Psychologist, 32*(12), 1013–1016.
Hinde, R. A. (1966). *Animal behavior: A synthesis between ethology and comparative psychology*. McGraw Hill.
Jablonka, E., & Lamb, M. (2020). *Inheritance systems and the extended evolutionary synthesis*. Cambridge University Press.
Kant, I. (1998). *Critique of pure reason*. Cambridge University Press.

Kudryavtseva, N. N. (2004). Lorenz was right! Or does aggressive energy accumulate? *Russian Journal of Genetics, 4*, 656–662.

Laland, K., & Brown, G. R. (2018). The social construction of human nature. In E. Hanon & T. Lewens (Eds.), *Why we disagree about human nature*. Oxford University Press.

Laland, K., Matthews, B., & Feldman, M. W. (2016). An introduction to niche construction theory. *Evolutionary Ecology, 30*, 191–202.

Lamm, E., & Jablonka, E. (2008). The nurture of nature: Hereditary plasticity in evolution. *Philosophical Psychology, 21*(3), 305–319.

Lehrman, D. S. (1953). A critique of Konrad Lorenz's theory of instinctive behavior. *The Quarterly Review of Biology, 28(4)*, 337–363.

Lehrman, D. S. (1970). Semantic and conceptual issues in the nature-nurture problem. In L. R. Aronson, D. S. Lehrman, E. Tobach, & J. S. Rosenblatt (Eds.), *Development and evolution of behavior* (pp. 17–52). San Francisco: Freeman.

Lorenz, K. (1941). Kants Lehre vom Apriorische im Licthe gegenwärtiger Biologie. *Blätter fur Deustche Philosophie, 15*, 94–125.

Lorenz, K. (1950). The comparative method in studying innate behavior patterns. *Symposia of the Society for Experimental Biology (Physiological Mechanisms in Animal Behavior), 4*, 221–254.

Lorenz, K. (1956). The objectivistic theory of instinct. In Foundation Singer-Polignac (Ed.), *L'instinct dans le comportement des animaux et de'l homme*. Masson.

Lorenz, K. (1963). *Das sogenannte Böse. Zur Naturgeschichte der Aggression*. Dr. G. Borotha-Schoeler Verlag.

Loucks, R. B. (1937). Reflexology and the psychobiological approach. *Psychological Review, 44*(4), 320–338.

Madsen, K. B. (1988). *A history of psychology in meta-scientific perspective*. North Holland.

Mameli, M. (2004). Nongenetic selection and nongenetic inheritance. *The British Journal for the Philosophy of Science, 55*(1).

Mameli, M. (2005). The inheritance of features. *Biology and Philosophy, 20*, 365–399.

Marler, P. (1999). Nature, nurture and the instinct to learn. In N.J. Adams & R.H. Slotow (Eds.), *Proceedings of the 22nd international ornithological congress, Durban* (pp. 2379–2393). BirdLife South Africa.

McCabe, B. J. (2019, May 29). Visual imprinting in birds: behavior, models, and neural mechanisms. *Frontiers in Physiology, 10*, 658.

McNally, R. J. (1987). Preparedness and phobias: A review. *Psychological Bulletin, 101*(2), 283–303.

McNally, R. J. (2016). The legacy of Seligman's "Phobias and Preparedness" (1971). *Behavior Therapy, 47*(5), 585–594.

Ongay, I. (2010). Desvanecimiento y reaparición de la idea de conducta en el marco de la Etología clásica: las perspectivas de Konrad Lorenz y de Niko Tinbergen. *El Catoblepas, 102*, 16.

Ongay, I. (2011). Ni innato ni aprendido. Capas básicas y capas metodológicas en las ciencias de la conducta. *El Catoblepas, 110*, 1.

Ongay, I. (2021). Educación y ciencias de la conducta. La etología como ciencia de la educación. *Eikasía. Revista de Filosofía, 99*, 393–420.

Ongay, I. (2022). B.F Skinner psicólogo/B.F. Skinner filósofo. In M. Porcel Medina & M. Pérez Álvarez (Eds.), *B.F Skinner revisitado*. Co-presencias.

Ongay, I. (2024). Cause and effect in biology, culture and the (extended) mind: a co-evolutionary approach. *Review of Evolutionary Political Economy, 5*(3), 445–471. Springer.

Pérez-Jara, J. (2022). Discontinuous materialism. In G. E. Romero, J. Pérez-Jara, & L. Camprubí (Eds.), *Contemporary materialism: Its ontology and epistemology. Synthese library* (Vol. 447). Springer. https://doi.org/10.1007/978-3-030-89488-7_3

Pérez-Jara, J. (2024a). The ontology of coevolution beyond economic systems. *Review of Evolutionary Political Economy, 5*, 425–444. https://doi.org/10.1007/s43253-024-00124-2

Pérez-Jara, J. (2024b). Quantum mechanics and inclusive materialism. *Philosophies, 9*(5), 140. https://doi.org/10.3390/philosophies9050140

Richards, R. J. (1974). The innate and the learned: the evolution of Konrad Lorenz's theory of instinct. *Philosophy of the Social Sciences, 4(2)*, 111–133.
Ryle, G. (1949). *The concept of mind*. Chicago University Press.
Schneirla, T. C. (1966). Behavioral development and comparative psychology. *The Quarterly Review of Biology, 41*, 283–302.
Seligman, M. E. (1970). On the generality of the laws of learning. *Psychological Review, 77*(5), 406–418.
Seligman, M. E. (1971). Phobias and preparedness. *Behavior Therapy, 2*, 307–320.
Shettleworth, S. J. (1972). Stimulus relevance in the control of drinking and conditioned fear responses in domestic chicks (*Gallus gallus*). *Journal of Comparative and Physiological Psychology, 80*(2), 175–198.
Skinner, B. F. (1938). *The behavior of organisms*. Appleton Century.
Skinner, B. F. (1948). Superstition in the pigeon. *Journal of Experimental Psychology, 38*, 168–172.
Skinner, B. F. (1956). A case history in scientific method. *American Psychologist, 11*(5), 221–233.
Skinner, B. F. (1957). *Verbal behavior*. Appleton Century.
Skinner, B. F. (1959). *Cumulative record*. Appleton Century.
Skinner, B. F. (1977). Herrnstein and the evolution of behaviorism. *American Psychologist, 32*(12), 1006–1012.
Sober, E. (1990). Contrastive empiricism. In S. theories (Ed.), *Wade Savage C*. University of Minnesota Press.
Soha, J. A. (2020). Nature, nurture and noise in bird songs ontogeny as determinants of phenotypic and functional variation among dialects. In H. Levine et al. (Eds.), *Phenotypic switching. Implications in biology and medicine*. Academic Press.
Stotz, K., & Griffiths, P. (2018). A developmental system account of human nature. In E. Hanon & T. Lewens (Eds.), *Why we disagree about human nature*. Oxford University Press.
Sun, S.-J., Narayan, V. P., Wang, Y., & Wasana, N. (2022). Digest: Nature and nurture: Influences of parental care and rearing environment on phenotypic plasticity in *Nicrophorus vespilloides. Evolution. International Journal of Organic Evolution, 76*(3), 681–684.
Thorndike, E. L. (1927). The law of effect. *The American Journal of Psychology, 39*, 212–222.
Tinbergen, N. (1954). The study of instinct. *British Journal for the Philosophy of Science, 5*(17), 72–76.
Vicedo, M. (2009). The father of ethology and the foster mother of ducks. Konrad Lorenz as an expert on motherhood. *Isis, 100*(2), 263–291.
Vicedo, M. (2013). The nature and nurture of love: from imprinting to attachment in cold-war America. *Bulletin of the History of Medicine, 88*(3), 584–585.
Viciana, H. (2021). Animal culture: But of which kind? *Studies in History and Philosophy of Science, 90*, 208–218.
Watson, J. B. (1913). Psychology as the behaviorist views it. *Psychological Review, 20*(2), 158–177.
Watson, J. B. (1930). *Behaviorism*. Chicago University Press.
Wilson, E. O. (1978). *On human nature*. Harvard University Press.

Chapter 8
Is Science an Ideology?

Gustavo E. Romero

Abstract The concept of ideology is central to understanding the many political, economic, social, and cultural processes that have taken place over the past two centuries in our societies. Yet the very concept of ideology remains a vague, openended, and much debated question. In this chapter I try to answer the question of whether science is a form of ideology or not from a philosophical point of view, taking a materialist approach. I begin by characterizing ideology as a complex, multifaceted concept. I then briefly discuss the material systems on which ideological movements operate, that is, societies and concrete human groups. I identify at least 11 different elements that seem to be present in most ideologies, and I compare these characteristics with those of contemporary science and technology. Although some superficial similarities can be identified, there are deep differences that make ideology completely different from science. With technology, however, the similarities are stronger. Ideologies are constantly evolving with technological advances, social changes, and even mere fashions. The current fragmentation of ideologies caused by the widespread use of new technologies and social networks has given rise to new phenomena of ideological diffusion that I think is very dangerous, especially for open societies. I discuss these processes in the context of the nature versus nurture debate, along with the question of whether we can get rid of ideologies.

8.1 Introduction

Ideology was ubiquitous throughout the twentieth century and still is today. Many millions of people have been killed in the name of ideologies. More millions have been imprisoned, persecuted, displaced, or tortured because of ideological conflicts.

G. E. Romero (✉)
Instituto Argentino de Radioastronomía (CONICET; CICPBA), Villa Elisa, Argentina

Facultad de Ciencias Astronómicas y Geofísicas, Universidad Nacional de La Plata, La Plata, Argentina

J. Pérez-Jara, Í. Ongay (eds.), *Beyond Nature and Nurture*, Synthese Library 497,
https://doi.org/10.1007/978-3-031-87715-5_8

The lifestyles of most people on this planet are determined or at least influenced by ideologies. Within the same society, families and friends are often divided by ideological differences. It seems almost impossible to understand today's world without some understanding of the many conflicting ideologies. And yet the concept of ideology itself remains woefully vague.

Entire libraries have been written on ideology and related topics. Most of these books examine specific ideologies, their history, characteristics, and impact on particular societies. Others compare different ideologies, or discuss conflicts between ideologies (or rather, conflicts between people and governments inspired by competing ideologies). Other books deal with the political, economic, and social aspects of ideologies. Some works map the world of contemporary ideologies, others rank them according to various criteria. Some of these books are written from a political perspective, others from an ideological perspective. Almost none discuss ideology from a philosophical perspective.

And yet, ideology is full of aspects that require philosophical investigation. A philosophical analysis of ideology is particularly appropriate because many ideologies are inspired by philosophical ideas. Others are inspired by scientific theories, and still others by pseudoscientific concepts.

The philosophy of ideology is more than a branch of political philosophy, for ideologies are not only political. They can also be biologist, economic, cultural, or even supernatural. Among the many philosophical questions that a philosophy of ideology can deal with, I mention the following:

- What is an ideology?
- What objects do ideologies deal with?
- How many kinds of ideology are there?
- Are ideologies testable?
- Do they evolve?
- What is the relationship between ideologies and society?
- Are there true ideologies?
- Is science some kind of ideology?
- Are we conditioned from our social environment to believe in some ideologies? Or are we, on the contrary, innately willing to accept some ideological attitudes?
- Do we need ideologies?

The variety of seemingly contradictory opinions that exist in the field of ideology studies makes a philosophical assessment desirable, one that can shed some new light on controversial issues and helps to dispel some of the vagueness that plagues the subject.

Let us consider, just as an example, the first question of our preliminary list above, the very basic issue of "what is an ideology?" Terry Eagleton (2007), for instance, offers a list of 16 competing meanings of the term 'ideology', collected from a wide variety of authors. Such a list, ranging from "a process of production of meanings, signs and values in social life" to "a socially necessary illusion" and "an action-oriented sets of beliefs", is far from exhaustive.

Certainly, not all definitions found in the literature are mutually compatible. Some point to only one specific aspect of ideologies. Others have a decidedly negative character. Still others are compatible with things that are certainly not ideologies (for example, we can imagine many sets of beliefs that are action-oriented but not ideological, such as those that guide a cook in preparing a particular menu).

Perhaps a better clue to the correct meaning of "ideology" can be obtained from the historical use of the word. The term seems to have been introduced by Antoine Destutt de Tracy (1754–1836). Following the spirit of the Enlightenment, soon after the French Revolution, he sought to found a general science of ideas. He tried to determine the process of forming ideas from observation and experience. But soon Napoleon and others began to use the word in a pejorative sense to describe a simplistic and idealized analysis of reality, totally divorced from facts, which tries to regulate people's lives with the excuse of improving them.

Marx introduced the word in social and political contexts with a different meaning in his classic work, written with Engels, *The German Ideology* (1976, originally written in 1846). Marx and Engels (1976) considered as ideological any set of political illusions produced by the social experiences of a class (briefly, according to Marx, a class in this context is a social group defined by its economic role, such as proprietors or workers). These illusions are used by the rulers, through the state, as an instrument of control and domination of the working class. Ideology, for Marx, was made up by the legal, political, religious, and philosophical principles proposed by the ruling classes with the aim of reinforcing the capitalist society. In other contexts, however, Marx also seems to conceive of ideology as a set of mental attitudes determined by the social environment. Thus, members of different classes are both directly and indirectly taught to think and behave in ways appropriate to their own class. With revolution and the suppression of classes, the ideology should disappear.

Lenin (1973, first published 1902), on the contrary, thought that a socialist ideology proper to the working classes was not only possible, but desirable. Such an ideology would help develop a working class consciousness and prevent this class from falling into trade unions. On the contrary, religion is an intentional creation of the ruling class for Lenin, a kind social tool whose purpose is to ensure the domination of the workers.

The Italian communist Antonio Gramsci (1891–1937) thought that ideology is consciously produced by intellectuals. Different ideologies conflict within the same society until one of them prevails. Engaging in the cultural war to achieve hegemony is the main task of the organic intellectual, that is, one closely connected with the class structure through some organization such as the communist party.

Karl Mannheim (1893–1947) realized that any social environment influences human thought, so a society with many different social groups and class environments will produce a multiplicity of ideologies. His "total" concept of ideology refers to the modes of thought and experience, the *Weltanschauung* or "worldview" of an age or group originating in a collective life situation.

Many social scientists and political analysts today use the term "ideology" in a purely descriptive way to refer to any discrete and relatively coherent system of beliefs that inform the social and political actions of a human group in a given society. Contrary to the Marxist conception, this view is essentially neutral with respect to the value of ideologies (see, for example, Seliger, 1976). Others adopt a more normative sense of the word, referring to ideas or beliefs that are in some way misleading, illusory, or one-sided, and that serve the interests of specific groups. There is a negative connotation associated with saying that something is 'ideological' in this sense. The critical conception of ideology typically holds that ideology is a way of using meaning to establish and sustain relations of domination (Thompson, 1990).

Before attempting to refine these first approximations to the concept of ideology, it should be noted that ideologies, whether in their neutral or negative sense, are always associated with groups and social classes. Perhaps, then, a good way to begin our inquiry into the nature of ideologies is to clarify the concepts of society, social group, class, social stability, and the like. This is what the next section is devoted to.

8.2 Some Concepts of Sociology

I maintain that social systems are material[1] objects, composed of organisms and the artifacts they produce (Bunge, 1979; Romero, 2018). Because they are material, social systems can change. They can grow, develop, collapse, disappear, and so on. They also interact with other social systems and with other concrete objects. Let us be more specific about these terms.

A **general material system** is a composed entity, i.e. a thing that is formed by other things. It is characterized by its *composition*, *environment*, *structure*, and *mechanism* (Bunge, 1979, 2003; Romero, 2022a).

The composition of the system is the set of its parts. The environment is the collection of things that interact with the system. The structure is the set of relationships (bonds or links) between the different components or parts, as well as with external objects. The first set of relations forms the *endostructure*, the second the *exostructure*. The total structure is the union of the two. Finally, the mechanism of the system is the collection of all its internal processes (a process is a series of lawful changes).

A **social system** is a concrete system composed of animals that (1) share an environment, (2) interact, (3) cooperate in some aspects and compete in others.

If the predominant animals in the group are humans, the social system is called the *human social system*. There are many types of human social systems, ranging from families to armies and societies. They can be natural and spontaneous, like a tribe or nation, or artificial, like a multinational company or a cooperative enterprise.

[1] The materialism I espouse is non-reductive, emergent, and systemic materialism (for further details, please see Romero, 2022a and Romero et al., 2022). It should not be confused with physicalism or other forms of eliminative materialism, nor with the so-called "new materialism" in the social sciences.

A **human society** (HS) is a human social system composed of four large subsystems: (1) biological, (2) economic, (3) political, and (4) cultural (Bunge, 2003).

The biological subsystem of a human society is composed by human beings (the dominant group) and the animals that depend on them; the structure is given by the biological relationships among all these individuals. The environment is formed by the territory and the ecosystem these people inhabit. The mechanisms are the biological processes experienced by the population.

The economic subsystem is composed of the economically active individuals (both producers and consumers) of society and the various means at their disposal. The corresponding structure is given by the relations of production, exchange and distribution of all kinds of goods among them. The environment includes the available material conditions, both internal and external. The mechanisms that operate in the subsystem are the economic processes of all kinds that occur in society.

The political subsystem is made up of those individuals who are dedicated to the management and control of common goods and activities related to the government and legislation of social actions. The corresponding structure consists of power, legal and administrative relations. The environment consists of the forces and conditions of power, both internal and external to society. The mechanisms are the actions of government and administration that take place, along with the reactions that these actions provoke from the rest of society.

Finally, the cultural subsystem is made up of the individuals who devote themselves to inventing, researching, teaching, discovering, planning, creating, representing, etc., as well as the various means at their disposal and the results of their work. The structure consists of the relationships between the various cultural actors and between them and the rest of the members of society. The environment is formed by the available material conditions, and the operating mechanisms are cultural processes of all kinds.

We can make now some postulates about social systems that seem plausible. For instance,

- Changes in a social system originate from (1) changes in its components, (2) changes in the links of its components, (3) interactions with the environment.
- Members of a social system cooperate in some respects and compete in others.
- A social system emerges if and only if its existence contributes to satisfying some needs or desires of at least one of its members.
- A social system disappears when the links among its components are dissolved.
- Changes in one subsystem of a society have impact on the other subsystems.

Regarding the latter postulate, I note that cultural changes, such as those initiated by new ideas, can affect the political and economic subsystems. These changes can be moderate or radical. Ideologies seem to operate in this sense, starting with certain groups of intellectuals and then extending to other parts of society.

A **social group** is a set of members of a society who have some common characteristics. For example, the group of the unemployed or the group of university students. We must remember that groups are not social systems. A population is divided into groups in order to study it or implement action plans. For example, the group of people over the age of 65 will be the first to be vaccinated in the event of a pandemic. Although the individuals who make up the groups are material, the groups themselves are not. You can vaccinate people, not groups. Groups are said to be vaccinated only when most or all of the members of the group are vaccinated.

Human groups are concepts abstracted from real people and used to think about various types of social systems. This does not mean that the groups are not objective, because objectivity is granted by the specification of a clear and explicit criterion for the formation of groups. The group, being a concept, can be well constructed, if the criteria are justified in a given context, or poorly defined, if the formation rule is not clear (for example, the "group of young people", or the "group of poor people", if the terms 'young' and 'poor' are not carefully defined).

There can also be arbitrary and subjective groups. For instance, the group of my friends at any given moment depends on my subjective evaluation of friendship at that moment.

A ***concrete*** **social group** is not a set but a collection of interacting individuals. Contrary to the social group, it is not a theoretical construction, but a material system. Specific social groups can achieve goals that are beyond the reach of their individuals. Some examples are work crews, police squads, gangs, sports teams, research teams, armies, musical bands, and mobs.

A **social class** is a set defined by a predicate on individuals belonging to a society. The members of a social class are supposed to strongly determine or be determined in some aspects by members of another human group. This relation of partial control or dominance is defined by Bunge (1979) as:

Dominance Relationship An individual x dominates an individual y in the aspect A, $(x > Ay)$, if and only if x determines the behavior of y in the aspect A.

More generally, we can say that social dominance refers to relationships wherein the goals of one individual prevail over the goals of another individual in a systematic manner (e.g. Sidanius and Pratto, 1999, 2004)

Note that according to the nature of the aspect involved in the dominance relationship, we can distinguish political, economic, religious, etc. classes. All these classes are conceptual constructs, not concrete social groups. Real individuals may be included in a certain class at a given moment t_1, but then at another moment t_2 they may fail to satisfy the defining relation of membership. In such a case we can say that a process of *social mobility* has occurred. Social mobility can be either ascending, in which case the individual becomes less constrained, or descending, in which case it is the other way around.

Some remarks:

- Classes, like groups, are not material.
- If classes are not material, there can be no such thing as "class struggle". Struggles are always between material systems.

- Classes cannot be self-aware because they don't have brains and therefore don't think.
- The concept of class is useful for analyzing the structures of a society, but classes cannot cause anything because they have no causal power. All real change in a society must occur because the behavior of real individuals changes.
- Although classes are not systems, some members of a class can group together and form a system, and their actions can have an impact on society. For example, unions or churches can bring together many members of a class and cause social events.

Some of the considerations of sociological concepts offered above give us clues about what ideology is. If social systems are real, concrete entities, then they are capable of change. They can be changed by external forces, such as conflicts with other social systems, or by changes in the environment, such as climate change or sudden disasters. Or they can change because of epidemics or other health problems that affect at least part of the biological subsystem. But social systems, and societies in particular, can also be changed by the actions of some of the intentional agents that make up the population. These actions can take various forms: voting, working, legislating, even violent revolution. And the actions of individuals arise from their mental configurations, their values and ideals, so that any cultural framework capable of promoting some actions and inhibiting others can be used to induce changes in a society in a particular direction. Ideologies seem to do this: they seem to be instruments for creating or preventing social change.

8.3 Some Neutral Concepts of Ideology

Several authors have proposed definitions of ideology in line with what was outlined above. Let us review a few of them.

Hamilton (1987) made an extensive examination of the literature on the concept of ideology, which led him to identify at least 27 definitional attributes to the term. He pondered these alleged attributes to ascertain their utility and coherence as definitional criteria for ideology finally proposing the following tentative definition:

> An ideology is a system of collectively held normative and reputedly factual ideas and beliefs and attitudes advocating a particular pattern of social relationships and arrangements, and/or aimed at justifying a particular pattern of conduct, which its proponents seek to promote, realize, pursue or maintain.

He notes that "This definition is coherent and sufficiently broad, yet sufficiently circumscribed, to meet the requirements of empirical application and research. It indicates the kind of ideas and beliefs that comprise ideologies rather than making claims about their causes, functions, or anything else that may or may not be empirically true about such ideas and beliefs." (Hamilton, 1987, p. 38). We see that in a broad sense this definition is in the line we have suggested.

Another author that offers a very neat definition is Lyman Tower Sargent (2009) in his classic book *Contemporary Political Ideologies: A Comparative Analysis*.

> An ideology is a system of values and beliefs regarding the various institutions and processes of society that is accepted as fact or truth by a group of people. An ideology provides the believer with a picture of the world both as it is and as it should be, and, in doing so, it organizes the tremendous complexity of the world into something fairly simple and understandable.

Here the values are explicitly mentioned, as is the fact that the adopted set of beliefs is considered true. People who have an ideology are believers, according to Tower Sargent. And they have an image of the kind of society they want to implement. They also have, as a starting point, a picture of the world as it is now. That picture may be simplistic, inaccurate, or just plain wrong. It is the difference between these two images, what is thought to be and what is believed should be, that drives the believer into action, likely guided by the value system proposed by the ideology. Here we see a great danger: if the believer's images of the world, present and future, are wrong, or the values adopted are misleading, the result of his or her actions may clash with the facts, with disastrous consequences.

Roger Eatwell definition takes note that ideologies not only can be used to promote or enforce changes in a society, but also can serve to stabilize it (Eatwell, 1993):

> A political ideology is a relatively coherent set of empirical and normative beliefs and thoughts, focusing on the problems of human nature, the process of history, and socio-political arrangements. It is usually related to a program of more specific immediate and short-run concerns. Depending on its relationship to the dominant value structure, an ideology can act as either a stabilizing or a radical force.

Finally, I mention the influential definition offered by Michael Freeden (2003), which is restricted to political ideologies:

> A political ideology is a set of ideas, beliefs, opinions, and values that
>
> 1. exhibits a recurring pattern,
> 2. are held by significant groups,
> 3. compete over providing and controlling plans for public policy,
> 4. do so with the aim of justifying, contesting, or changing the social and political arrangements and processes of a political community.

Contrary to the classical Marxist vision, Freeden emphasizes that there can be several ideologies in the same society and that they compete to control public policies. It could be argued that total ideologies compete to impose the general worldview of the society that hosts them. Freeden also stresses the importance of language control in that fight:

> Ideologies compete over the control of political language as well as competing over plans for public policy; indeed, their competition over plans for public policy is primarily conducted through their competition over the control of political language.

We can extend Freeden's definition beyond the realm of politics in line with our previous characterization of a society as follows:

Definition An ideology is a body of ideas, beliefs, opinions, and values such that

1. exhibit a recurring pattern,
2. is held by the members of one or more social groups,
3. individuals in each group work together and compete with other groups to gain control of public policy plans,
4. they do so with the aim of justifying, contesting, or changing the state and direction of the biological, economic, socio-political, and cultural processes of a community.

In short, in a broad sense, an ideology is more or less a collection of beliefs and values, not necessarily true or coherent, held by a group of individuals in a human society and used as a tool to gain control over that society.

8.4 Essential Features of Any Specific Ideology

Although the definitions given above offer a broad characterization of the concept of ideology, we can still ask ourselves 'what characteristics are essential to call a certain body of beliefs ideology?' In order to clarify this, I propose to consider any specific ideology, political or not, as a multidimensional field of beliefs. We can represent such a field as a 11-dimensional conceptual space generated by the following basis (see Bunge, 1985 for a similar approach):

$$\mathcal{I} =< C, S, D, G, B, A, P, V, I, O, M > \tag{8.1}$$

where,

C is the community of believers (a subset of which are militants) in $\mathcal{I}$.
S is the society that hosts C, and on which the members of C act to modify it in some aspect.
D is the domain of objects, real or imagined, that are studied, revered, or manipulated by members of C.
G is the general worldview adopted by the members of C.
B is the background knowledge that members of C take for granted regarding the objects of their interest (for example, historical or economic facts).
A is the total set of statements about objects in D that members of C hold true. A subset A_{core} of this set A forms the central core of beliefs of the ideological field to which believers are unwilling to change without renouncing their ideology.
P is the problem or set of problems, conceptual or empirical, that the members of C face.
V is the value system shared by the members of C.
I is the set of ideals to which the members of C aspire.
O is the set of concrete objectives of the members of C with which they hope to realize the ideals I.

M are the methods that members of C adopt to achieve their objectives O.

I will now make a few remarks about this minimal characterization of ideology as a field of belief. First, I must say that it is a tentative scheme, and certainly open to improvement. There may be other essential features of ideology that escape our analysis. However, new elements can easily be added to our definition as additional dimensions, once they have been identified through comparative research on actual ideologies. The different components of the expression (8.1) can be interpreted as discrete sets, the elements of which will depend on the specific ideology. The first set, C, has as elements the members of a community of believers in the ideology. I use the word 'community' and not just 'group' because some interaction is expected between people who have the same ideology, especially between those who are militants, that is, those who care about spreading the ideology.

People who have an ideology will only exist in a society in which they try to act. Hence the dimension S. Different types of societies will have different ideologies. The action of the individuals of C on S is possible because, as we have seen, a society is a material entity that exists in space and time and is capable of evolution. It is this evolution that the members of C wish to control or at least influence.

The scope of an ideology is given by the domain D of objects that concern the believers. If these objects are social or political, the ideology will be sociopolitical. But, as we will discuss in the next section, it can be broader or narrower domains.

People who hold an ideology will share a worldview, or at least elements common to several worldviews, however simple. This will give their beliefs the minimum coherence necessary for argument, or at least for propaganda. For example, adherents of an ideology may view the world as a material system in which human beings are free to act, or they may believe that God created the world and we must act according to his designs, or that human beings are inherently evil and therefore must be controlled and regulated down to the minutest detail if they are to live in society without killing each other, etc. Of course, the worldview could be much more sophisticated, informed by science and technology.

If the adherents of an ideology want to change some aspect of society, they must have, or pretend to have, some knowledge of the relevant aspects in order to devise a course of action. For example, an economist who is ideologically identified with a left-wing ideology may think he or she knows that inflation is always caused by speculative behavior on the part of manufacturers, merchants, and distributors of goods. Then, given the right power, he or she might try to impose and enforce price controls in supermarkets and other stores. Another liberal economist might think that his colleague's actions are nonsense because the government's practice of having the central bank print money to finance public spending is the main cause of inflation. So, for the second economist, reducing the deficit and stopping printing money are the first steps to suppressing inflation. Without some knowledge or presumed knowledge of the causes of inflation, there is nothing to be done in either case. If the position of economists is ideologically oriented, and their decisions and choices are based on ideas belonging to the hard core $A_{\rm core}$ of beliefs of their respective ideologies, it will be very difficult, if not impossible, for them to change their minds in the face of evidence of failure and take a different course of action. If there were not a hard and irreducible core of ideas that are not negotiable, we would not be talking about ideology, we would be talking about economics. It is easy to find examples that extend the existence of a core of canonical truths to all kinds of ideologies.

Every ideology faces problems that it claims to solve. Problems of power, job opportunities, gender issues, freedom, etc. The extent of these problems depends on the nature of the ideology and the characteristics of the society. To solve them, members of C rely on their background knowledge B and a set of methods M that are considered acceptable and effective. They are also guided by their values V and ideals I. Values may be part of an elaborate axiology or simply a set of morals, depending on the sophistication of the ideological field. Ideals are conceptual models in which various values are exemplified. The purpose of ideals is to facilitate the visualization of goals and to motivate action. The ideal itself is not a goal that we can hope to achieve, but it serves to push our struggle in a direction that can lead us to our goals. Therefore, all ideologies contain problems, methods, values, and ideals. They are the driving forces and self-imposed constraints necessary to achieve goals. The latter are also essential: people guided by ideologies are goal-oriented. Everything else is a means to achieve the changes they want to implement. Perhaps the shortest way to characterize a particular ideology is to list its core beliefs and goals.

8.5 Types of Ideologies

From the characterization of ideology offered in the previous section, we see that what makes an ideological field different from other fields of activity in human societies is that ideology is based on beliefs and adopts specific arrangements of values and methods with the ultimate goal of achieving some influence on the evolution of a society or a concrete social group. Depending on the scope of the ideology, that is, according to its domain D, we can differentiate different types of ideologies (see also Bunge, 1985 and Freeden, 2003).

- **Macro-ideologies**: they are concerned with all or several of the subsystems of the society. Examples: liberalism, socialism, totalitarianism, etc.
- **Micro-ideologies**: they refer only to a limited class of issues within the society. Examples: nationalism, pacifism, feminism, gender ideology, etc.
- **Super-ideologies**: they refer to all areas of human existence, not only social ones. Examples: Catholicism, Islamism, Marxism (in some of its various forms), etc.

If an ideology of any kind does not change in the face of evidence adverse to it, we say that this ideology is a *fundamentalism*. Many religious, political, and cultural ideologies fall into this class.

8.6 Science vs Ideology

We arrive now to the central question of this chapter: is science just another ideology? Many people seem to think so. I strongly disagree. But before discussing the relationship between science and ideology, it is useful to characterize science. I will use an approach similar to the one I used earlier for the concept of ideology. I will first offer a general outline and then a multidimensional analysis. Science is the result of highly complex human activity, and any simple attempt to define it by a single salient characteristic tends to degenerate into a mere caricature.

Science is the result of a systematic human effort to gain true knowledge about the world. It is not the only way to gain human knowledge; we can learn many things just by observing, practicing, reading, etc. Science differs from these and other ways of acquiring knowledge in that it is systematic and its results are subject to a variety of controls. It is also a progressive activity in the sense that scientific knowledge increases with research. There are several indicators of scientific progress, including improvements in the ability to predict events and increases in human ability to manipulate the environment (through science-based technology).

Unlike other forms of knowledge acquisition, *science produces conceptual representations of the world that are articulated in theories and models* (Romero, 2018). A theory is a logically organized set of statements, endowed with a specific interpretation, that refer to objects of the same class. If we introduce a set of statements P, a set of semantic instructions to interpret them S, and a domain (reference class) R, then a theory is represented by the quadruple (Bunge, 1967; Romero, 2018):

$$T =< P, S, R, \vdash >, \tag{8.2}$$

where $\vdash$ is the logical entailment operation. So a theory is a context that is closed under deduction: every statement in a theory is either a premise or a deductive consequence of a set of premises. The premises are called axioms and the consequences are theorems.

Certainly, theories in the making are seldom presented in this way. It is the task of the philosophers of science to render them into such a format to investigate their structure, ontology, and deep meaning. The working scientists usually do not care about this.

Contrary to a popular belief, the theorems of a theory never can be directly tested (except for coherence). Any evaluation of a theory against empirical evidence must be implemented through individual statements produced through a *model.* Models are obtained from a number of theories $(T_1, T_2, \ldots, T_n)$ and sets of specific assumptions $(A_1, A_2, \ldots, A_m)$ that describe concrete situations. Symbolically,

$$(T_1 \wedge T_2 \wedge \ldots \wedge T_n) \cup (A_1 \wedge A_2 \wedge \ldots \wedge A_m) \vdash M. \tag{8.3}$$

The model represents a collection of processes occurring in a specific circumstance. When we go from general theories to models the reference class shrinks enormously.

General theories, unlike models, are not expected to make concrete predictions unless they are considerably enriched with special assumptions and data. We test the theories by *consistency analysis* (both internal and with the total network of theories) and by *empirical evaluation of models* obtained from the theories with specific assumptions and data on applications to individual cases. These evaluations are made by comparing predictions (statements) of models with data. An empirical datum is not a fact, but a proposition that informs about a fact and that is acquired with the help of empirical operations (experiments or observations). We always compare propositions with propositions, never propositions with facts. Since propositions are conceptual objects, they are loaded with theory. The fact itself, on the other hand, is independent of theory (Bunge, 1983; Romero, 2018).

It is very difficult to rule out a theory by the predictions of a model, because any observation or experiment involves many theories and assumptions about the specific situation to which the model is applied. Any of the theories or any of the ancillary data could be the cause of the lack of confirmation. Similarly, positive claims of confirmation of a theory should be treated with caution for the same reason: false positives can be caused by a variety of sources. That is why multiple independent tests are needed. We need to think about the validity of a given theory in a given domain across many models in quite different situations and with different experimental setups before we can draw conclusions about the validity of the theory in question.

We can apply a multidimensional approach to science, as we did with the concept of ideology. The result is always tentative and perfectible, but I think it takes into account most traits that are characteristic of scientific research. Science itself can be defined as a set of research fields where each research field R comprise the following items (Romero, 2018):

- C: A community of researchers specialized in the field R.
- S: A society that houses the activities of those individuals in C.
- D: A domain of material or conceptual objects to investigate and study in the field R.
- G: A general and basic philosophy shared by the members of C.
- F: Set of formal languages used by researchers in their field.
- B: A background of previous scientific knowledge.
- Q: A collection of problems of questions about the domain D.
- A: A collection of goals of the members of C with respect to D.
- M: A specific set of scientific methods, techniques, error analysis approaches, and heuristic recipes that are used to solve the problems in Q.
- E: An ethic common to the members of C and a set specific moral norms for the field R.

Then, a research field R is formally represented by:

$$R =< C, S, D, G, F, B, P, A, M, E > . \qquad (8.4)$$

The elements of each component change over time, hence these components are sets only at a fixed moment, otherwise they are collections, not sets. The field of research evolves according to the evolution of its components. At a given instant, science can be defined as the set of all research fields at that instant:

$$\text{Sci} = \{R_1, R_2, \ldots, R_n\}. \quad (8.5)$$

This type of characterization is similar to that proposed by Bunge (1983), but I have included an ethic component that in my opinion is essential for any research activity.

Now, a few remarks. Many of the components of (8.4) are similar to those of our characterization of ideology. There are, however, important differences. Scientific research, the activity of the members of C, consists of discovering, posing, examining, and if possible solving problems. Not every problem is a scientific problem: scientific problems, Q are those that are posed against a scientific background, B, and are studied with scientific means, M, and with the primary objective of increasing our knowledge about the world and the mechanisms that operate in it (A).

The result of research is articulated in theories and models that intend to represent as correctly as possible some aspect of the world. Theories of basic sciences (a) are made up of hypotheses, including regularities (trends or laws) about classes of entities that are assumed to exist, (b) are capable of producing more or less accurate predictions based on specific models, and (c) contain no value judgments about reality or action programs aimed at changing it.

On the other hand, an ideology usually does not result from research (although some may be inspired by it) nor does it change with its results: up to now, ideologies have been fairly dogmatic beliefs and resistant to scientific innovations. An ideology can change but only in details. If an ideological "ism" were to radically change, it would cease to be that "ism."

Major changes in an ideology are usually introduced by some charismatic leader and resisted by other charismatic leaders, rather than being the consequence of rigorous investigation and admission of mistakes. The result is that ideologies, instead of progressing like the sciences, fragment. As an example, let us consider the many divisions of communism: Marxism, Marxism-Leninism, Stalinism, Maoism, Trotskyism, etc. The methodological component and the verifiable element are mostly absent from ideologies. The ideologues will apply, in many occasions, the ideological package prescribed for each case regardless of the situation, to later blame the failure on whatever is at hand at the time.

Several authors, in the last decades, have proposed that science is a form of ideology, since it would actually deal with power and domination. The attitude of considering science as ideology perhaps goes back to Heidegger's obscure analysis of technology (Heidegger, 1954, English translation 1977). According to Heidegger (1) technology "is not an instrument", it is a way of understanding the world; (2) technology "is not a human activity" but develops beyond human control; and (3) technology is "the greatest danger", risking seeing the world only through technological thinking. There is a confusion here between science and technology on which I will talk about later.

The view of science as a form of power and an ideology is more clearly stated by Marcuse (1964, 1968). According to Marcuse, "domination perpetuates and extends itself not only through technology, but as technology, and the latter provides the great legitimation of the expanding political power which absorbs all spheres of culture." He adds, "science, by virtue of its own methods and concepts, has projected and promoted a universe in which the domination of nature has remained linked to the domination of man—a link which tends to be fatal to this universe as a whole. Nature, scientifically comprehended and mastered, reappears in the technical apparatus of production and destruction which sustains and improves the life of individuals while subordinating them to the masters of the apparatus. Thus, the rational hierarchy merges with the social one." The question for Marcuse is not whether technology has played any role in the promised emancipation of humanity. The question for him is simply whether technological advance leads to more repression and domination or not. He fears that "the very concept of technological reason is perhaps ideological". Since industrial technology facilitates the exploitation of the working class by the bourgeoisie, the instrumental manipulation of nature in the natural sciences is bound to be guided by bourgeois ideology. Science has become ideology.

According to Habermas, the central error in Marcuse's formulation of the problem is that he has retained a concept of ideology appropriate to an era long past (Habermas, 1970, see also 1971). For Marx, ideology was tied to a defined social class. Such a view is now untenable. Habermas thinks that classes have not disappeared in contemporary society, but have been irreversibly integrated. When Habermas calls for the abandonment of the critique of technology and science as ideology, he is not trying to imply that they are neutral; he recognizes that they have become forms of domination as well as emancipation from deprivation and hard work, so he sees them in an ambivalent light.

Other authors who challenge the neutrality of science and defend a science loaded with ideology are Thomas Kuhn,[2] Paul Feyerabend, and Bruno Latour. These well-known and widely debated views have many similarities with Foucault's thesis that the knowledge/power relationship becomes characteristic of modern society.

Proposals to identify science and ideology arise not only from the sociology of science, but also from the camp of ideology itself, especially from environmentalism and feminism.

Ecology's critique of science and technology asserts that ethical neutrality with respect to the results of scientific research is not justified. Green ideologues challenge the idea that any form of knowledge can be divorced from its consequences. If science has formed an alliance with the dominant forces that contribute to the destruction of the environment, then its neutral character is not such, and science

[2] At least in his classic book *The Structure of Scientific Revolutions*, University of Chicago Press, 1962. In later works he moderated his original views.

must be fought. Science, we are told, has become an ideological tool of those who exploit the planet for their own short-term gain.

Radical feminism sees science as a way of knowing burdened with presuppositions and permeated by the interests of male domination over women. Modern science is not objective, but a vehicle for Western masculine values (Mies and Shiva, 1993). The privilege of determining what counts as scientific knowledge and how it is used has been controlled by men and, for most of history, limited to men, so that science is a kind of male ideology.

Some of these observations miss the point by confusing science with ideology, or with the attitudes of individual scientists who hold ideological positions, or by confusing science with technology. When we compare our characterizations of science and ideology, we can see deep differences beyond the superficial similarities. It is true, however, that basic science involves values. As we have seen, each field of research is composed of individuals who share a certain ethic, and the ethic is based on a common axiology, a set of values accepted by the researchers. These values, however, are not universal; they refer to the research methods used to gain knowledge about the objects in the field of study. They imply things like the rejection and condemnation of plagiarism and fraud, working conditions in laboratories, proper citation behavior, and many other things. Of course, researchers themselves may have ideological positions on many issues, but such ideology should not pervade their work. Ideological content, when it infiltrates scientific results, is often detected and purged by the many checks and balances available to modern science.

The confusion of science and technology, so common among Heiddeger, Marcuse, and many postmodern critics, can be avoided by an adequate characterization of both activities. Technology is related to our ability to manipulate our environment. Not all technology is necessarily based on science. Technology predates science. Science-based technology can be characterized as a human activity that aims to design, develop, build, and control artifacts using knowledge gained through science (Romero, 2018). An artifact is something artificial that can be controlled and used for specific purposes. Artifacts are not just mechanical. They can be electronic, thermodynamic, biological, or cultural, depending on the proposed goals.[3] Science-based technology includes not only the many fields of engineering, but also medicine, didactics, normative epidemiology, economics, law, and all disciplines of social planning.

As we did with ideology and science, we can distinguish several components in a technological field T_i (Romero, 2018):

- C_i: A community of technologists.
- S: A society that welcomes those individuals in C_i.
- D_i: A set of material or conceptual things that T_i deals with.
- F_i: A set of formal theories used by the members of C_i.

[3] For "conceptual artifacts" see Romero (2018), pp. 94–96, and Romero (2022b).

- E_i: Set of scientific theories and data used by the members of C_i.
- P_i: A set of specific practical problems.
- A: The total technological knowledge available to those in C_i.
- O_i: A set of final technological goals.
- M_i: A specific set of norms and methodological instructions.
- V: A value system adopted by the members of C_i.

Each scientific technology is a science-informed activity aimed at solving practical problems. Although closely related to science through F_i and E_i, it differs radically from science in its methods and goals. There is, however, a strong virtuous link between basic science and scientific technology: scientific results motivate questions whose answers demand new technologies, and then the results and empirical explorations obtained with these new technologies make it possible to formulate new questions and problems. that cause a mutually reinforcing progress.

Technology is not without values. Technologists in any field have values regarding their rules of action in certain activities, such as animal experimentation, environmental contamination, or weapons production. In contrast to basic science, where the general ethical neutrality is clearer, discussions about values and ethical implications are possible and desirable in technology. Ideally, such debate should be rational and science-based, but it is often ideologically motivated.

The moral responsibility of technologists in their efforts to produce technological artifacts and solutions to the problems entrusted to them should not obscure the more important responsibility of politicians, bureaucrats, or executives who demand, finance, and decide on the use of the products of scientific technology, from vaccines to weapons (see Doorn and van de Poel, 2012 and references therein).

I conclude that more than science being a kind of ideology, ideology is a kind of social pseudo-technology.

8.7 Ideological Fashions and Packages

Why do we believe in ideologies? Ideologies offer a kind of chart or map of our social and political environment (Freeden, 2003). Ideologies give us mental representations that seem to guide us by means of identifiable patterns in the midst of what otherwise appears to be a mere chaos of events beyond our control. Ideologies offer us ways to make sense of what we experience in society. They promise security, a sense of belonging, purpose, and change for the future. More importantly, they help us organize our values and give us motivation to act.

Ideologies, however, do not objectively represent external reality. Diverse and sometimes conflicting ideologies compete for our attention, offering radically different versions of the social world. Often they are not even internally coherent or consistent with the scientific worldview. It is not surprising, then, that many people end up adopting ideological positions in various combinations, with no justification other than that these positions are popular in a particular time and place. Or perhaps they simply follow a charismatic person who holds those views and expresses them in a way they can relate to. Or maybe it is simply because there is no other option and the world seems too strange, chaotic, and threatening otherwise.

On the other hand, the spread and diffusion of certain ideologies, as well as the decline of others, is a complex phenomenon that obeys a combination of causes and circumstances of an economic, social, and historical nature. Think, for example, of the decline of liberalism and the concomitant spread of totalitarianism in Europe and Asia in the first half of the twentieth century, and especially after the Great Depression. Or the spread of leftist ideologies in South America at the beginning of this century, coinciding with the rise in international commodity prices caused by the growth of China. The complex combination of external, internal, environmental, and psychological events necessary to bring about such sociopolitical changes is not yet fully understood. Fashion may not even be aware of some of these processes.

The universalization of the use of social networks at the beginning of this century facilitated the dissemination of simple ideas and combinations of ideological positions, forming ideological kits ready to be adopted by social groups with similar backgrounds. As mentioned by Pérez-Jara and Camprubí (2022): "These *packs* structure collective identities by gathering themes and narratives which lack internal logical connections between them. Moreover, what comes in each pack varies with the cultural landscapes." These ready-to-use ideological kits assembled for immediate gratification need not even be coherent. Pérez-Jara and Camprubí add:

> The specific contents of ideological packs thus depend on the historical and cultural conditions behind their construction and maintenance. And yet, once they are functioning, these menus provide a significant degree of predictability of a person's ideological nebula just by attending to one or a couple of isolated ideological items.

In Argentina, for example, most people who now identify themselves as leftists and progressives will also be nationalists, something unimaginable before the advent of Peronism, when socialists were internationalists (Sebreli, 2002). It is also very likely that the self-proclaimed progressive will support dictatorships such as those in Venezuela and Nicaragua and sympathize with the regimes in China and Iran, while strongly advocating the use of inclusive language and special legislation in favor of transgender people and other sexual minorities. Minorities that would be outlawed and persecuted in most of the countries with which these "progressives" ideologically identify. The lack of logical connection between these and other positions seems to go unnoticed or unimportant to whoever adopts the particular ideological package. Surprisingly, many of these people will assert the importance of science in society. Similarly, many self-identified liberals may support anti-abortion laws, be nationalistic, support the death penalty, or oppose gun control or vaccination. Similar ideological kits are popular in Brazil and other countries in the region. It seems clear that reason plays little or no role in the adoption of these ideological packages. It is the same lack of rational support that probably makes it possible for the believer to undergo a sudden "conversion" and change sides, something that is occasionally observed. Pérez-Jara and Camprubí (2022) remark:

> At a cultural level, social groups usually change opinion not because they are really convinced by the logical force of the arguments of their opposite enemies. Rather, there are other cultural factors at work, such as the changes of position of the public charismatic figures that those social groups have already bestowed legitimacy. But sacred cows can also be sacrificed and canceled if they step too far into fields which their audiences are convinced to be inherently evil or polluted.

One of the results of these attitudes is a marked polarization where each side considers the other the source of all evil. Another result is to facilitate the sudden emergence of new charismatic leaders who can embody with effective dramaturgy the basic ideological kit of preference in a specific social group (Alexander and Pérez-Jara, 2021).

8.8 Why People Believe in Ideology? Nature vs. Nurture

In the previous section, we mentioned how the cultural environment influences people's ideological choices. But that seems to be only part of the story. The source of ideological differences can also be traced in part to physiological, genetic, cognitive, and neural patterns.

Twin studies consistently find that political orientation has a heritable component (Martin et al., 1986; Alford et al., 2005). More recent genetic studies show a statistically significant association between self-reported political ideology and the 7R variant of the dopamine D4 receptor (DRD4) gene (see Dawes and Fowler, 2009, Settle et al., 2010). These studies suggest that the gene creates a disposition, resulting in personality traits that lead an individual to seek cultural environments that, in turn, incline some people toward innovation and liberal ideologies and others toward conservatism.

In a much larger sample of 1771 Han Chinese university students in Singapore, Ebstein et al. (2015) observed a significant effect of association between the DRD4 exon III variable number of tandem repeats and ideological attitude. Subjects with two copies of the 4-repeat allele (4R/4R) were significantly more conservative. These results provide further evidence for a role of the DRD4 gene variants in contributing to individual differences in ideological attitude. Although these and other similar studies should be extended to larger and more diversified samples to achieve stronger statistical support, the preliminary results are extremely encouraging. In particular, transcultural studies are highly desirable to separate environmental effects, that might introduce some systematic biases. For a recent review see Méndez (2017).

Physiological responses also seem to trace the ideological dichotomy between liberals and conservatives. When exposed to negative images, for example, people with a leftist ideological orientation report a smaller increase in sympathetic nervous system activation, indicated by changes in electrodermal activity (Oxley et al., 2008). A strong relationship has also been observed between physiological responses to unpleasant images and conservatism (Dodd et al., 2012), in the sense that people of this orientation are more sensitive to crude and repulsive visual stimuli.

Many of the correlations of physiological responses with ideological identifications have been corroborated at a deeper neurological level by functional analysis of brain activity using MRI (fMRI). Recently, Ahn et al. (2014) applied machine learning techniques to fMRI data to test the hypothesis that brain responses to emotionally evocative images predict broad ideological orientation. Disgusting images (for example, a mutilated body) generate neural responses that are highly predictive of the ideological group they belong to. Specifically, machine learning analysis enabled the identification of the subject's ideological orientation from whole-brain blood oxygen level distribution (BOLD) patterns during imaging exposure. The hemodynamic response of the conservative group had a steeper slope and a higher peak than that of the liberal group.

These results are independently supported by analysis of the correlations found in brain-injured patients (Nam et al., 2021). People with frontal lesions show a preponderance of more conservative (or less liberal) beliefs than those with or without anterior temporal lobe lesions. Additional studies predict ideology by extent of damage, yielding evidence that greater damage to the dorsolateral prefrontal cortex, but not to the amygdala, is associated with a higher incidence of conservatism. These last results suggest that emotional reactions are stronger in conservative people, reinforcing what was found by Ahn et al. (2014).

The general picture is also consistent with the neurophysiological studies carried out by Nam et al. (2018), who found that a larger bilateral amygdala volume is positively correlated with the tendency to believe that the existing social order is legitimate and desirable, that is, with a conservative position (see also Kanai et al., 2011, and Kim et al., 2020 for a functional connectivity analysis).

It is important to note that the studies cited above do not pertain to the political positions typically designated as "leftist" and "rightist". The former term is frequently employed to describe individuals who prioritize the collective over the individual, advocate for equality, support environmental protection, and promote expanded educational opportunities. This encompasses the provision of social safety nets for those who require them, among other measures. In contrast, those who identify themselves as "rightist" tend to espouse an individualistic worldview, wherein the individual is placed above the community. Additionally, they typically espouse a limited government approach, which they believe is conducive to individual freedom and personal property rights, as well as competition and other related concepts.

This political dichotomy between "left" and "right" has been correctly identified as a problematic concept by Lewis and Lewis (2023). They have proposed that the concept of fixed dichotomies, namely that there exists an inherent polarity between "left" and "right," and that individuals can be situated along a continuum between these two extremes, is a fallacious assumption. The assumption that the political left is more collectivist and the political right is more individualist has been challenged by sociological and historical findings. For example, Adolf Hitler is regarded as a collectivist, but he is usually labeled as an extreme right individual. Also, it is frequently assumed that those on the right wing of the political spectrum espouse a belief in limited government. However, an examination of historical evidence suggests that this may not be the case. For example, Mussolini's political philosophy was defined by a robust centralization of power within the state, with the state itself becoming the primary object of collective action. Certainly, Mussolini and fascism are regarded as right-wing, as is militarism. Nevertheless, it could be argued that militarism is inherently communal in nature. One might also consider whether militarism is, in fact, statist and whether it is not indicative of a larger, more centralized government. From a political standpoint, the nuances are not as straightforward as they may appear.

The aforementioned neurological and genetic studies appear to indicate that there are fundamental human tendencies that influence how individuals respond to specific social circumstances. These tendencies appear to be shaped by genetic and unconscious neurological processes. Additionally, these tendencies manifest as a predisposition to make choices that are not necessarily aligned with the political stereotypes or semantic labels associated with a particular society or attitude, with the standard usage of left and right. Rather, the tendency seems to show a predisposition to better adapt to groups that embrace or reject change, novelty, order, and pluralism.

All these findings and others reported in the current literature (see Jost et al., 2014 and Krastev et al., 2016 for reviews) invite us to revise the traditional view that ideological positions are the product of rational, conscious, and socialized thought. The adoption of ideological stances seems to be, instead, more related to an emotional process intimately linked to complex neural dispositions. Neither a unique product of nature nor of nurture, ideology seems to emerge rather when our dispositions find the appropriate cultural and material conditions for their development.

8.9 Are Ideologies Necessary?

Our own time, since the early nineteenth century, has been called the "age of ideology" because of the ubiquity, prevalence, and importance of these belief systems (Watkins, 1964). The death or decline of ideologies, however, has often been claimed since the concept of ideology itself emerged at the end of the eighteenth century. After World War II, Aron (1955) and later Bell (1960) and Lipset (1960) formulated the "end of ideology" thesis. According to this thesis, as a result of the gigantic struggle of ideologies that took place in the middle of the twentieth century, a struggle that cost millions of lives, destroyed entire nations, and caused indescribable suffering, both right-wing and left-wing ideologies had been discredited. As a result, the power of ideologies to motivate and mobilize people would have been exhausted, at least in the West.

After the collapse of the Soviet Union, this thesis was further developed and modified by Fukuyama (2006), who argued that liberalism was the ultimate victor in the ideological wars of the twentieth century. Even in countries where a liberal society had not yet been fully realized, liberalism would be seen as the only acceptable ideological view.

Not surprisingly, the end of ideology thesis has been widely contested (for an overview of the debates in this controversy, see Brick, 2013). Articles titled "The End of the End of Ideology" soon began to proliferate. Hodges (1967) and later Jost (2006) reviewed the evidence against the claim that ideologies are in decline and concluded that ideology is very much alive. Palmer (1994) suggests that ideologies change their tone and mode of operation rather than end.

As we have seen in Sect. 8.7, current ideological opinion, at least among large groups of people defined by common interests, similar cultural backgrounds, and connected by media and social networks, often emerges in the form of a variety of "ready-to-eat" menus or ideological packages. Because each package contains a set of positions that may not be logically related or even contradictory, its acceptance depends more on social interactions than on cognitive evaluation. The result is an atomization of the great conflicts of the past and an expansion of the ideological battlefield to include hitherto unsuspected issues. For example, refusing to wear a mask for sanitary reasons becomes an ideological statement because this act is perceived as part of an ideological package that defines a particular group. These associations can lead many people to make completely unreasonable decisions, decisions that affect their lives and the society in which they live, sometimes with catastrophic consequences.

Given that ideology can so easily evolve and change, driven by unprecedented technological change, fostered by our innate neural dispositions, can we hope to get rid of it in the future, at least its most damaging effects? Should ideology disappear, if possible, from our cultural landscape?

Ideologies undoubtedly are powerful forces that shape our societies. As Palmer (1994) says: "They are instrumental in recruiting the early enthusiasms of the more schooled and aware citizens, the intellectuals. They give rise to simplified slogans that encapsulate in a popularly attractive fashion the main concerns and thus recruit the greater numbers needed for effective political action. They catalyze the adoption of policies either to conserve or to change social, economic and political institutions." These are important social functions, instrumental in the mobilization of the different political, social, economic and cultural agents. Ideology has the power to motivate people to act, something that in the past could only be achieved through religion, force, or more rarely, reason. The problem is when the ideological body of beliefs is in disagreement with well-known facts, includes meaningless or incompatible propositions, or the value system is outdated or inconsistent with the proposed goals.

Ideological fanaticism can be mitigated by scientific and technical knowledge, which helps us to see the inconsistencies of thought and the incompatibility of belief with evidence (Lane, 1966). Education in the habits of critical thinking, on the other hand, makes it more difficult to accept ideological packages that are incoherent or lack clear meaning. Cognitive activity and learning can activate neural circuits that serve to inhibit dispositions that originate in less plastic subcortical architectures. I don't know of any more powerful tools for destroying fanaticism and unreason than science and fact-based philosophy, with the cognitive habits they produce.

Ideology is not inevitable, though it is highly functional in gaining and maintaining power in modern technical societies. When it becomes dysfunctional, we must fight it. Our very existence may depend on it. In the past, and often in the present, this has often been done by violence. At least in open societies, the struggle should begin much earlier, trying not to get to the point where violence seems inevitable.

Is it possible to have a scientific ideology, i.e. an ideology with a body of beliefs that is informed by science and free of inconsistencies? Bunge (1985) thinks so. I doubt it. True science has nothing to do with beliefs. It is about research, conjecture, evidence, and testing. Science lacks the degree of conviction necessary to move the masses. And the values of science refer only to the researchers, how they relate to their methods and protocols, and how they interact with the society that hosts them. In this case, I think we should look to philosophy for help. A philosophy informed by the best science of the day. In political philosophy, in the philosophy of sociology, in axiology, and in ethics.

Perhaps the best thing to do is to finally put aside ideologies, if we can educate ourselves enough, and simply agree on our goals as a society, and then look for the optimal means to achieve them within each subsystem of society: political, economic, biological, and cultural.

A scientifically and philosophically informed policy will always be better than an ideologized one.

8.10 Summary and Conclusions

Ideology is an ill-defined concept that is sometimes used very loosely. In general, the term denotes a set of beliefs, values, methods, and goals that concrete human groups consider necessary, or at least desirable, to achieve lasting changes in various aspects of modern human societies. In this article I have identified at least 11 essential components or "dimensions" that are present in every ideology, that is, elements that must be manifested if a body of ideas is to be called an ideology.

To claim that there is an ideology, there must be a community of people who have produced, believe in, and are committed to the basic assumptions of the ideological corpus. These people should be in contact with a particular society that they are trying to change in some way. Different ideologies deal with different objects or referents of their discourse in each society. Certain groups of people, mobilized by ideals, try to change some aspect of a society. They can succeed because both the militants and society are material entities, and material things are capable of change. The tools they have at their disposal to achieve their goals and solve the problems that concern them are a variety of methods, some background knowledge, a set of values, a worldview, however sketchy, and a core of ideas that they consider inalienable. This core of ideas may or may not be informed by current science, and may or may not form a coherent body.

Ideology is created by human beings to affect other human beings in a society. Ideologies are not produced by social classes, because social classes are not material entities, but conceptual ones. Only human beings, concrete human groups and societies are material systems. And only material objects and systems can change their states and interact (see Romero, 2022a for an updated review of systemic materialism).

Ideologies can be divided into three broad groups according to the scope of action they seek: (1) macro-ideologies, (2) micro-ideologies, and (3) super-ideologies. The first group includes sociopolitical ideologies that seek to change or control societies through their major subsystems: political, economic, and cultural. If an ideology focuses on only one specific subsystem, it may be appropriate to call it an "intermediate ideology. The second group deals with a more limited range of issues, from nationalism to feminism. There are a large number of micro-ideologies in the world today. Finally, super-ideologies deal with every aspect of human life. They even offer an interpretation or intended insight into the entire universe. In this group we find religions that affect human affairs, such as Catholicism or Islam, as well as philosophically motivated macro-ideologies that adopt a metaphysical worldview, as seen in certain totalitarianisms.

In addition to these three main types of ideologies, we also find some ideologies that appeal to syncretism. They adopt a group of ideas from different ideologies depending on the circumstances in order to defend their goals. These ideologies, like conservatism and progressivism, are also very flexible in their methods and programs.

I have not considered capitalism and democracy to be ideologies because they are an economic system and a way of organizing governments. There are many ways to implement both, and none of these ways brings together the multiple dimensions that characterize ideologies. Likewise, populism and terrorism are not ideologies. They are methods of gaining and maintaining power.

Nor is science another kind of ideology. Despite some superficial similarities, there are profound differences. The goals of ideology are more like those of technology than those of science. And every ideological movement has at its core a set of beliefs and propositions that are non-negotiable. Science, on the other hand, is always in flux. All scientific assumptions and presuppositions are always considered provisional and open to revision. Values in science, moreover, concern only issues related to scientific practices and do not extend to the full range of human activity. Ultimately, science seeks true representations of the world, while ideologies seek to command, control, and direct certain kinds of human action to achieve certain human ends.

Ideology is highly mutable as technology advances. This is not surprising, since ideology seems to have emerged as a way of coping with the complications of modern industrial societies, where technology plays a fundamental role in determining social structure. With the advent of radio and television in the twentieth century, the major ideologies were widely disseminated, leading to major conflicts as the main contenders defended substantially opposite world views. In the twenty-first century, social networks have played a crucial role in the fragmentation and simplification of ideologies, creating ideological packages or kits. These packages consist of combinations of slogans and simple maxims. Ideologies are developed by intellectuals, but ideological packages are produced by politicians and populists for quick use and consumption. The goal of these packages is to provide a ready-to-use guide that encourages behavior and justifies the opinions of their followers. The inconsistencies that these ideological packages often contain are reflected in the irrational behavior of those who adhere to them. One of the most alarming results of the massive proliferation of these ideological menus is the extreme political polarization observed in many societies. Such polarization has become the most powerful force against rational discussion and perhaps the greatest internal threat to open societies in this century.

Since its appearance on the political, social, and cultural landscape in the nineteenth century, ideology has been behind great confrontations and wars, social upheavals and riots, terror and totalitarian control, the destruction of goods and freedoms, the division of families and the separation of friends. But it has also been a source of great progress, advancing civil reforms, improving working conditions, resisting tyranny and oppression, defending the rights of minorities, promoting industrialization, and much more. Its ever-changing legacy remains controversial. Can we do better without ideologies? Perhaps. But in the meantime, we would do well to try to prevent the spread of incoherent and scientifically uninformed ideological packages by encouraging people to think for themselves, to be guided by science and reason, and to leave dogma and prejudice behind.

Our brains are wired to believe. We have innate predispositions that favor the attitudes that underlie the major ideological camps. And yet such predispositions work only under certain conditions that can be controlled. And even under such conditions, our brains are plastic enough to inhibit a wide range of primitive impulses. We must and we can, for the sake of our civilization.

Acknowledgments I thank Javier Pérez Jara for useful suggestions and stimulating comments, as well as for his invitation to contribute to this volume. I also thank Óscar Teixidó, Matt Suárez Holze, and especially Santiago Perez Bergliaffa for helpful remarks.

References

Ahn, W. Y., Kishida, K. T., Gu, X., Lohrenz, T., Harvey, A., Alford, J. R., Smith, K. B., Yaffe, G., Hibbing, J. R., Dayan, P., & Montague, P. R. (2014). Nonpolitical images evoke neural predictors of political ideology. *Current Biology, 24*(22), 2693–2699. https://doi.org/10.1016/j.cub.2014.09.050.

Alexander, J., & Pérez-Jara, J. (2021). Mito y rito en el corona-drama internacional. Jeffrey Alexander en diálogo con Javier Pérez Jara. In *Sociedad entre pandemias* (pp. 257–258). Fundación Gaspar Casal.

Alford, J., Funk, C., & Hibbing, J. (2005). Are political orientations genetically transmitted? *American Political Science Review, 99*(2), 153–167. https://doi.org/10.1017/S0003055405051579.

Aron, R. (1955). The end of the ideological age?. *The opium of the intellectuals* (pp. 305–24). Trans. Terence Kilmartin. Norton (2nd ed., 1962).

Bell, D. (1960). *The end of ideology*. Free Press.
Brick, H. (2013). The end of ideology thesis. In M. Freeden, L. T. Sargent, & M. Stears (Eds.), *The Oxford handbook of political ideologies* (p. 90). Oxford University Press.
Bunge, M. (1967). *Foundations of physics*. Springer-Verlag.
Bunge, M. (1979). *Treatise on basic philosophy, Vol.4: Ontology II: A world of systems*. Kluwer.
Bunge, M. (1983). *Treatise on basic philosophy, Vol.6: Epistemology and methodology II: Understanding the world*. Kluwer.
Bunge, M. (1985). *Seudociencia e Ideología*. Alianza Editorial.
Bunge, M. (2003). *Emergence and convergence*. University of Toronto Press.
Dawes, C. T., & Fowler, J. H. (2009). Partisanship, voting, and the dopamine D2 receptor gene. *The Journal of Politics, 71*(3), 1157–1171.
Dodd, M. D., et al. (2012). The political left rolls with the good and the political right confronts the bad: Connecting physiology and cognition to preferences. *Philosophical Transactions of the Royal Society B: Biological Sciences, 367*, 640–649.
Doorn, N., & van de Poel, I. (2012). Editors' overview: Moral responsibility in technology and engineering. *Science and Engineering Ethics, 18*, 1–11. https://doi.org/10.1007/s11948-011-9285-z.
Eagleton, T. (2007). *Ideology: An introduction*. New and updated ed. Verso.
Eatwell, R. (1993). Ideologies: Approaches and trends. In R. Eatwell & A. Wright (Eds.), *Contemporary political ideologies*. Pinter Publishers.
Ebstein, R. P., Monakhov, M. V., Lu, Y., Jiang, Y., Lai, P. S., & Chew, S. H. (2015). Association between the dopamine D4 receptor gene exon III variable number of tandem repeats and political attitudes in female Han Chinese. *Proceedings of the Royal Society B, 282*, 20151360. https://doi.org/10.1098/rspb.2015.1360.
Freeden, M. (2003). *Ideology: A very short introduction*. Oxford University Press.
Fukuyama, F. (2006). *The end of history and the last man*. Free Press (Original work published 1992).
Habermas, J. (1970). Technology and science as 'ideology'. In *Toward a rational society*. Beacon Press.
Habermas, J. (1971). *Toward a rational society: Student protest, science and politics*. Heinemann.
Hamilton, M. B. (1987). The elements of the concept of ideology. *Political Studies, XXXV*, 18–38.
Heidegger, M. (1954). Die Frage nach der Technik. *Vorträge und Aufsätze*. Neske Verlag.
Hodges, D. C. (1967). The end of "The End of Ideology". *American Journal of Economics and Sociology, 26*, 135–146. https://doi.org/10.1111/j.1536-7150.1967.tb00996.x.
Jost, J. T. (2006). The end of the end of ideology. *American Psychologist, 61*(7), 651–670. https://doi.org/10.1037/0003-066X.61.7.651.
Jost, J. T., Nam, H. H., Amodio, D. M., & Van Bavel, J. J. (2014). Political neuroscience. *Political Psychology, 35*, 3–42. https://doi.org/10.1111/pops.12162.
Kanai, R., Feilden, T., Firth, C., & Rees, G. (2011). Political orientations are correlated with brain structure in young adults. *Current Biology, 21*, 677–680. https://doi.org/10.1016/j.cub.2011.03.017.
Kim, T., Hur, J. W., Kwak, S., et al. (2020). Intrinsic functional connectivity of blue and red brains: Neurobiological evidence of different stress resilience between political attitudes. *Scientific Reports, 10*, 15877. https://doi.org/10.1038/s41598-020-72980-x.
Krastev, S., McGuire, J. T., McNeney, D., Kable, J. W., Stolle, D., Gidengil, E., & Fellows, L. K. (2016). Do political and economic choices rely on common neural substrates? A systematic review of the emerging neuropolitics literature. *Frontiers in Psychology, 7*, 264. https://doi.org/10.3389/fpsyg.2016.00264.
Lane, R. E. (1966). The decline of politics and ideology in a knowledgeable society. *American Sociological Review, XXXI*, 649–662.
Lenin, V. I. (1973). *What is to be done?* Progress.
Lewis, V., & Lewis, H. (2023). *The myth of left and right: How the political spectrum misleads and harms America*. Oxford University Press.

Lipset, S. M. (1960). The end of ideology? In S. M. Lipset (Ed.), *Political man: The social bases of politics* (pp. 403–417). Doubleday.
Marcuse, H. (1964). *One dimensional man*. Beacon Press.
Marcuse, H. (1968). *Negations: Essays in critical theory*. Penguin Press.
Martin, N. G., et al. (1986). Transmission of social attitudes. *Proceedings of the National Academy of Sciences of the United States of America, 83*, 4364–4368.
Marx, K., & Engels, F. (1976). *The German ideology*. Vol. 5 of K. Marx & F. Engels, *Collected works* (50 vols.). International Publishers.
Méndez, M. F. (2017). *The Journal of Neuropsychiatry and Clinical Neurosciences, 29*(2), 86–94.
Mies, M., & Shiva, V. (1993). *Ecofeminism*. Fernwood Publications.
Nam, H. H., Jost, J. T., Kaggen, L., et al. (2018). Amygdala structure and the tendency to regard the social system as legitimate and desirable. *Nature Human Behaviour, 2*, 133–138. https://doi.org/10.1038/s41562-017-0248-5.
Nam, H. H., Jost, J. T., Meager, M. R., & Van Bavel, J. J. (2021). Toward a neuropsychology of political orientation: Exploring ideology in patients with frontal and midbrain lesions. *Philosophical Transactions of the Royal Society B, 376*, 20200137. https://doi.org/10.1098/rstb.2020.0137.
Oxley, D., et al. (2008). Political attitudes vary with physiological traits. *Science, 321*, 1667–1670.
Palmer, L. M. (1994). The end of the end of ideology. *History of European Ideas, 19*, 709–713. https://doi.org/10.1016/0191-6599(94)90054-X.
Pérez-Jara, J., & Camprubí, L. (2022). *Science and apocalypse in Bertrand Russell: A cultural sociology*. Lexington Books.
Romero, G. E. (2018). *Scientific philosophy*. Springer.
Romero, G. E. (2022a). Systemic materialism. In G. E. Romero, J. Pérez-Jara, & L. Camprubí (Eds.), *Contemporary materialism. Its ontology and epistemology* (pp. 79–107). Springer.
Romero, G. E. (2022b). Mathematics do not refer to material entities. In G. E. Romero, J. Pérez-Jara, & L. Camprubí (Eds.), *Contemporary materialism. Its ontology and epistemology* (pp. 324–327). Springer.
Romero, G. E., Pérez-Jara, J., & Camprubí, L. (Eds.). (2022). *Contemporary materialism. Its ontology and epistemology*. Springer.
Sebreli, J. J. (2002). *Crítica de las Ideas Políticas Argentinas*. Editorial Sudamericana.
Seliger, M. (1976). *Ideology and politics*. George Allen and Unwin.
Settle, J. E., et al. (2010). Friendships moderate an association between a dopamine gene variant and political ideology. *The Journal of Politics, 72*, 1189–1198.
Sidanius, J., & Pratto, F. (1999). *Social dominance: An intergroup theory of social hierarchy and oppression*. Cambridge University Press. ISBN 978-0-521-62290-5.
Sidanius, J., & Pratto, F. (2004). Social dominance theory: A new synthesis. In J. T. Jost (Ed.), *Political psychology: Key readings in political psychology*. Psychology Press.
Thompson, J. B. (1990). *Ideology and modern culture: Critical social theory in the era of mass communication*. Polity Press.
Tower Sargent, L. (2009). *Contemporary political ideologies: A comparative analysis* (14th ed.). Wadsworth.
Watkins, F. (1964). *The age of ideology - Political thought, 1750 to the present*. Prentice-Hall.

Chapter 9
Nature, Nurture, and Nous: The Role of Reflection and Reason in the Nature-Nurture Debate

Liane Gabora and Cosette Robertson

Abstract Biological and social factors (i.e., nature and nurture) interact to give rise to specific human attributes; but they alone are insufficient to explain our most uniquely human attributes: our creative ideas, artifacts, and cultures, and the diverse worldviews that underlie them. We posit that these attributes stem from the capacity for an autonomous, generative stream of thought; it relies on nous. The word 'nous' comes from the Greeks, and refers to intellect, understanding, thought, or reason. Our abstract thoughts, ideas, and unique creative styles, and how they crystallize into distinct ways of seeing, being, and contributing to the world may seem harder to pin down or study scientifically than innate drives (nature) and socially learnt (nurture). We show how these attributes can be modelled using autocatalytic networks. Originally used to model the origin of life, autocatalytic networks provide an abstract formal framework with which to model the emergence and growth of networks; not just the networks of catalytic molecules at the core of biological evolution, but also the psychological structures at the core of cultural evolution. Applied to culture, they allow us to model the cognitive steps involved in how knowledge obtained through social or individual learning (modeled as foodset items) 'catalyzes' the generation of new ideas and perspectives (foodset derived items), and to track lineages of cultural adaptation and change. We outline how they have bene used to model significant transitions in human cultural prehistory, as well as cognitive development in the mind of a child, and cultural discontinuities caused by cross-domain transfer.

Keywords Autocatalytic networks · Cognition · Creativity · Evolution · Culture · Concepts · Development · Nature · Nurture

---

L. Gabora (✉)
Department of Psychology, University of British Columbia - Okanagan Campus, Kelowna, BC, Canada
e-mail: liane.gabora@ubc.ca

C. Robertson
Fipke Centre for Innovative Research, University of British Columbia - Okanagan Campus, Kelowna, BC, Canada

J. Pérez-Jara, Í. Ongay (eds.), *Beyond Nature and Nurture*, Synthese Library 497,
https://doi.org/10.1007/978-3-031-87715-5_9

9.1 Introduction

It is often said that it isn't nature or nurture that matters, so much as the interaction between them. Yet even 'the interaction between them' doesn't capture what makes humans different from other species, and from each other; something extra is required. For example, say you buy a new house. This decision is fuelled in part by the biological drives for shelter, i.e., influenced by *nature*. The kind of house you buy is also influenced by what features your family and friends taught you to look for in a house; i.e., it is also influenced by *nurture*. However, what if the new house inspires a novel? No one taught that novel to you; you came up with it yourself. Although it is indirectly affected by nurture—by creative writing exercises in school, what novels you've read and enjoyed, and so forth—your novel goes above and beyond what you've acquired through social learning. It can't be accounted for by nature; indeed, while you are immersed in the writing of the novel, you may forget to eat, or ignore your children, i.e., ignore biological imperatives. Nor is the novel any kind of straightforwardly computable function of the *interaction* between nature and nurture, though both nature and nurture had an indirect impact on the decision to write the novel and the final form it took. Each step along the way as you figured out the plot, developed the characters and setting, and manifested these ideas as passages of text and dialogue, your mental contents become less firmly tethered to the influences of nature and nurture, and more deeply rooted in forces of your own making. Nature and nurture are vital sources of knowledge that often provide initial points of departure for a novel or work of art or music, but they are insufficient to explain the form such works achieve. Nature and nurture alone cannot account for how we alter, make sense of, or discard learned or innate knowledge, nor the complex thought processes by which we generate *new* knowledge. They cannot account for the overall configuration of an individual's mental model of reality, or *worldview*. To explain this, and the fruits of its ideation and self-expression, we need something extra.

In this chapter, we show how this 'something extra', beyond what can be accounted for by nature, nurture, or direct interactions between them, can nevertheless be explained and understood in scientific terms, using a relatively new kind of abstract formal framework, referred to as *autocatalytic networks* (Kauffman, 1993). The term *autocatalytic* refers to the fact that interactions amongst initially distinct *parts* can give rise to (i.e., *autonomously catalyze* the emergence of) a new *whole*. As we shall see, autocatalytic networks provide a means of distinguishing the contributions of nature (i.e., innate traits), the contributions of nurture (i.e., environmental influences), and the contributions of that 'something extra,' *nous*. The word 'nous' in the title of this chapter comes from the Greeks, and refers to intellect, understanding, thought, or reason. We posit that the ability to make a novel contribution to human culture relies on the capacity for an autonomous, generative stream of thought; it relies on nous.

9.2 Nous: The Glue that Binds Innate and Socially Transmitted Knowledge

Like humans, other species possess innate knowledge (nature) and knowledge obtained directly from the environment (nurture), which they use to elicit appropriate responses to stimuli. However, these stimulus-response are relatively fixed. Even when they exhibit creative responses that get copied by others, any alteration of those responses is accidental; there is little in the way of cumulative, adaptive change. In contrast, humans[1] are flexible; we tailor responses to our personal tastes and the specific situation we're in, and we cumulatively build on each other's ideas. For example, the invention of the wheel led to the cart, which in turn led to the automobile. However, this kind of strategic cumulative change can only occur in a mind in which elements of knowledge are not discrete and distinct from one another (as in fixed action patterns) but integrated into webs of understanding, such that any one is potentially accessible to, and influenced by, or combined with, any other. Only then can, for example, the concept of a cart be combined with the concept of an internal combustion engine to generate the automobile. Nous is the glue or connective tissue that binds together innate knowledge (nature) and knowledge obtained directly from the environment (nurture). Humans routinely engage in streams of thought that explore possible interrelationships between different elements of their knowledge they possess. These streams of thought continuously modify and update the configuration of our worldviews.

It makes sense, then, that to formal model and understand the nature of nous, and how it gives rise to an integrated understanding of the world that can adapt existing ideas to new circumstances, a network model is required.

9.3 Modeling What Emerges at the Nature/Nurture Interface

Building a scientific model of nous can help us understand it by providing a means to examine variables and their interrelationships, test hypotheses, and serving as a framework for communication and elaboration. Since nous is the glue that *connects* innate knowledge and environmentally sourced knowledge into the unique structure that is an individual's worldview, the obvious route to modeling nous is using *network science:* the study of complex networks (such as cognitive networks, semantic networks, social networks, or computer networks). Network science has roots in graph theory. Elements of the network are represented by *nodes* (or *vertices*), and the connections between them are represented by *links* (or *edges*). However, to model how nous—streams of reason and creative thought—weave

[1] Perhaps other species elicit flexible responses too; the point here is not that humans are different from any other species; it is that there is a distinction to be made here.

instincts and acquired knowledge into a worldview, we need a specific kind of network, one that can model how new elements form through interactions between existing elements, and how networks come about in the first place.

9.3.1 Autocatalytic Network Theory

A promising avenue is to use autocatalytic networks. ***Autocatalytic network theory*** grew out of studies of the statistical properties of random graphs consisting of points randomly connected by edges. As the ratio of edges to points increases, connected points join to form clusters, and as the size of the largest cluster increases, so does the probability of a phase transition resulting in a single giant connected cluster, i.e., an integrated whole (Erdös & Rényi, 1960). The point at which this happens is referred to as the *percolation threshold.*

Autocatalytic networks were first used to develop the hypothesis that life began, not as a single self-replicating molecule, but as a set of catalytic molecules that crossed the percolation threshold to a self-organizing, self-sustaining whole (Kauffman, 1993). The theory postulates that the first living thing, and all subsequent life forms, take the form of an autocatalytic networks. Autocatalytic networks have subsequently been used to model phase transitions underlying the origin of complex cognition (Gabora, 1998; Gabora et al., 2022; Gabora & Steel, 2017, 2020a, 2020b, 2022; Ganesh & Gabora, 2022a, 2022b). Autocatalytic network theory is simply an abstract mathematical framework with multiple applications; indeed, it has been suggested that they may hold the key to understanding the origins of *any* evolutionary process be it biological or cultural (Gabora & Steel, 2021). Autocatalytic networks model, not just network structure and dynamics, but how networks *emerge* through the formation of connections amongst existing elements, and how they *grow* as external elements are incorporated.

9.3.1.1 Key Concepts

A key development that made autocatalytic network theory useful for modeling nous was the differentiation between network elements sourced from the environment, and network elements generated anew within the system. Elements sourced from the environment are referred to as the ***foodset***, since they serve as raw materials for the generation of new elements, while elements generated within the system are referred to as ***foodset-derived*** elements, since they are constructed from other elements (either foodset elements, or previously generated foodset-derived elements) (Hordijk et al., 2010). This distinction enables us to study under what conditions foodset items become woven together by foodset-derived items into integrated wholes of various types. The process by which a new foodset-derived item is generated from one or more foodset items is referred to as a ***reaction***.

Another key idea in autocatalytic network theory is that of ***catalysis***. The generation of a particular new foodset-derived element may be galvanized, or sped up, by a specific existing element (which may be either a foodset or foodset-derived element). Thus, in autocatalytic networks, nodes are not just passive transmitters of activation; they actively trigger, or *catalyze* the synthesis of novel (foodset-derived) nodes from existing ones. An element serves as a ***catalyst*** if it is involved in a reaction but is not, itself, altered by the reaction.

9.3.1.2 Cognitive Autocatalytic Networks

While the initial application of autocatalytic networks to the origin of life involved catalytic molecules forming chemical reaction networks, in cognitive applications they are networks of knowledge and memories, composed of chunks of procedural or declarative information, scripts and schemas, concrete episodes, and abstract concepts. (*Concepts* are mental constructs such as CAT or FREEDOM that enable us to interpret new situations in terms of similar previous ones.) For simplicity, all these different kinds of mental contents may be referred to as *mental representations.*

Elements obtained through both nature and nurture serve as foodset elements, whereas elements obtained through 'nous' (e.g., insights or ideas that are the outcome of reason or creative thought) serve as foodset-derived elements. Thus, the foodset includes not just innate knowledge, but also socially transmitted knowledge, as well as information acquired directly from the environment through individual learning (e.g., from a book, or observations of nature) so long as it is not modified or reinterpreted in the process. For example, an inborn fear of snakes constitutes an *innate foodset item*. Learning from a teacher the scientific name for the bird results in a *socially learned foodset item* because pre-existing information is obtained from the environment through social means. Discovering for oneself that lightning tends to be followed by thunder results in an *individually learned foodset item;* the information itself does not change form simply because the individual now knows it. An example of a *foodset-derived item* is the generation of a poem or song about lightning; here, new information is generated by the individual in question, through internal thought processes, i.e., through nous. (In reality, the distinction between foodset and foodset-derived items is not always black and white, but we will ignore that subtlety for now.) The distinct roles played by foodset and foodset-derived items in RAF models is consistent with cognitive differences arising when learning is through (a) instruction or imitation, versus (b) opportunities to experiment or explore, or direct challenges that prompt students to learn for themselves (e.g., Bonawitz et al., 2011; Chen & Kalyuga, 2020).

It is common to treat reason and creative thought—i.e., nous—as a form of individual learning (e.g., Boyette, 2016). However, it is necessary to distinguish between these two processes to describe and model the unfolding of cultural lineages. By distinguishing individual learning, wherein *pre-existing* (foodset) information is obtained from the environment by non-social means, from nous, wherein *new* (foodset-derived) information is generated, it becomes possible to specify who contributed to any given lineage of cultural innovation, and how. Another reason to distinguish the two is that creative thought is quite different in character from either social or individual learning; it often involves reiterated reflection, in which a problem or task is considered from multiple perspectives, and one's conception of the solution or output transitions from ill-defined (or 'half-baked') to well-defined (Scotney et al., 2020). Creative thought can also solidify a sense of identity (Karwowski & Kaufman, 2017). It is what brought to human culture everything from myths about storm gods, to thunderous rain sticks, to the invention of electrical circuits.

Note that an item can be part of the foodset of one network, while the identical item may be foodset-derived in another network; what matters is whether it materialized in its present form in the network in question. (For example, if you invent a new game, that's a foodset-derived item in your mind, but if you tell your friends about it, it's a foodset item in their minds, because the friends didn't think it up for themselves.)

In cognitive applications of autocatalytic networks, insights or ideas obtained through nous are said to be *catalyzed* by needs, goals, knowledge gaps, or external stimuli, and (in keeping with autocatalytic network convention), the process by which this happens is referred to as a *reaction*. In a cognitive context, *reactions* are mental operations such as induction, conceptual restructuring, concept combination, or metaphor. Just as reactions between molecules generate new molecules, interactions between mental representations generate new mental representations, which, in turn, enable new interactions. Just as a catalyst makes a certain reaction more likely to occur, a question, desire, or external stimulus can trigger or 'catalyze' a new idea, or perspective. For example, you *might* just happen to write a poem about lasagna, but it is more likely that this would be catalyzed by a stimulus such as seeing or smelling lasagna, or a memory of a particular instance of lasagna, or the drive state of hunger. The *reactants* in this cognitive reaction are words and their referents, and the *product* is the poem. The poem is foodset-derived because it is composed of foodset items put together in a novel way, resulting in something new.

The products of a cognitive reaction are often more complex than reactants, but not necessarily; for example, a foodset-derived item might come about through the realization that two seemingly distinct categories are both instances of the same more abstract category (as often happens in mathematics). The *reactivity* of a concept or idea refers to its capacity to stimulate or catalyze recursive sequences of conceptual change. Ideas may become highly reactive by aligning with needs, drives or desires, thereby leveraging the power of biological urges to further cultural goals.

9.3.1.3 Mathematical Development of the Theory

Autocatalytic network theory has been mathematically extended and generalized as ***Reflexively Autocatalytic Foodset Generated networks***, also referred to as RAF networks, or simply, **RAFs** (Hordijk et al., 2010, 2011; Hordijk & Steel, 2004, 2015, 2016; Steel, 2000; Steel et al., 2019, 2020; Xavier et al., 2020). The approach offers an established formal framework for modeling how the origins of self-organizing structures that are sufficiently complex to reproduce and *evolve*, whether they be organisms undergoing biological evolution, or novelty-generating minds driving cultural evolution.

Technically, a RAF is thus a non-empty subset of 'reactions' that satisfies the following properties:

- *Reflexively autocatalytic*: each reaction is catalyzed by at least one element type that is either produced by the network or present in the foodset.
- *F-generated*: all 'reactants' (i.e., elements undergoing change) can be generated from the foodset through a series of reactions in the network.

The union of all RAFs in a system forms a single unique maximal RAF referred to as the maxRAF, and any other RAFs subsumed by it are referred to as subRAFs. RAFs can be hierarchically structured, i.e., the maxRAF may consist of multiple levels of subRAFs. A subRAF becomes an *irreducible* RAF—or simply, an *irrRAF*—when it cannot be reduced any further without losing its RAF status. Table 9.1 provides definitions of key terms in RAF network theory and how they apply to cognitive RAF networks. The terminology reflects its initial application to biochemistry, but there is nothing about the theory that is specific to that setting; it is a general mathematical framework that is equally applicable to cognition.

Figure 9.1 is a RAF network diagram illustrating how innate knowledge, as well as information acquired from the environment through social or individual learning, which collectively constitute the *food set*, serve as raw materials for new information generated *by* the individual, i.e., *foodset-derived knowledge*. These foodset items (in conjunction with innate knowledge) participate in a reaction, and through catalysis, a new foodset-derived item is generated *by the individual.* For example, this could schematically illustrate the use of innate, individually learnt, and socially acquired information in the creative generation of a cartoon. A second RAF network, representing knowledge from a different knowledge domain—e.g., music—is formed, resulting in a sound track for the cartoon, followed by a co-RAF that connects the sound track to the cartoon as the art and music are adjusted to align. The figure also shows the formation of a domain-general (DG) maxRAF that subsumes the other RAFs, which are subRAFs of the maxRAF. Note that if we omit the distinction between the three sources for foodset items (innate, acquired through individual learning, or acquired through social learning) the figure could serve equally well to illustrate, at an abstract schematic level, the kinds of interactions amongst abiotic molecules, postulated to underlie the origin of life.

9.3.1.4 Strengths of the RAF Approach

We now summarize the features of RAF networks that make them particularly suited to modeling nous: thoughts and ideas that emerge at the nature-nurture interface but extend beyond what could be classified as nature or nurture.

Table 9.1 Alphabetically listed definitions of terms in autocatalytic network theory

Term	Definition
Catalyst	Element that triggers or speeds up a reaction but that is not itself altered by the reaction (Kauffman, 1993). In a cognitive context, it may be a stimulus, thought, or drive state that triggers a specific cognitive change (Gabora & Steel, 2017)
Co-RAF	Does not constitute a RAF by itself, but can be co-opted by an existing RAF to form a larger RAF (Steel et al., 2013)
Foodset-derived item	Element that is constructed from scratch during a reaction (Hordijk et al., 2010). In a cognitive context, it is a mental representation that the individual thought of on their own (Gabora & Steel, 2017)
Foodset item	Element that is not constructed from scratch but is available in environment (Hordijk et al., 2010). In a cognitive context, it is a mental representation that was either innate, or obtained from the environment, e.g., through social learning (Gabora & Steel, 2017)
IrrRAF	RAF that cannot be reduced further without losing its RAF status (Hordijk et al., 2012)
MaxRAF	The union of all RAFs in a system that forms a single unique maximal RAF (Hordijk et al., 2012)
Persistent RAF	A RAF that does not depend on any particular element from the external environment to persist (Gabora & Steel, 2020a, 2020b)
Reactants	Raw materials (either foodset items or foodset-derived items) that undergo change in a particular reaction; in a cognitive context, they may be concepts, beliefs, and so forth (Gabora & Steel, 2017)
Percolation threshold	Point at which there is a phase transition such that network elements form a single giant connected cluster, i.e., an integrated whole
Products	New elements that result from a reaction; in a cognitive context, the products may be ideas (Gabora & Steel, 2017) or perspectives (Ganesh & Gabora, 2022b)
Reaction	Process wherein catalyst triggers transformation of reactants into products; in a cognitive context, 'reaction' refers to cognitive change due to a mental operation such as deduction, analogical transfer, or concept combination (Gabora & Steel, 2017)
Reactivity	Capacity to catalyze reactions. In a cognitive context, the reactivity of an idea refers to its capacity to stimulate conceptual change (Gabora & Steel, 2020a)
SubRAF	Subset of a maxRAF that is itself a (smaller) RAF (Hordijk et al., 2012)
Transient RAF	A RAF that does depend on a particular catalyst from the external environment to persist (Gabora & Steel, 2020a, 2020b)

Modeling Endogenous Change The RAF framework is consistent with the goal of understanding, not just network structure, but also how networks emerge and dynamically restructure themselves in response to internal and external pressures, i.e., generative network growth, including phase transitions to structures with dramatically new properties (Steel et al., 2019). They can model the transition from disconnected parts to a network that is self-sustaining and self-organizing (Hordijk & Steel, 2004; Mossel & Steel, 2005), and able to self-replicate and evolve (though

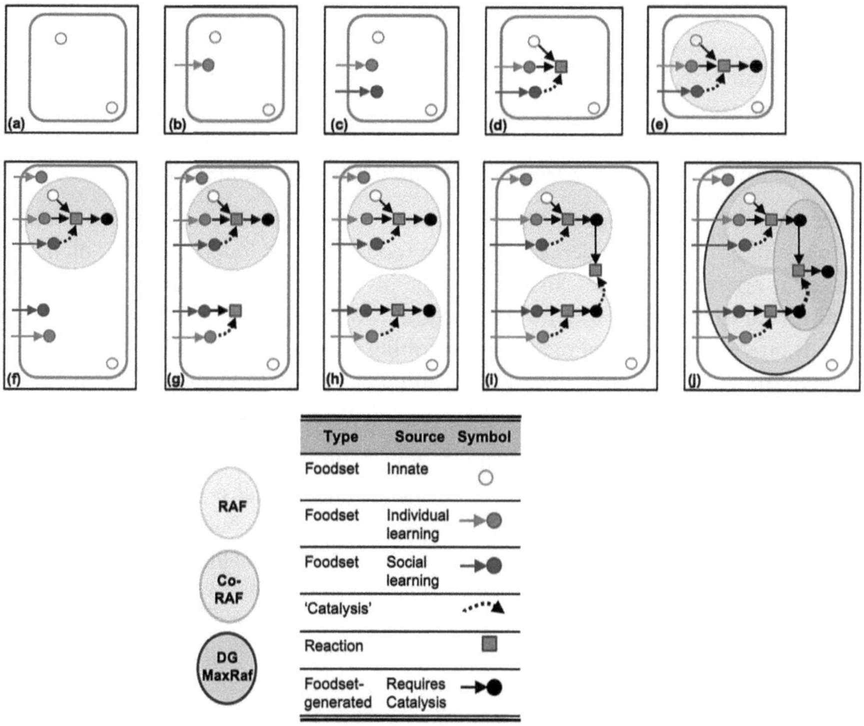

Fig. 9.1 An example of the transition from non-RAF (**a–d**), to RAF (**e**), to multiple RAFs (**f–h**), to their integration into a domain-general maxRAF (**i, j**). In panel (**a**) the individual's mind (brown rounded rectangle) initially contains only innate knowledge (white dots) which constitute foodset items. In panels (**b**) and (**c**), more foodset items are acquired through individual learning (green dot) and social learning (red dot), respectively. Panel (**d**) illustrates a socially learnt foodset item catalyzing (dotted arrow) a reaction (blue square) between other foodset items, resulting in a foodset-derived item (black dot), and a first RAF (yellow oval). Panels (**f–h**) depict a similar process resulting in the formation of second RAF. Panel (**i**) depicts the formation of a coRAF (orange oval) network that is reliant on the continued existence of the earlier two RAFs. Panel (**j**) shows the domain general (DG) maxRAF (purple oval) that encompasses each of the other RAFs as subRAFs. The innate item on the lower right of all panels, and the one acquired through individual learning on the upper left of panels (**f**) through (**j**) do not assimilate into a RAF

in a haphazard manner, without reliance on a self-assembly code) (Gabora & Steel, 2021; Hordijk & Steel, 2015).

Potential to Scale Up The generalized RAF setting is conducive to the development of efficient (polynomial-time) algorithms for questions that are computationally intractable (NP-hard) (Steel et al., 2019). Since RAFs can be described as the (nonempty) subsets of the 'reactions' that are the fixed points of an (efficiently computable) interior map operating on subsets of these 'reactions,' it is possible to efficiently identify and classify RAFs within large systems such as cognitive systems (Steel, 2023).

Semantic Grounding When applied to cognition, the demarcation into foodset versus foodset-derived elements provides a natural means of grounding abstract concepts (such as FREEDOM, a foodset-derived item) in direct experiences (i.e., foodset items such as experiences of freedom, or the lack of it, in all their rich concrete details). One can trace the ancestry of how a foodset-derived concept emerged through series of 'reactions' originating in foodset knowledge.

Tracking Ideas Not only can one follow such lineages within individuals, but also between individuals. The demarcation into foodset versus foodset-derived elements makes it possible to trace existing innovations back through each individual that contributed to them in a way that demonstrates what each individual's contribution was. It provides a means of precisely describing how new ideas and cultural outputs emerge from previous ones, and tracking cumulative change in cultural lineages step by step (Gabora & Steel, 2020a, 2020b, 2022; Ganesh & Gabora, 2022a).

Problem Solving and Resolving Uncertainty A problem or source of uncertainty (e.g., a dog owner's inability to communicate with her deaf dog) is modeled as an item that is disconnected from the maxRAF containing the bulk of the individual's mental representations. The disconnect alerts one to relevant stimuli (e.g., while lying next to the dog with her phone in her pocket, the dog-owner notices that the dog's head turns when her phone vibrates). Such a stimulus may catalyze a solution, described as a catalyzed reaction between foodset items (e.g., 'phone vibration' and 'dog communication') resulting in a foodset-derived item (e.g., the dog-owner invents an app-connected dog collar that vibrates in different patterns to convey messages to dogs). The problem (e.g., inability to communicate with a deaf dog) is now integrated into the worldview by way of this new foodset-derived item.

9.4 Nous Enabled Emergence of a Second Level of Autocatalytic Structure

The hypothesis that cognitive networks constitute a *second* level of autocatalytic structure (the first being the structure that confers life to a set of abiogenic molecules) is consistent with the observation that, like living organisms, cognitive networks are self-sustaining, self-organizing, and self-reproducing (Gabora, 2004; Maturana & Varela, 1973; Varela et al., 1991). The self-sustaining nature of a cognitive network is evident in the tendency to reduce cognitive dissonance, resolve inconsistencies, and preserve or revise existing schemas in the face of new information (Antrobus et al., 1970; Festinger, 1957; Klinger & Cox, 1987). Although the contents of a cognitive network change over time, it maintains integrity as a relatively coherent whole. Its capacity to solving problems through the generation of new connections (as in the example above) is evidence of its spontaneously self-organizing nature. Cognitive networks *self-replicate* in a piecemeal manner through social learning, and our propensity to share stories and perspectives, thereby influencing the organization of each others' cognitive networks. Finally, cognitive networks *evolve* in the sense of exhibiting cumulative adaptive, open-ended change over time.

9.4.1 Modeling the Origins of Nous

We now look at how cognitive RAFs have been used to model the transition from minds that consist of disconnected islands of innate or 'nurtured' knowledge and memories, to minds that autonomously generate the conceptual glue that integrates these mental representations into a *worldview*, beginning with some of the earliest evidence for nous.

It has been proposed that the transition from Oldowan to Acheulean tool technology 1.76 mya was precipitated by the onset of representational redescription, which enabled the forging of associations, and the emergence of hierarchically structured concepts, making it possible to shift between levels of abstraction to carry out complex tasks (Gabora & Smith, 2018). Unlike more primitive Oldowan stone tools, the Acheulian hand axe required not only the capacity to envision and bring into being something that did not yet exist, but hierarchically structured thought and action, and the generation of new mental representations: the concepts EDGING, THINNING, SHAPING, and a meta-concept, ACHEULIAN TOOL. The RAF model lays out what socially or individually learned foodset items were needed to serve as reactants in the generation of these concepts, and the drive state that catalyzed their representational restructuring (Gabora & Steel, 2020a). Early cognitive RAF structure consisted of *transient RAFs* that persisted only in the presence of particular catalysts (such as specific external stimuli). There was no conceptual redundancy or reinforcement in place ensuring the persistence of these new mental representations, no alternative routes by which they could be generated. Nevertheless, this constituted a critical step toward what has been referred to as *conceptual closure,* characterized by the emergence of a persistent autocatalytic cognitive structure that encompasses the bulk of the individual's mental representations (Gabora, 2000, 2004).

The RAF framework also provides a promising avenue for unraveling one of the greatest archaeological mysteries: that of why development of the Acheulian hand axe was followed by over a million years of cultural stasis. The rate of representational redescription at that time may not have risen above the critical percolation threshold required to form an integrated whole, so streams of abstract thought were short, and few in number.

9.4.2 Modeling the Transition to Persistent Nous and Cognitive 'Modernity'

The Paleolithic marked the appearance of art, science, and religion, and people who were behaviorally and cognitively 'modern,' i.e., they lived and thought in a way that feels recognizably human. Synthesizing research from psychology, anthropology, and cognitive science, it was suggested that this was brought on by one or more genetic mutations that allowed thought to be spontaneously tailored to the situation by modulating the degree of (1) divergence (versus convergence), (2) abstractness (versus concreteness), and (3) context-specificity (Gabora & Steel, 2022; Winslow & Gabora, 2022). This in turn marked a phase transition culminating in the capacity for a emergence of a persistent, unified RAF network that bridged previously compartmentalized knowledge and experience, and collectively formed a stable, hierarchical maxRAF (Gabora & Steel, 2020b). The capacity to tailor one's mode of thought to the kind of experience one is having is referred to as contextual focus (Gabora, 2003). Once this was in place, a problem could be reflected upon from different perspectives until a solution was achieved, described as the formation of a new subRAF of the maxRAF. Onset of this capacity enabled our human ancestors to make sense of and connect seemingly disparate information or experiences. This in turn made it possible to creatively adapt ideas to ever-new situations in a genuinely human way, and marked the beginnings of cumulative, adaptive, open-ended cultural evolution.

9.4.3 Using RAFs to Construct Cultural Lineages

The creative products of nous are further stabilized as other individuals develop and build on early ideas. The fact that the RAF approach tags mental representations with their source, i.e., whether they were (1) innate, (2) acquired through social learning or individual learning (of *pre-existing* information, i.e., elements of the foodset), or (3) the result of creative thought (resulting in *new* information, i.e., foodset-derived elements) makes it possible to distinguish the contribution of these three different sources on the emerging cognitive networks of individuals and social groups.

Conventional approaches to artifact networks categorize artifacts on the basis of predefined superficial attributes (such as whether a projectile point has flat or fluted edges), and use a Darwinian framework to model the chronological evolution of and interrelationships amongst artifacts (e.g., O'Brien et al., 2001). In contrast, taking a network-based approach, artifacts are classified not just on the basis of objective superficial attributes but also on the basis of what makes them *meaningful to humans*, including symbolic and utilitarian functions and affordances (e.g., despite dissimilar material properties, both GUN and KNIFE are connected to WEAPON by their function), as well as the underlying cognitive processes required to transition from one artifact to another. In analyses of both PaleoIndian projectile points and Baltic psalteries (which both have a well-documented ethnographic and archaeological record) the network approach recovered previously unacknowledged patterns of historical relationship that were more congruent with geographical distribution and temporal data than what was obtained with other approaches (Gabora et al., 2011; Veloz et al., 2012). The RAF approach lends itself to analyses of cultural lineages that incorporate not just predefined superficial attributes but also the (sometimes spontaneously arising) affordances that make an artifact meaningful to humans by incorporating how needs and desires 'catalyze' modifications to raw materials or pre-existing artifacts (the 'reactants'), resulting in 'products' specific to that need or desire.

Cultural discontinuities, which present a formidable challenge for modeling cultural lineages (Kolodny et al., 2015), can also accommodated by the RAF framework. The underlying psychological processes behind cultural outputs are not always evident from examination of the outputs themselves (Heyes, 2018; Osiurak & Reynaud, 2020). Therefore, cultural lineages cannot be documented merely through the analysis of cultural outputs, it is necessary to incorporate the cognitive networks and psychological processes that spawn cultural innovations. Cultural discontinuities have their origin in cognitive processes and social, emotional, and personality factors that constrain and enable the exchange of ideas. In the RAF approach, collaborative creativity and cultural discontinuities are modeled in terms of underlying cognitive processes such as analogy and cross-domain transfer resulting in intermittent bursts and cascades of cognitive 'reactions' within and between individuals (Gabora & Steel, 2017, 2020a, 2020b, 2022; Ganesh & Gabora, 2022a, 2022b).

9.4.4 RAFs Applied to Cognitive Development

RAF networks can be used to develop a step-by-step process model of conceptual change, i.e., of the process by which a child acquires the capacity for nous and develops an integrated internal model of the world. A young child's mental life consists largely of episodes of lived experience comprised of disconnected foodset items. The child begins using abstract thought (nous) to make connections and reconcile inconsistencies, and this is described as the catalysis of foodset-derived items by external stimuli as shown in Fig. 9.2a. Because children are reliant on such external cues to elicit recall of items in memory, this results in transient RAFs. As the child matures and their knowledge expands, this reliance on external cues decreases, and persistent RAF structure emerges as illustrated in Fig. 9.2b, c.

This process was studied using data on children's mental models of the shape of the earth (Vosniadou & Brewer, 1992) and modeled using the RAF framework (Gabora et al., 2022). We modeled different trajectories by which children's conceptualize of the planet changes—from flat earth, through different intermediate steps, to spherical earth—as well as the impact of factors such as pretend play on this process. As RAFs increase in size and number, they begin to merge and form a maxRAF that bridges previously compartmentalized knowledge, and which constrains and enables the scaffolding of new conceptual structure. Once most conceptual structure is subsumed by the maxRAF, the child can reliably frame new knowledge and experiences in terms of previous knowledge and experiences, engage in recursive representational redescription, i.e., abstract thought, and adapt existing

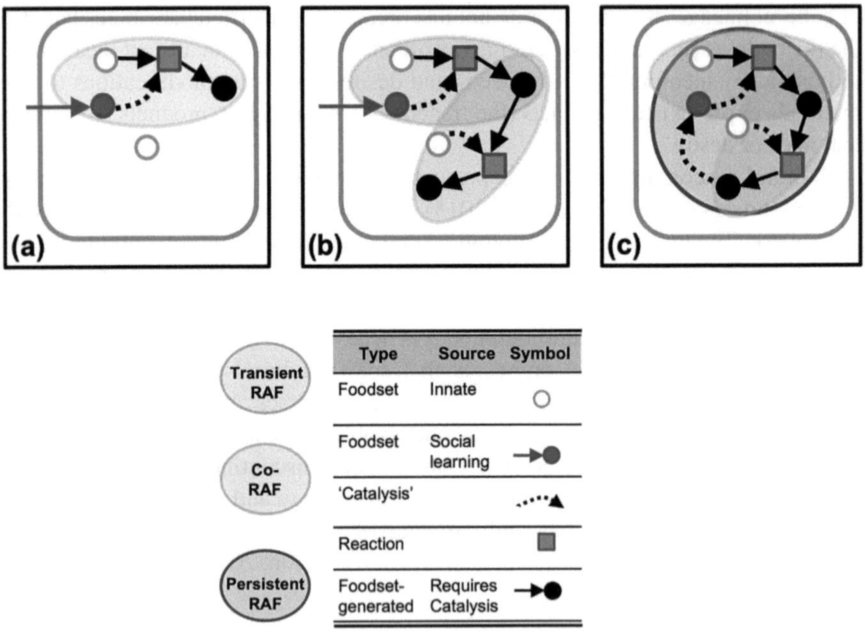

Fig. 9.2 Panel (**a**) depicts a *transient RAF,* so-called because it relies on and exists only in the presence of a particular stimulus or 'catalyst' (red arrow). (**b**) Endogenous growth of RAF structure through the formation of a second RAF (as well as a max-RAF that encompasses both, which for clarity is not shown). (**c**) Since the product (lower black dot) now catalyzes the two-reaction process that generates it, the transient RAF has become a *persistent RAF* that persists in the absence of the stimulus

ideas to their own perspectives. Individual differences in reliance on foodset versus foodset-derived information sources may culminate in different kinds of cognitive networks and concomitant learning trajectories, which ultimately affect the kind and magnitude of the individual's impact on society and the world.

9.4.5 RAF Models of Social Groups

Cognitive restructuring often occurs in a collaborative context. When people work together in teams, they have a richer repertoire of perspectives from which to view, and refine, an idea. The RAF framework has been used to model the emergence of conceptual structure through dyadic and group interaction. Although the configuration of an individual's maxRAF is unique to that person, RAF structure can extend *across* individuals (Gabora et al., 2022; Gabora & Steel, 2022). When one individual communicates a 'conceptual reaction' to colleagues, their minds may catalyze further elaboration of the idea, such that each person's foodset-derived 'product' ideas serve as 'catalysts' and 'reactants' for 'reactions' in the minds of others.

9.4.6 *Individual Creative Styles*

As people's worldviews emerge at the interface between nature and nurture, the more they attempt to express feelings and ideas that have not yet been articulated, the greater the probability that their worldviews unfold in unique ways that deviate from the status quo. This may then be reflected in what they put out in the world; many or most creators develop a style that is recognizably theirs. The upshot is that the foodset-derived sources of information that arise from nous vary more widely from one person to another than either (1) 'innate' sources of information —i.e., nature (e.g., most people have some degree of fear of heights), or (2) socially learnt sources of information—i.e., nurture (e.g., we share scripts such as 'what to do in a restaurant').

This is related to the concept of a writer starting out imitating the styles of authors they admire, but eventually finding their 'own voice,' a phenomenon that extends beyond creative writing to other creative domains. In a study in which creative writing students who had studied together for years and could correctly identify above chance which of their peers wrote particular creative writing excerpts could *also* correctly identify above chance which of their peers created which work of art (Gabora et al., 2012). Evidently, knowledge derived from nous enables conceptual structure to coalesce into a unique configuration that can even be detected across different avenues for self-expression.

9.4.7 *Future Directions*

We have laid out a framework for characterizing and modeling 'nous': the reasoning and creative thought processes that emerge at the interface between nature and nurture. Though rooted in innate drives and social learning, the outputs of nous become quite distinct from them, and we showed how the RAF framework is well suited to modeling this.

The framework lends itself to testable hypotheses. Because innate knowledge is biologically 'expensive,' it may be subject to a high mutation rate, so that it can be expunged if it is not 'earning its keep.' This leads to the speculative prediction that a mutation that eliminates a particular innate knowledge item in a particular individual may stimulate the development of RAFs that replace that innate knowledge with foodset-derived knowledge, which is more explicit than innate knowledge, thus more amenable to analysis and modification. This is consistent with Einstein's famous claim that because he lacked the intuitive understanding of space and time that came naturally to others he was forced to acquire a deeper understanding of these concepts (Einstein, 1949). It is also consistent with findings that intellectual giftedness can co-occur with learning disorders (Toffalini et al., 2017). Future RAF models will benefit from empirical research into how innate foodset items constrain and enable the assimilation of socially transmitted foodset items, and of the conditions under which children generate new ideas (foodset-derived items).

This leads to another question that could be tested with the RAF framework: is someone who acquires a particular piece of information by learning it for themselves (i.e., in their mind it is a foodset-derived item) more likely to generalize that representation and apply it to other domains than someone who acquires it through social learning (i.e., in their mind it is a foodset item)? This hypothesis is reminiscent of those explored in developmental studies of generalization and causal reasoning (Gopnik et al., 2000; Gopnik, 2001). Research on learning and memory suggests that self-generated knowledge and autonomous learning are conducive to the encoding of information, and its later retrieval in novel situations (e.g., Deci & Ryan, 2008; MacLeod et al., 2012; McCabe, 2015). The RAF approach could be useful for identifying gaps in knowledge, and exploring how cognitive development, social environment, and the form and content of socially transmitted information impact RAF formation. (For example, narratives that spread quickly through cognitive contagion may be describable as more compact RAFs.)

RAFs have been applied to the therapeutic setting to model how distorted client self-perception can undergo changes during client-therapist interaction (Ganesh & Gabora, 2022a, 2022b), and future directions could explore the inverse, i.e., how negative biases develop in the first place. We note that creativity—while frequently innovative, exciting, or a source of pleasure—can potentially be detrimental (e.g., Andreasen, 1987, 2008; Kaun, 1991; Strong et al., 2006). A specific RAF integrating with the maxRAF could promote restructuring that is harmful to an individual's worldview. For example, assimilation of criticism could catalyze feelings of unworthiness. The RAF framework has also been used to model how the same events can be conceptualized in different ways (Gabora et al., 2022; Ganesh & Gabora, 2022a, 2022b), and could shed light on why one person may be prone to a certain psychopathology while another person with a comparable 'foodset' is not.

Finally, and more speculatively, has been hypothesized that self-hood requires a structure that is self-organizing and self-preserving, i.e., the kind of structure that is described by a RAF network (Gabora & Bach, 2023). In this view, until AIs can use nous processes to restructure their cognitive networks, they may remain merely tools, as opposed to autonomous agents.

9.5 Summary and Conclusions

The fruits of human culture and civilization cannot be directly attributed to genetically coded traits attributed from our ancestors (nature), nor to lessons we learn and experiences we have (nurture). Generative thought processes, while internal, are not purely biological (recall how a writer may neglect food or children while writing a novel); nor purely social (e.g., while socially transmitted knowledge and expectations influence the form of a novel, their integration and reorganization into an original story may be accomplished in isolation). Cognitive networks begin restructuring themselves as soon as we are born, and this process continues throughout our lives, shaping our uniquely configured worldviews. Human achievements are the product of our creative self-expression and intellect, of *nous:* our capacity to generate ideas and think them through, modifying and refining them in light of different constraints and perspectives. This process is as much responsible for who we are, individually and collectively, as nature and nurture. Thus, methods for modeling and understanding this are indispensable to understanding what makes us uniquely human. The framework of autocatalytic networks provides a means of describing and modeling the distinction between nature, nurture, and nous.

References

Andreasen, N. (1987). Creativity and mental illness: Prevalence rates in writers and their first-degree relatives. *The American Journal of Psychiatry, 144*, 1288–1292. https://doi.org/10.1176/ajp.144.10.1288

Andreasen, N. (2008). The relationship between creativity and mood disorders. *Dialogues in Clinical Neuroscience, 10*(2), 251–255. https://doi.org/10.31887/DCNS.2008.10.2/ncandreasen

Antrobus, J., Singer, J., Goldstein, S., & Fortgang, M. (1970). Mindwandering and cognitive structure. *Transactions of the New York Academy of Science, 32*, 242–252.

Bonawitz, E., Shafto, P., Gweon, H., Goodman, N., Spelke, E., & Schulz, L. (2011). The double-edged sword of pedagogy: Instruction limits spontaneous exploration and discovery. *Cognition, 120*(3), 322–330.

Boyette, A. H. (2016). Children's play and culture learning in an egalitarian foraging society. *Child development, 87*(3), 759–769.

Chen, O., & Kalyuga, S. (2020). Exploring factors influencing the effectiveness of explicit instruction first and problem-solving first approaches. *European Journal of Psychology of Education, 35*, 607–624.

Deci, E., & Ryan, R. (2008). Self-determination theory: A macrotheory of human motivation, development, and health. *Canadian Psychology/Psychologie Canadienne, 49*(3), 182–185. https://doi.org/10.1037/a0012801

Einstein, A. (1949). *Albert Einstein, philosopher-scientist* (Vol. 1). Library of Living Philosophers.

Erdös, P., & Rényi, A. (1960). On the evolution of random graphs. *Publication of the Mathematical Institute of the Hungarian Academy of Sciences, 5*(1), 17–60.

Festinger, L. (1957). *A theory of cognitive dissonance*. Stanford University Press.

Gabora, L. (1998). Weaving, bending, patching, mending the fabric of reality: A cognitive science perspective on worldview inconsistency. *Foundations of Science, 3*, 395–428. https://doi.org/10.1023/A:1009646612330

Gabora, L. (2003). Contextual focus: A cognitive explanation for the cultural transition of the Middle/Upper Paleolithic. In R. Alterman, & D. Hirsch (Eds.), *Proceedings of the 25th Annual Meeting of the Cognitive Science Society* (pp. 432–437). Hillsdale: Lawrence Erlbaum Associates.

Gabora, L. (2004). Ideas are not replicators but minds are. *Biology and Philosophy, 19*, 127–143.

Gabora, L. M. (2000). Conceptual closure: How memories are woven into an interconnected worldview. *Annals of the New York Academy of Sciences, 901*(1), 42–53.

Gabora, L. (2006). Self-other organization: Why early life did not evolve through natural selection. *Journal of Theoretical Biology, 241*(3), 443–450. https://doi.org/10.1016/j.jtbi.2005.12.007

Gabora, L. (2019). Creativity: Linchpin in the quest for a viable theory of cultural evolution. *Current Opinion in Behavioral and Brain Sciences, 27*, 77–83. https://doi.org/10.1016/j.cobeha.2018.09.013

Gabora, L., & Bach, J. (2023). A path to generative artificial selves. In *Proceedings of the 22nd Portuguese conference on artificial intelligence.* Springer.

Gabora, L., Beckage, N., & Steel, M. (2022). An autocatalytic network model of conceptual change. *Topics in Cognitive Science, 14*(1), 163–188. https://doi.org/10.1111/tops.12583

Gabora, L., Leijnen, S., Veloz, T., & Lipo, C. (2011). A non-phylogenetic conceptual network architecture for organizing classes of material artifacts into cultural lineages. In L. Carlson, C. Hőlscher, & T. F. Shipley (Eds.), *Proceedings of 33rd annual meeting of the Cognitive Science Society* (pp. 2923–2928). Cognitive Science Society.

Gabora, L., O'Connor, B., & Ranjan, A. (2012). The recognizability of individual creative styles within and across domains. *Psychology of Aesthetics, Creativity, and the Arts, 6*(4), 351–360.

Gabora, L., & Smith, C. (2018). Exploring the psychological basis for transitions in the archaeological record. *Journal of Anthropological Sciences, 96*, 1–26. https://doi.org/10.48550/arXiv.1811.10431

Gabora, L., & Steel, M. (2017). Autocatalytic networks in cognition and the origin of culture. *Journal of Theoretical Biology, 431*, 87–95.

Gabora, L., & Steel, M. (2020a). Modeling a cognitive transition at the origin of cultural evolution using autocatalytic networks. *Cognitive Science, 44*(9), e12878. https://doi.org/10.48550/arXiv.2007.06442

Gabora, L., & Steel, M. (2020b). A model of the transition to behavioral and cognitive modernity using reflexively autocatalytic networks. *Journal of the Royal Society Interface, 17*(171). https://doi.org/10.1101/2020.07.10.185512

Gabora, L., & Steel, M. (2021). An evolutionary process without variation and selection. *Journal of the Royal Society Interface, 18*(180), 20210334. https://doi.org/10.1098/rsif.2021.0334

Gabora, L., & Steel, M. (2022). From uncertainty to insight: An autocatalytic framework. In R. Beghetto & G. Jaeger (Eds.), *Uncertainty: A catalyst for creativity, learning and development* (pp. 125–158). Springer. https://doi.org/10.1007/978-3-030-98729-9_8

Ganesh, K., & Gabora, L. (2022a). A dynamic autocatalytic network model of therapeutic change. *Entropy, 24*(4), 547. https://doi.org/10.3390/e24040547

Ganesh, K., & Gabora, L. (2022b). Modeling discontinuous cultural evolution: The impact of cross-domain transfer. *Frontiers in Psychology—Theoretical and Philosophical Psychology, 13*. https://doi.org/10.3389/fpsyg.2022.786072

Gopnik, A. (2001). Theories, language, and culture: Whorf without wincing. In *Language acquisition and conceptual development* (pp. 45–69). Cambridge University Press. https://doi.org/10.1017/CBO9780511620669.004

Gopnik, A., Meltzoff, A. N., & Kuhl, P. K. (2000). *The scientist in the crib: What early learning tells us about the mind.* William Morrow Paperbacks.

Heyes, C. (2018). *Cognitive gadgets: The cultural evolution of thinking*. Harvard University Press.

Hordijk, W., Hein, J., & Steel, M. (2010). Autocatalytic sets and the origin of life. *Entropy, 12*(7), 1733–1742. https://doi.org/10.3390/e12071733

Hordijk, W., Kauffman, S. A., & Steel, M. (2011). Required levels of catalysis for emergence of autocatalytic sets in models of chemical reaction systems. *International Journal of Molecular Science, 12*(5), 3085–3101. https://doi.org/10.3390/ijms12053085

Hordijk, W., & Steel, M. (2004). Detecting autocatalytic, self-sustaining sets in chemical reaction systems. *Journal of Theoretical Biology, 227*(4), 451–461. https://doi.org/10.1016/j.jtbi.2003.11.020

Hordijk, W., & Steel, M. (2015). Autocatalytic sets and boundaries. *Journal of Systems Chemistry, 6*(1). https://doi.org/10.1186/s13322-014-0006-2

Hordijk, W., & Steel, M. (2016). Chasing the tail: The emergence of autocatalytic networks. *Biosystems, 152*, 1–10. https://doi.org/10.1016/j.biosystems.2016.12.002

Hordijk, W., Steel, M., & Kauffman, S. (2012). The structure of autocatalytic sets: Evolvability, enablement, and emergence. *Acta Biotheoretica, 60*(4), 379–392. https://doi.org/10.1007/s10441-012-9165-1

Karwowski, M., & Kaufman, J. C. (Eds.). (2017). *The creative self: Effect of beliefs, self-efficacy, mindset, and identity*. Academic Press.

Kauffman, S. A. (1993). *The origins of order: Self-organization and selection in evolution*. Oxford University Press.

Kaun, D. (1991). Writers die young: The impact of work and leisure on longevity. *Journal of Economic Psychology, 12*(2), 381–399. https://doi.org/10.1016/0167-4870(91)90021-K

Klinger, E., & Cox, W. (1987). Dimensions of thought flow in everyday life. *Imagination, Cognition, and Personality, 7*, 105–128.

Kolodny, O., Creanza, N., & Feldman, M. (2015). Evolution in leaps: The punctuated accumulation and loss of cultural innovations. *Proceedings of the National Academy of Sciences, 112*(49). https://doi.org/10.1073/pnas.1520492112

MacLeod, C., Pottruff, M., Forrin, N., & Masson, M. (2012). The next generation: The value of reminding. *Memory & Cognition, 40*, 693–702. https://doi.org/10.3758/s13421-012-0182-8

Maturana, H., & Varela, F. (1973). Autopoiesis and cognition: The realization of the living. In R. S. Cohen & M. W. Wartofsky (Eds.), *Boston studies in the philosophy of science* (Vol. 42). Reidel.

McCabe, J. (2015). Learning the brain in introductory psychology: Examining the generation effect for mnemonics and examples. *Teaching of Psychology, 42*(3), 203–210. https://doi.org/10.1177/0098628315587617

Mossel, E., & Steel, M. (2005). Random biochemical networks and the probability of self-sustaining autocatalysis. *Journal of Theoretical Biology, 233*, 327–336. https://doi.org/10.1016/j.jtbi.2004.10.011

O'Brien, M., Darwent, J., & Lyman, R. (2001). Cladistics is useful for reconstructing archaeological phylogenies: Paleoindian points from the Southeastern United States. *Journal of Archaeological Science, 28*, 1115–1136.

Osiurak, F., & Reynaud, E. (2020). The elephant in the room: What matters cognitively in cumulative technological culture. *Behavioral and Brain Sciences, 43*, e156. https://doi.org/10.1017/S0140525X19003236

Scotney, V., Schwartz, J., Carbert, C., Adam Saab, A., & Gabora, L. (2020). The form of a 'half-baked' creative idea: Empirical explorations into the structure of ill-defined mental representations. *Acta Psychologica, 202*, 102983. https://doi.org/10.1016/j.actpsy.2019.102981

Steel, M. (2000). The emergence of a self-catalysing structure in abstract origin-of-life models. *Applied Mathematics Letters, 13*(3), 91–95. https://doi.org/10.1016/s0893-9659(99)00191-3

Steel, M. (2023). Interior operators and their relationship to autocatalytic networks. https://arxiv.org/abs/2305.16714

Steel, M., Hordijk, W., & Smith, J. (2013). Minimal autocatalytic networks. *Journal of Theoretical Biology, 332*, 96–107. https://doi.org/10.1016/j.jtbi.2013.04.032

Steel, M., Hordijk, W., & Xavier, J. C. (2019). Autocatalytic networks in biology: Structural theory and algorithms. *Journal of the Royal Society Interface, 16*(151). https://doi.org/10.1098/rsif.2018.0808

Steel, M., Xavier, J., & Huson, D. (2020). Autocatalytic networks in biology: Structural theory and algorithms. *Journal of the Royal Society Interface, 17*, 20200488.

Strong, C. M., Nowakowska, C., Santosa, C. M., Wang, P., Kraemer, H., & Ketter, T. (2006). Temperament–creativity relationships in mood disorder patients, healthy controls and highly creative individuals. *Journal of Affective Disorders, 100*(1), 41–48. https://doi.org/10.1016/j.jad.2006.10.015

Toffalini, E., Pezzuti, L., & Cornoldi, C. (2017). Einstein and dyslexia: Is giftedness more frequent in children with a specific learning disorder than in typically developing children? *Intelligence, 62*, 175–179. https://doi.org/10.1016/j.intell.2017.04.006

Varela, F., Thompson, E., & Rosch, E. (1991). *The embodied mind*. MIT Press.

Veloz, T., Temkin, I., & Gabora, L. (2012). A conceptual network-based approach to inferring the cultural evolutionary history of the Baltic psaltery. In N. Miyake, D. Peebles, & R. P. Cooper

(Eds.), *Proceedings of the 34th annual meeting of the Cognitive Science Society* (pp. 2487–2492). Cognitive Science Society. http://arxiv.org/abs/1310.0737

Vosniadou, S., & Brewer, W. (1992). Mental models of the earth: A study of conceptual change in childhood. *Cognitive Psychology, 24*(4), 535–585. https://doi.org/10.1016/0010-0285(92)90018-W

Winslow, M., & Gabora, L. (2022). Beyond two modes of thought: A quantum model of how three cognitive variables yield conceptual change. *Frontiers in Psychology (Section: Quantitative Psychology and Measurement), 9054*(46). https://doi.org/10.3390/e24040547

Xavier, J., Hordijk, W., Kauffman, S., Steel, M., & Martin, W. (2020). Autocatalytic chemical networks at the origin of metabolism. *Proceedings of the Royal Society of London. Series B, Biological Sciences, 287*, 20192377.

Part II
Case Studies in the Nature vs. Nurture Debate

Chapter 10
Naturalism of Mother's Milk and Its Historical Change in Japan: Modernization of *Nature* as Normative Concept

Shinji Kajitani

Abstract "It is natural for children to be fed with mother's milk". This seems to be biologically as well as historically right. As shown in such cases, naturalness often has normative connotation and becomes the basis of belief that something is true and good. However, a detailed analysis of the content, the reasons and the background in which the word "nature" is used reveals that it has varied dramatically over time. This article will explore its historical change in Japan in terms of mother's milk or breastfeeding, from the early modern to the contemporary period, mainly by investigating the description of child-rearing books from each era. The historical variability of "nature" is paradoxically due to the fact that it is always supposed to be unchanging and serves as the "ground" of truth and goodness. In every age, our lives are beset by various difficulties and problems, and we repeatedly return to "nature" to get out of our predicament. However, what is "natural" cannot be determined from the beginning. It is only in the social and historical context that its meaning finally becomes concrete in the light of the problems we face. Whatever we consider "natural," the culture of human society always has something contrary to it. This is especially true today, when our lives are so far removed from naturalness. In such an age, "nature" becomes a strong support for people, especially in areas such as child-rearing, where it is difficult to find reliable guidelines.

Keywords Mother's milk · Breastfeeding · Japan · Nature · Norm · Modern · Child

---

S. Kajitani (✉)
Center for Philosophy, Tokyo University, Tokyo, Japan
e-mail: kajitani@fusehime.c.u-tokyo.ac.jp

J. Pérez-Jara, Í. Ongay (eds.), *Beyond Nature and Nurture*, Synthese Library 497,
https://doi.org/10.1007/978-3-031-87715-5_10

"It is natural for children to be fed with mother's milk"[1]— This may seem to be an unquestionably correct statement. This is not the case only today. The "naturalness" of mother's milk has been repeatedly advocated throughout history. It has been presented as if it were a truth with universal validity that transcends the differences of time and culture.

Certainly, there seems to be some validity to such claims. It is a *biological* fact that women (females) give birth to children and thereafter produce milk for them. Moreover, it is a *historical* fact that humans have breastfed their children since the beginning of time everywhere in the world. If this is true, there is no question that this thesis has been repeated throughout history. However, a closer look reveals that the meaning of "nature" has varied considerably from period to period, and has its own socially and historically specific substance in each era.

In order to consider this, we must distinguish two major meanings of the word "nature". Today, the word "nature" often refers to the "domain" contrasted with "culture" or "human activity", as in the terms "natural environment" and "natural phenomena". On the other hand, "nature" related to mother's milk, which is the subject of this paper, differs from the first meaning in that it refers to the "state" of things. For example, it implies a kind of "norm", a "state of things" as they should be, as seen in the expressions "natural behavior", "natural death" or "natural law". This is especially evident in the antonym "unnatural" as is said like "unnatural death" or "unnatural smile". What is "natural" is right and good, while "unnatural" is wrong and bad.

Today, giving birth and raising a child is still associated with this normative naturalness. Natural childbirth is desirable, and mothers and children are expected to share the same room after birth, while disposable diapers and powdered milk are considered unnatural. Even before the modern era, "nature" was often used to describe childbirth and child-rearing, and the fact that "nature" is a kind of norm meaning "the way things originally should be" has not changed. However, a detailed analysis of the content, the reasons and the background in which the word "nature" is used reveals that it has changed dramatically over time. Perhaps that is why, in the context of each era, the word "nature" has always had the nuance of being true and good.

This article will use the term "naturalism" for the belief that something is true and good on the basis of its being "natural," and will explore its historical change in Japan, and the discussion is divided into three parts: (1) the early modern period (Edo period), (2) the modern period (from the Meiji era to the postwar period), and (3) the contemporary period (from the postwar period to the present, especially after 1960s–70s, after the so-called "period of high growth"). The investigation is mainly based on child-rearing books of each era. The division above is of course rough and just for convenience. What was common in the earlier period would carry over

[1] Today, feeding with mother's milk is usually called "breastfeeding", but because breastfeeding was not by mothers, but can be also by other women such as wet nurses, especially in the pre-modern period (or even until the 1950s in Japan), this paper prefers to the word mother's milk.

for some time into the later period, and conversely, what becomes dominant in the later period would have already appeared earlier. Nevertheless, it can be said that the main aspects of naturalism of mother's milk have shifted in that way.

10.1 The Early Modern Period (Edo Period: 1603–1868)

10.1.1 The Caution Against Wet Nurses and the Idea of Qi and Blood

In the Edo period, the discourse on the "naturalness" of mother's milk was directed first and foremost to hiring wet nurses. In Japan, child-rearing books began to be published in the mid-Edo period, around 1700. Gyuzan Kotsuki's *Shoni Hitsuyo Sodategusa* (*Necessary Advices for Small Children*) (1703), the most comprehensive and systematic book on child-rearing that became the standard during the Edo period, states in the chapter "Giving Milk to a Newborn Child" as follows: "Even in noble or wealthy families, it is most advisable for a mother to give her own milk to her baby if she is healthy and produces a lot of milk."[2] Genryo Hirano, who published a comprehensive medical manual for the home *Byoka Suchi* (*Encyclopedia for Medical Treatment at Home*) (1832), also wrote: "The child in the womb shares the mother's flesh and blood, and milk is made from mother's blood and is destined to be given to her child. Therefore, if we wish for the healthy growth of the child, it is best to nourish it with her own milk. This the way in accordance with the natural order of things."[3]

The focus here is not put on the care of the child in general, but only on feeding the child. Why, then, was breastfeeding by mother supposed to lead to the healthy growth of the child? Why was it thought that a wet nurse inhibits this? At the root of this question lies a view of the body that was peculiar to pre-modern Japan or East Asia.

In pre-modern East Asia including Japan, everything in this world is thought to be made from *qi*, something like vital energy which spreads throughout the universe. So are body and mind of human. These are not heterogeneous dualistic entities, but in principle homogeneous monistic ones. Therefore, body and mind directly affect and interact each other through *qi*, and that is also the case between persons.

Generally speaking, a mother's mental and physical disposition is transmitted to her child through the *qi*[4] and blood (a sort of *qi*) circulating in the body. Therefore, through milk, or *qi* and blood from the breast, the temperament and disposition,

---

[2] Katsuki (2016), 47.

[3] Hirano (2006), 187.

[4] *Qi* is the fundamental ontological concept of the world and human beings in East Asian cultures. It is a sort of vital fluid force which prevails throughout the world and permeates all things. Blood is a liquid form of *qi* in the human body, while *qi* in a narrow sense is a pneumatic form as breath

mental and physical state, and character of breastfeeder (mother or wet nurse) have a great influence on the child's mind and body.

For example, in Katsuki's *Necessary Advices for Small Children*, there are such instructions for breastfeeder: when she eats something hot food in summer, the milk also becomes hot, and the child vomits if he drinks it; if she eats something cold in winter, the milk also becomes cold and the child will cough or has diarrhea if he drinks it; if she breastfeeds after she has gotten angry, the child will get upset and has a shock or epilepsy; if she breastfeeds after a meal, the child may cry violently and have a seizure; if she breastfeeds before she has calmed down from sadness or joy, the child will drool and cough, etc.[5]

In this way, the breastfeeder must not only be healthy, but also moderate and temperate, able to care about the state of her own body as well as that of the child, and act quietly according to any situation. In theory, it does not make any difference whether mother or wet nurse breastfeeds. In the case of wet nurse, however, there are more special circumstances. They were usually from lower class and poor in comparison with their employing families. According to Katsuki, such women were "cynical, jealous, arrogant, and prone to anger."[6] In the past, inferior class were associated with inferior and worse constitution of the person, and were considered to be quite undisciplined and untrustworthy. In fact, as wet nurses were much more paid than other servants, poor mothers often became wet nurses even if they had to leave or abandon their own children for others.[7] Therefore, it was believed that if wet nurses breastfed, such inferior qualities of body and mind would enter the child and cause serious harm.

When people said in the middle Edo period that it was "natural" for a mother to breastfeed her own child, it implied the "purity" of the blood relationship, i.e., that it should not violate class differences. Moreover, they believed that the healthy development of the child was also closely related to the hierarchical class order. But that is not contrary to nature. Just as there are distinctions between heaven as above and earth as below in nature, "so the class order is natural" in that there is an order of high and low classes among human beings, and it was thought to be the correct state of affairs when these distinctions were maintained without confusion.

However, there were two other difficulties even if the mother breastfed her baby. The first is the "taidoku (poison from the womb)" as the main cause of child disease, the second is that "shonyu" (mother's first milk) or colostrum which was considered poisonous. And "nature" was also spoken of related to these two factors.

---

and emotion. In the premodern period, the health and illness of the human body and spirit were regarded as dependent on the circulation of *qi* and blood.

[5] Cf. Katsuki (2016), 54–56.

[6] Katsuki (2016), 51f. Cf. Hirano (2006), 192.

[7] Cf. Sawayama (2017), 41, 116.

10.1.2 The Change of the Relationship Between Fetal Poison and Mother's First Milk

"Taidoku", a poison from the womb, which could be translated as "fetal poison", is a black, sticky substance that a newborn child has in mouth when he or she comes out of the mother's womb. According to today's medicine, this is merely the rest of amniotic fluid which the fetus swallowed in the womb, while it was considered highly toxic in the Edo period. People thought that if the child swallowed it, or kept it inside the body, it could cause all kinds of diseases, especially smallpox, which was the most feared disease of the time. Therefore, treatment for removal of this substance immediately after the birth of a child was extremely important. The first thing to do was to wipe the inside of the mouth with a cloth, and then to expel anything that had entered the body with cathartic or emetic drugs. The child must not be breastfed until the complete expulsion of the poison is confirmed by the disappearance of black stools. (today they is called "meconium")

This was a rather harsh ordeal for a newborn's delicate life, but in those days, as far as the etiological theory was concerned, this was the treatment for the child's lifelong health. This practice seems "unnatural" to us today, who take for granted the bond between mother and child from the very beginning, both physically and mentally, by having the child suckle on the breast immediately after birth. Conversely, however, even if we consider the connection between mother and child through milk to be "natural", it does not mean that it is self-evident from the beginning.

Moreover, the problem was not only on the side of the child's body. The mother's milk itself, which is supposed to mediate the natural connection between mother and child, also had its own difficulty: before the modern era, colostrum, the first milk produced by the mother, was considered harmful to the child's body. In fact, colostrum is thicker and yellower than the milk that comes after normal breastfeeding, and even today's scientific knowledge suggests that it has a different composition. It is not only in Japan, but in many cultures around the world, whether traditional or modern, that colostrum was discarded as bad or irrelevant.[8]

However, the notion that colostrum was poisonous began to lose its validity in the nineteenth century in Japan. At least among physicians who wrote books on child-rearing and obstetrics, colostrum had a detoxifying effect, and this effect was explained by associating it with "nature". For example, Jusei Hirano wrote, "Mother's first milk has the natural effect of removing fetal poison taken in by the child from the womb, and is better than medicine."[9] Ryusai Kuwata, from the end of the Edo period, wrote in his *Aiiku Sadan* (*Talk over Tea about Child-rearing*) (1853), "Mother's first milk has the effect of improving bowel movements and removes fetal poison well. This is a natural function to detoxication."[10]

[8] Cf. Palmer (1988), 25.

[9] Hirano (2006), 202f.

[10] Kuwata (1853), 23b.

With this change in the notion of colostrum, the fear of fetal poison seems to gradually disappear. In Kuwata's etiology, this is no longer emphasized. This view of colostrum as a natural cathartic became more common during the Meiji period. Later, after around the 1920s, it was regarded as a highly nutritious drink for newborn children. After World War II, it came into the limelight as a substance that contains immune substances and will contribute to protecting children from disease, and became the first thing to be given to children after birth.

As colostrum was positively evaluated and the theory of fetal poison diminished in influence, it became "natural" for mother and child to be bonded by milk from the beginning. Conversely, the mother-child relationship through milk could not have been easily created until this traditional medical theory had been cast aside.

10.2 The Modern Period (from 1868 to the 1950s)

10.2.1 Emergence of the "Preferred Milk" Hierarchy

Even in the Meiji period (1868–1912), there remained a criticism of the practice of entrusting breastfeeding to wet nurse, but it was mainly because of the healthy development of the child, and not because of the class of the wet nurse. The views on fetal poison and colostrum also did not differ much from those of the late Edo period.

However, the naturalism of mother's milk entered the new phase: after the statement that feeding with mother's milk is the best way to raise a child because it is natural, many child-rearing books give detailed explanation in the following order: if breastfeeding by mother is difficult, then by wet nurse; if this is difficult, then cow's milk (or animal milk); and after that, artificial milk such as condensed milk or powdered milk. The dichotomy of mother or wet nurse recedes, and instead, the "preferred milk" hierarchy, that is, the sequence of mother's milk–wet nurse's milk–cow's milk–artificial milk appeared. Moreover, the milk of mother and wet nurse was lumped together as "human milk," and "cow's milk" and others were categorized in "animal milk".[11]

Kan Adachi, director of the Army Medical College, refers to feeding with human milk as a "natural nursing method," while that with cow's milk and artificial milk are called "artificial nursing method".[12] The important indicator of whether feeding is "natural" or "unnatural" was no longer the difference in class among humans, as in the Edo period, but the difference in species. The difference in the quality of milk between biological mothers and wet nurses became the same in terms of human milk, and a much greater difference was found from animal milk.

[11] Cf. Shimoda (1907), p. 206; Shindo (1894), 18.

[12] Cf. Adachi (1907), 78.

Perhaps related to this, the criteria for the selection of wet nurses are generally the same as in the Edo period, but there seems to be no longer a sense of caution that the mental and physical condition of the wet nurses would be transmitted to the child through the milk. In the Meiji era, the severe changes in the mental and physical condition of breastfeeders were only associated with deterioration of the quality of milk.

According to Michiyoshi Mishima, Japan's first pediatrician trained in Western medicine, the intense anger, hatred, surprise, sadness, or joy of a nursing woman "changes the nature of the milk, reduces its quantity, and causes very terrible harm,"[13] so she should not breastfeed at such times. Whether mothers or wet nurses, they should always be calm in mind and body.[14] (Today, it is said that the quantity of milk depends on the emotional state of the mother, not even the quality.)

In this way, during the Meiji Era, people became less wary of wet nurses and concerned only with the deterioration of milk quality. This is probably because the differences between mothers and wet nurses were no longer highlighted. Even though the superiority of breastfeeding by the mother remained unchanged, as long as the child was fed with human milk, it was called "natural" and this was contrasted with the unnaturalness of giving cow (or animal) milk.

10.2.2 Orientation Toward the Western and Modern

The superiority of mother over wet nurse and that of human milk over cow's milk was not merely supported by its "naturalness". In the Meiji era, statistics were also introduced with Western sciences. Many child-rearing books use statistics to show the value and correctness of breastfeeding. When people read them, they should have believed that it is too frightening to feed a child with cow's milk or artificial milk. Therefore, there should have been no doubt that "natural" breastfeeding is better. Moreover, the difficulties of feeding with cow's milk or powdered milk are explained in all books. There are many things that must be cared about, such as obtaining and storing fresh, good quality milk, hygienic use of baby bottles, and so on.

However, we cannot see much wariness in most child-rearing books. They rather give the impression as if it were encouraged, explaining in detail the manual for feeding with milk. In fact, Terumaro Kato, for example, recommends mixed feeding from the beginning, saying that it is better to get the baby used to milk and rubber nipples in case the mother becomes ill or otherwise unable to breastfeed.[15] The authors of these books clearly warned of the dangers of cow's milk and artificial milk on the one hand, while on the other hand, they were very positive about them.

[13] Cf. Miura (1899), 104, 108.

[14] Cf. Shimoda (1907), 202.

[15] Kato (1912), 18–25.

What is important here is the orientation toward the Western and the Modern that was strong at this time, especially among the wealthy.

This kind of orientation is also evident in the new method of breastfeeding called "regular breastfeeding". The instructions for this can be found in most child-rearing books that I have seen. Today, "unrestricted breastfeeding," in which the child is allowed to take as much as and whenever he/she wants, is said to have been the original Japanese tradition.[16] This new method was strongly recommended in Japan during the Meiji era as part of the trend toward Westernization. This was done, above all, to foster the development of disciplined human beings. Michiyoshi Mishima described breastfeeding as "the beginning of life rules" and said: "The first week after childbirth is really the beginning of life for the baby. From this time on, you should have your child follow strict rules. Once accustomed to an irregular rhythm, it would be the cause of a lifetime indiscipline."[17] While it is commonly believed today that regular breastfeeding interferes with "natural breastfeeding" and is not good for the healthy development of the child, we can see that at that time, the exact opposite was the case. Rather, it was considered a threat to the health of the child that one feeds the child whenever he or she wants.

When regular breastfeeding came to be strongly demanded in the Meiji era, a completely different framework was introduced, not only for child-rearing, but also for a wide range of views on illness, the body, and the human condition. A change of this magnitude could not have occurred without an orientation toward the Western or the Modern. It is on this basis that the discourse on naturalness of mother's milk developed in the modern Japan.

10.3 Contemporary Period (the 1960s–Today)

10.3.1 Anti-science Based on Science

By some time after the war, the early modern elements of the "nature" of mother's milk had almost disappeared. This is largely due to changes in society as a whole. There are two main factors: first, with the advent of postwar democracy and mass society, the custom of hiring wet nurses faded away, and the readership of child-rearing books became the general public. Second, the spread of artificial milk meant that cow's milk was no longer used as a milk substitute. As a result of these two events, the hierarchy of milk (mother's milk–wet nurse's milk–cow's milk–artificial milk) that characterized the early modern Japan collapsed, and at the same time, the difference between human and animal was de-emphasized. And as the two in-between terms were dropped, a new dichotomy emerged: mother's milk–artificial

[16] Cf. Tsuneyoshi and Boocock (1997), 30f.; Yamamoto (1983), 61.

[17] Miura (1889), 5a; (1899), 109.

milk. Nowadays, "natural" means "not artificial", but simply "flesh and blood," and breastfeeding became just the same as mother's milk.

The scheme itself has certainly become simpler. Artificial milk itself, however, is a tremendously complex entity. Behind its production lie natural science, industrial technology, capitalism, and the social system that supports them. Moreover, when feeding the baby, the important factors are not only the equipment such as bottles and rubber nipples, but also the hygiene management and the quality of the water used to dilute the milk, which also involves various scientific knowledge, technologies, industries, social systems, and living environments. Mother's milk now has to face such a huge and unpredictable system. Since a mother's milk is supported by these conditions, breast milk itself is a part of the system. Herein lies the difficulty of the modern naturalism of mother's milk.

Although it had already begun in the Meiji period, it was after World War II that people began to stress the importance of breast milk on the basis of more rigorous and detailed scientific knowledge. In that context, there are mainly two following arguments: firstly, feeding with mother's milk is regarded as a biological necessity, and secondly, it is tried to prove scientifically how beneficial mother's milk is for the child and the mother.

In demonstrating the biological necessity of breastfeeding, the animal nature of human beings, which was distanced during the Meiji period, is rather positively evaluated. In other words, humans evolved as mammals, and mammals are designed to raise their own children on their own milk. Therefore, it is best for mammals, including humans, to grow up on their own milk, which follows the natural order of things. Besides, scientific research proved the value of mother's milk as an artificially irreplaceable food in the 1970s based on research in immunology, nutrition, and psychology.

On the other hand, as the research on breastfeeding progressed during this period, the quality of powdered milk also improved. As a result, the percentage of breastfeeding increased, but the use of artificial milk did not decrease that much. In other words, science became a means of proving that breastfeeding is in accordance with the "natural order of things," as well as a means of producing and justifying artificial alternatives to it. Today, when science has become the "court of truth," the naturalism of mother's milk is self-contradictory, relying on science and resisting science at the same time.

10.3.2 Dilemmas in Industrial Capitalism

Historically speaking, since pediatrics developed as a medicine for abandoned children, the main issues were dealing with infectious diseases and nutritional support for children who could not be breastfed. Milk substitutes were needed for this purpose, initially cow's milk and eventually shifting to the more convenient condensed milk and powdered milk. Thus, from the modern era onward, the more child-rearing became involved in medicine (especially pediatrics), the more artificial milk was preferred to breast milk. In Japan, this trend accelerated as hospital births became more common during the so-called high-growth period in 1950s and 60s.

Moreover, artificial milk differs from cow's milk in that it is a highly industrialized product. Therefore, its production and use contain various scientific knowledge and technology, capitalist systems, and social environments. This extremely complex intertwining of many factors was probably difficult to foresee in the beginning, and even today it is not brought to attention in our daily life. It is therefore necessary to give an overview of the historical background of artificial milk and its spread.

The control of quantity and time is a characteristic not only of medicine but of modern society in general. In industrial capitalism as well, the control of production and labor became necessary with mechanization, and quantification and regularization in the whole society. Regular breastfeeding was also in line with these trends of the times, and artificial milk is better suited for managing children's nutrition than breast milk which is not known how much is given.[18] Especially because many children died of malnutrition, it was widely used as a "scientific" method to determine how much milk to feed and how much weight to gain. On the other hand, development of dairy technology made it possible to produce powdered milk from the whey left over from milk production. Thus, medical need for artificial nutrition and the new industry of artificial milk production made a "fortunate encounter" and spread across the world through markets and hospitals.[19]

Another important factor to remember is the change in labor patterns brought about by industrial capitalism. Work in factories and plantations separated the home from the workplace. When mothers were working, their children were separated from the mother and often not breastfed. Even if the children were taken to the factory, feeding time was often severely restricted.[20]

After the war, artificial milk became available to the general public and spread throughout society. Therefore, even though the dangers of artificial milk were pointed out, there must have been more admiration than caution toward artificial milk. Artificial milk was modern and perceived as a symbol of women's liberation. The term "I cannot breastfeed much" was used as a kind of status symbol, and the so-called "intellectuals" did not breastfeed their children, but rather fed them with powdered milk. Conversely, breastfeeding was associated with an image of being old-fashioned or poor, and was regarded as a practice of people of low cultural status.[21]

This was, in part, inevitable. In time, the spread of artificial nutrition and the decline in infant mortality occurred in parallel. This must have given the impression that the infant mortality rate fell because of artificial nutrition. In reality, the decline in infant mortality and morbidity was the result of a combination of many factors, including improvements in living standards and public health, health care system,

[18] Palmer (1988), 22f.; Yamamoto (1983), 1–7.

[19] Palmer (1988), 7ff.

[20] Palmer (1988), 118ff., 129ff.

[21] Kamata (1990), 217.

establishment of insurance systems, and the widespread use of immunizations and antibiotics. Artificial milk embodies a conjunction of these various achievements, and it is a historical product realized during the high-growth period in Japan.

10.3.3 The Pitfall of Nostalgia and Idealism

In discussing the naturalism of mother's milk in postwar Japan, it is important to note that it emerged after the high-growth period. During this period, Japan had fully modernized and joined the ranks of advanced nations. There, one can find a nostalgia for the good old days of Japan, a desire to return to Japanese traditions in opposition to the modern and Western culture. Therefore, the endeavor to restore "natural" breast milk in opposition to artificial milk was also an attempt to recover the "lost past".

According to Tatsuo Matsumura, an enthusiastic breastfeeding advocate from the 1970s, all mothers used to produce breast milk to raise their children, while today, breast milk is no longer produced, and children are raised artificially.[22] He describes these changes in infant nursing as a "loss of motherhood" and "women's motivation for natural child-related behaviors such as pregnancy, breastfeeding, and childcare is diminishing."[23] Sociologist Koji Yokoyama also notes that "the nature of breastfeeding that used to be a part of women's lives, such as letting a baby suckle in between farm work or holding it in one hand while walking, has disappeared in modern culture."[24] The word "nature" is here used to describe and glorify the old ways of breastfeeding.

However, this is only nostalgia and has no historical basis. There must have been many mothers in the past who were too busy working to breastfeed, and not every mother had an abundant supply of milk. That is why people prayed to gods to produce breast milk. Even today, it is said that emotional stress can cause breast milk to fail, and women's position in the past was far more oppressive than it is today.

These authors with nostalgia are not interested in what the facts are. There is a far more complex reality behind their beliefs. To overcome it, it is important for them to find the "nature" of breastfeeding that has been glorified in the past and to uphold the ideal of breastfeeding. This was especially true for mothers raising their children in isolation in the urban society. In the past, women such as mothers or relatives of a new mother were living together or nearby and helped her in raising new born child.

However, after the period of rapid economic growth, urbanization and increase of nuclear families made this difficult. Under such circumstances, many doctors and child-rearing books that mothers read at home suggested that if breast milk could

[22] Cf. Matsumura (1972), 116.

[23] Cf. Matsumura (1972), 200.

[24] Yokoyama (1986), 33.

not be produced, powdered milk would be sufficient, and because of this, mothers tended to use artificial nutrition (including mixed feeding).

It is no wonder that in such a situation, mothers were required to have firm determination and conviction if they were eager to breastfeed. Matsumura states that breastfeeding is possible for every mother, if one really tries with a firm will.[25] Of course, in reality, it cannot be true. Only housewives can devote their entire energies to breastfeeding. In addition, they must have a variety of household appliances available, so that they do not have to spend much effort and time on household chores. Moreover, someone such as her mother, sisters, and husband must be present to support the mother in the mental and physical sense.

However, only those in privileged and fortunate circumstances would be able to breastfeed in this way, with the exception of a few mothers whose milk supply is originally abundant. After all, the fact that breast milk "always comes out" is not so much a natural state as an ideal state that is impossible to achieve in reality. It is, so to speak, similar to the scientific spirit of setting up various conditions in an isolated laboratory and trying to control the results. Moreover, all of these are based on the intertwining of today's social system, science and technology, industrial capitalism, and other various factors. Today, breastfeeding is possible only under modern social and historical conditions, and in that sense, it is in no way "natural", as it has always been considered throughout history. What naturalism of mother's milk is against is the very condition of our existence today on which it now depends.

10.4 Conclusion

As discussed above, the concept of "nature" in breastfeeding or mother's milk has not remained unchanged throughout history, nor is it a simple concept with a unified meaning at any given time. This historical and social variability of "nature" is perhaps paradoxically due to the fact that it is always supposed to be unchanging and serves as the "ground" of truth and goodness. In every age, our life is beset by various difficulties and problems, and we repeatedly return to "nature" to get out of our predicament. It is always seen as the "original value" or "original state" to which we should return. However, what is "natural" cannot be determined from the beginning. It is only in the social and historical context that its meaning finally becomes concrete in the light of the problems we face.

This is probably why it is rarely recognized that "nature" is never simple, but has a complex content and history. Even when scholars and experts use this word, it is often presented as if it were an uncontroversially correct opinion, without questioning its basis or substance. However, no matter what the time period, stating that something is "natural" does not mean it is the most correct or good. Whatever we consider "natural," the culture of human society always has something

[25] Cf. Matsumura (1972), 118–138.

contrary to it. In this sense, human beings are essentially "anti-natural". This is especially true today, when both society and human beings are so far removed from naturalness. In such an age, "nature" has the power to become a source of support for people, especially in areas such as child-rearing, where it is difficult to find reliable guidelines. When we speak of "nature" in difficult conditions, we need to ask ourselves again what we intend and what we are seeking.

References

Adachi, K. (1907). *Tsuzoku Ikuji Shinwa (Popular New Theory of Child-rearing)*. Nanko Shoten.

Hirano, J. (2006). *Byoka Suchi* (*Encyclopedia for Medical Treatment at Home*). H. Kosodo & A. Nakamura (Eds.). Nozan Gyoson Bunka Kyokai

Kamata, H. (1990). *Nihonjin no Koumi, Kosodate: Ima – Mukashi (Childbirth and Child-rearing of Japanese: Today and Past)*. Keiso Shobo.

Kato, T. (1912). *Ikuji Ho (Method of Child-rearing)* (revised expanded edition). Shufu no Tomo sha.

Katsuki, G. (2016). *Shoni Hitsuyo Sodategusa* (*Necessary Advices for Small Children*). Transliterated by Setsuko Nakamura. Nosangyoson Bunka Kyokai.

Kuwata, R. (1853). *Aiiku Sadan (Talk over Tea about Child-rearing)*. Izumoji Bunjiro.

Matsumura, T. (1972). *Bonyu Shugi—Anata no Kodomo ha Ushi dehanai (Breastfeedism—Your Child is not a "Calf")*. Kobunsha.

Miura, M. (1889). *Haha no Tsutome* (*Obligations of Mother*). Maruzen.

Miura, M. (1899). *Haha no Tsutome* (*Obligations of Mother*) (11th expanded edition). Hakubunkan.

Palmer, G. (1988). *The Politics of Breastfeeding*. Pandora Press.

Sawayama, M. (2017). *Edo no Chichi to Kodomo (Breast Milk and Child)*. Yoshikawa Kobunkan.

Shimoda, U. (1907). *Joshi no Eisei (Hygiene for Women)*. Toyamabo.

Shindo, G. (1894). *Ikuji Hikkei—Chichi no Tomo (Necessary to Child-rearing—Friend of Breastfeeding)*. Hakubunkan.

Tsuneyoshi, R., & Boocock, S. S. (Eds.). (1997). *Ikuji no Kokusai Hikaku—Kodomo to Shakai to Oyatachi (International Comparison of Child-rearing—Children, Society, and Parents)*. Nihon Hosho Kyokai.

Yamamoto, K. (1983). *Bonyu (Mother's Milk)*. Iwanami Shoten.

Yokoyama, K. (1986). *Kosodate no Shakaishi (Social History of Child-rearing)*. Keiso Shobo.

Chapter 11
Fathers' Involvement in Childcare: A Historical View on Nature and Nurture in Edo and Meiji Japan

Atsuko Fujino

Abstract It is often highlighted that from the mid to late Edo period, fathers in almost all social classes were earnestly involved in childcare. However, the modernization and industrialization of the Meiji period positioned Japanese fathers not as caregivers, but as economic providers. The resulting absence of fathers from home continues to be one of the crucial contemporary problems in Japan. This paper examines why Japanese fathers in the Edo period were more engaged as caregivers and how their involvement in childcare became inhibited with historical change. This examination leads to an identification of important sociocultural factors influencing fathers' involvement in childcare. Four key factors facilitating fathers' involvement in childcare were identified. The first factor is the social environment in which fathers pay attention to the survival and maturity of their offspring. The second factor is the ideological belief that embraces everyone as co-parents. The third factor is a method of childrearing emphasizing the importance of fathers' physical interaction with children from an early age. The fourth factor is the existence of means to share norms and values surrounding fathers' childcare. Understanding these sociocultural factors in the light of current research, it can be said that the Edo period is a unique age in which paternal behavior emerged in almost all fathers. The findings suggest that fathers' involvement in childcare is "nature," "nurture," and their dynamic reciprocity.

Keywords Fathers' childcare · Edo period · Meiji period · Body–mind unity

---

The original version of the chapter has been revised. A correction to this chapter can be found at https://doi.org/10.1007/978-3-031-87715-5_17

A. Fujino (✉)
Faculty of Sociology, Kyoto Sangyo University, Kyoto, Japan
e-mail: afujino@cc.kyoto-su.ac.jp

J. Pérez-Jara, Í. Ongay (eds.), *Beyond Nature and Nurture*, Synthese Library 497,
https://doi.org/10.1007/978-3-031-87715-5_11

11.1 Introduction

In most countries, social policies such as paternity and parental leave have been promoted to encourage fathers' involvement in childcare. Similarly, Japan revised the *Child Care and Family Care Leave Law* to promote fathers' caregiving in early childhood.

Despite the increased involvement of fathers in childcare due to such policies, the Japan Cabinet Office's "2020 White Paper on Gender Equality" reports that mothers spend 4.6 times as much time with their preschool-aged children as fathers do.[1] This report states that Japan's gender gap in childcare is much greater than that of the United States and European countries. While the image of fathers as caregivers has become popular within Japanese society, the long absence of fathers from their homes remains a crucial problem.

This study focuses on Japanese fathers' childcare involvement from the mid-seventeenth to mid-nineteenth century, known as the Edo period, when Japanese society was stable, and the influence of foreign countries was minimal. It has been highlighted that fathers across nearly all social classes were earnestly involved in childcare from the mid-Edo period onward (Koizumi, 2020; Koyama, 2002; Nakae, 2003; Ohta, 2017). Certain Westerners who visited Japan in the early Meiji period, just after the Edo period ended, recorded in their writings that they were deeply impressed with the closeness in the father-child relationship (Matsuyama, 2020; Nakae, 2003 pp.7–13).

However, with the onset of modernization and industrialization in the Meiji period, Japanese fathers were positioned not as caregivers, but as economic providers. Consequently, by the mid-Meiji period, childcare completely shifted from men to women (Arichi, 1986). Thereafter, Japanese society started viewing the absence of mothers due to full-time work as a problem for child development, whereas little attention began being paid to children growing up in fathers' absence from home. Until recently, the emphasis was on fathers' roles as breadwinners to support their children's lives financially. Thus, the fathers' roles and the father-child dyad have drastically changed over time in Japan.

Such a phenomenon has been seen in other countries as well. Fathers' roles have varied historically, with the social construct of fatherhood shaped through public discourse and popular literature (Gregory & Milner, 2011; LaRossa, 1988). Similarly, feminist scholars have pointed out that the role of mothers has varied historically, and was constructed through dynamic social interactions and relationships (Arendell, 2000). As such, could fatherhood be assumed to be a *tabula rasa* position?

Meanwhile, the parental investment theory based on a sociobiological and evolutionary psychology perspective assumes that men's role of a caregiver occurred as an evolutionary process, as a result of the comparison between the benefits and costs

[1] The Japanese data shown by the Cabinet Office's "White Paper on Gender Equality 2020" is as of 2017.

for the sake of men's reproductive success. (Boyette & Gettler, 2019; Geary, 2007; Hrdy, 2007; Kaplan & Lancaster, 2003). In other words, fatherhood is a universal characteristic of human fathers.

This theory explains how men's involvement with non-biological and biological offspring emerged. Compared to other mammals, a human child is demanding, requiring mothers to spend more time and energy in taking care of their children. Humans are also unique in terms of raising multiple children of different ages at the same time. Therefore, it is difficult for mothers to care for their children alone; they need someone else to help them. If a man commits to a mother and her child, he will elevate his chance of being chosen as the father of her next child. Simultaneously, he risks reducing the number of his offspring, owing to the loss of mating opportunities. Hence, he chooses to protect or care for his own offspring to ensure their survival and prosperity in terms of his reproductive success. Thus, this theory highlights the importance of fathers' care for their offspring and their concerns about the survival and maturity of human infants.

Numerous studies focusing on fatherhood behavior have developed in the field of neurobiology over the last 20 years, demonstrating that as men become fathers, they undergo neurobiological changes such as hormonal changes (Feldman et al., 2010; Gordon et al., 2010). In other words, men's physiology changes when they are involved in the physical care of their babies for the formation of the father-child bond. This is considered a biological factor encouraging men's childcare involvement.

Nevertheless, as Japanese history supports, fathers' commitment to their children varies dramatically across society, culture, and historical periods. Hence, fathers' involvement in childcare is determined by not only biological factors, but also sociocultural ones, including the social system, ideology, and culture (Hrdy, 2009; Roopnarine & Yildirim, 2019).

Additionally, the sociobiological and psychological evolutionary perspective states that fathers' commitment to childcare is affected by the continual interaction between biological and sociocultural factors (Roopnarine & Yildirim, 2019). Biological factors provoking paternal behavior interact with sociocultural factors motivating fathers toward involvement in childcare, producing complex interactions between and among factors over time (Lewis & Lamb, 2007). In summary, current perspectives see men's childcare involvement to be both "nature" and "nurture/culture," and their dynamic reciprocity.

This chapter discusses the sociocultural factors that gave rise to the father's behavior as a caregiver, from Japan's Edo to Meiji periods. We focus on the social institution and ethos—thoughts or ideologies—peculiar to these ages in Japan, investigating the following questions: Why were Japanese fathers in the Edo period more engaged as caregivers? Why were they subsequently less involved in childcare? and what sociocultural factors promoted fathers' childcare behavior?

Moreover, we discuss the relationship between the sociocultural factors identified historically and the biological factors assumed by current research in the sociobiological evolutionary and neurobiological fields. Ultimately, we consider whether the findings obtained from the historical perspective support the current assumption—that men's childcare involvement is a combination of "nature," "nurture/culture," and their dynamic reciprocity.

11.2 Sociocultural Factors of Fatherhood Since the Mid-Edo Period

11.2.1 Social Institutions and the Effect of Media

As stated before, fathers across nearly all social classes were keenly involved in childcare from the mid-Edo period onward. This was mainly because they had a crucial duty to contribute to the maintenance and continuance of "the family" (*ie* in Japanese).[2] The *ie* in this period is defined as a kinship group, existing perpetually from the ancestor to the descendant, which comprises the family head of the male direct line, his wife, children, and parents (Ariga, 1954). The family is conceived as an institution where family property, affairs, name, and social status are passed on to one's descendants. In other words, the family in this period was a public entity.

The Edo period began in 1603 when the emperor appointed Ieyasu Tokugawa, a member of the warrior (or *samurai*) class, as the *shogun*, the de facto ruler of the country. The Edo period spanned the next 264 years, until Yoshinobu Tokugawa, the fifteenth *shogun*, returned rulership to the emperor in 1867. During this period, the *Tokugawa* family inherited the position of *shogun*. This was a period when the warrior class, constituting about seven percent of the population, monopolized government posts.

Edo society was characterized as a hierarchical social structure based on the feudal system. The Edo shogunate was administered under the master–servant relationship between the *shogun* and the *daimyo* (high-ranking warriors), who received their lands from the *shogun* and dominated there. Moreover, four feudal classes were distinguished: warriors at the top, followed, in order, by peasants, artisans, and merchants.

Edo society was also noted for its social stability. The period before Edo was called *sengoku*, when wars took place on an almost constant basis, partly because of the conflicts among relatives over territorial succession. Hence, in addition to the establishment of a rigid social structure, the Edo shogunate undertook many other measures on behalf of the warrior class to ensure political stability. The shogunate adopted the Chu-His school of Confucianism, teaching the orderly interrelationship between the cosmos and human society as an official doctrine. Moreover, it established the rule of primogeniture with the eldest son for succession and inheritance, based on the Confucian norm.[3]

---

[2] Koyama (2002), Nakae (2003), Ohta (2017), and Koizumi (2020) have pointed this out.

[3] The eldest son succeeded in the original family. The younger sons were allowed to establish new families of their own branches. At the same time, however, adoption was common in this period as a means of preserving the family line (Moore, 1970).

Consequent to these measures, the maintenance and continuance of the family became the warrior class's greatest concern. Such a social situation prompted warrior-class fathers to be more committed to their children. Thereafter, the attitudes and practices of the warrior class permeated within the rest of the society for two reasons.

The first reason is demographic and economic factors. According to estimates by historians, Japan's total population was 17 million in the seventeenth century, 31.29 million in the eighteenth century, and 30.69 million in the nineteenth century (Saito & Takashima, 2017). Thus, the population increased from the seventeenth to the eighteenth century, whereas it stagnated or slightly declined from the eighteenth to the nineteenth century. The population stagnation in the latter half of the period was mainly due to famine and infanticide as a strategy for surviving natural calamities (Kalland & Pedersen, 1984).

In other words, commoner household sizes shrank to adjust to the severe economic conditions: from a large family with many collateral relatives, there was a transition to a small family with a male direct line, his wife, children, and parents (Hirai, 2016). Such a family functioned efficiently as a small agricultural or commercial entity during times of economic hardship, particularly if the family's maintenance and succession were done well. This type of family imitated the families of the warrior class; in this way, the commoner class came to have similar concerns and began concentrating on their childrearing.

The second reason was the wide dissemination of childcare and childrearing information through books and articles.[4] While it is difficult to estimate how many such documents actually circulated among the people, the extant documents written during the Edo period exceeded 60 (Kojima, 1986, p.318). A more thorough search could reveal a far greater number (Koizumi, 2020, p. 357).

The literacy rate in the Edo period was high. All warriors were literate as it was essential for them to study Confucius's teachings written in Chinese characters. Even the commoner class literacy rate was assumed to be rather high (Tsujimoto, 2000). Based on the number of private educational institutions (*terakoya* in Japanese), an American researcher estimated the literacy rate to be 40–50% for men, and 10–15% for women (Dore 1976, 2010). Another researcher estimated the male literacy rate to be above 60% in the great cities of Tokyo, Kyoto, and Osaka in the mid-eighteenth century, and the female literacy rate to be 40% (Tsujimoto, 2000). While the literacy rate of this period differed according to social status and region, male literacy remained higher. In this case, men would have had more access to the writings on childrearing.

On the one hand, the expansion of fathers' general interest in childrearing might have led to the increase in published material on the topic, on the other hand, the model of the warrior-class father portrayed in these books was widely conveyed to the fathers of the commoner class. The reason for this is given in an anecdotal

[4] The development of printing technology underpinned mass production of print books from the mid-seventeenth century onward (Iriguchi, 2019).

experience of the first shogun, Ieyasu, which is also referred to as the origin of "childrearing books" in the Edo period.

The warrior class had a custom of writing the family precepts (*kakun* in Japanese) for their descendants, stemming from before the *Edo* period.[5] The childrearing books stemming from the *kakun* include advice on childcare and childrearing (Ohta, 2017). Ieyasu failed in raising his children; he was unable to make his eldest son his successor and had him commit suicide, *seppuku*. Thus, to prevent his third son, *Hidetada*, who became his successor, from repeating his mistakes, he wrote a *kakun* and sent it to *Hidetada*'s wife and servants. Confucians and warriors wrote this didactic story in their books or articles for the general public, which became popular among fathers in all social classes, and are considered to be the beginning of "childrearing books." Ieyasu's lessons came to be regarded as childrearing guidelines within society (Nakae, 2003, pp. 15–31).[6]

Against this background, the information and advice on childcare were directed mainly at fathers, who were responsible for childrearing. However, some books were written for mothers, not in *kanji* (Chinese characters), but in *kana* (Japanese phonetic script). The books for male readers tended to discuss raising boys, while those for female readers tended to explain about wifely duties and raising girls (Koizumi, 2020, pp. 155–162), as fathers played a major role in educating their sons and mothers in training their daughters (Fuess, 1997: p.384). Raising boys properly was extremely important because they were eventually responsible for maintaining family continuity as the family head, either in their own family or after establishing a new family. Raising a girl properly was no less important, for she would get married in another family and help her husband as a wife and mother.

Representative childrearing books for male readers are Soko Yamaga's *Yamaga Gorui* (*Sayings of Yamaga*), written between 1663 and 1665; Shihei Hayashi's *Fu-kei kun* (*Moral Treatise on Fathers and Elder Brothers*), written in 1786, and Koki Nakamura's *Fu-shi kun* (*Moral Treatise on Fathers and Children*), written in 1811. Representative childrearing books for female readers are Tekisai Nakamura's *Himekagami (A Mirror of Young Women)*, written in 1661, and Johaku Namura's *Onna Chohoki* (*A Useful Book for Women*), written in 1692.

Thus, although the parental role varied with the gender of the children, fathers surely provided their children with childcare and education. The Edo period saw both fathers and mothers being keenly involved with their children, as a way of strengthening their future roles.

[5] The family precepts in the warrior class had been handed down since the thirteenth century. They were originally a means of transmission of didactic advice on a way of living from the head of the warrior's family to their descendants (Ohta, 2017, p. 197).

[6] According to Confucianism, the father–son relationship is cited as one of the most important sources of morals among human relationships. This is the reason why many Confucian scholars were authors of childrearing books.

11.2.2 The Ethos Underpinning Childcare and Childrearing

This section explores the ethos of the Edo period— people's thoughts, values, beliefs, or ideologies underpinning fathers' childcare and childrearing.

Confucian thought was undoubtedly the most widespread, given that Ieyasu, the first *shogun*, had officialized Confucianism as a doctrine.[7] While Western influence in this period was minimal, it is conceivable that the maternal principle influenced by Buddhism, which was a dominant religion in Edo period, was at the root of people's thoughts. As many scholars, including Kawai (1997), have stated, a strong maternal principle has been dominant in Japanese society throughout the ages. Even though a patriarchal family system was politically and socially established based on Confucianism as a paternal principle in this period, the maternal principle underlying people's thoughts complemented this family system (Hirayama, 2002, p. 134).

According to Kawai (1997), the paternal principle produces an ideology of separation, whereas the maternal principle generates a cosmology to embrace everything. Hence, the intimate relationship between a father and child in the Edo period can be conceived as a Confucian fatherhood affected by the maternal principle. At the same time, fathers cared for children cooperating with their spouse although women were under the men's subordination. Confucian norms emphasized the difference between men and women in their statuses and roles within the hierarchical social structure. However, the maternal principle, provided more egalitarian ideas, sometimes making their gender roles ambiguous.

Furthermore, the perspective of "mind–body unity" was peculiar to Japan during this period, influenced by the Zen philosophy of Buddhism, Confucianism, and Shintoism, generally known as an ethos of the warriors (DeMarco, 2015). According to Ichikawa (1993), the Japanese regarded the body as a highly integrated concept made up of several elements such as natural, spiritual, and social existence. A term from ancient Japan, "*mi*," denotes "body" but does not include the nuance of mind–body dichotomy. Ichikawa (1993) inferred the Japanese perspective of "body" in a broad variety of meanings and usages.

This perspective can also be cited as a factor for fathers' involvement in childcare in the Edo period. Fathers positively interacted with their children, who could not yet speak, using their bodies. The childrearing books in the Edo period also stressed the importance of fathers' involvement with their children from birth. Thus, the Edo period is referred to as the era of the "discovery of infancy," in contrast with Philippe Ariès' "discovery of childhood" (Ohta, 2017, p. 202).

---

[7] Within Confucianism, it was neo-Confucianism (i.e., the Cheng-zhu school) that Ieyasu specifically officialized. In the eighteenth century, there was an increase in the number of scholars who adhered to Yomei-gaku (Yangmingism), which was not an official school of Confucianism, as well as in the number of Kokugaku (the study of Japanese classical literature) scholars interested in studying ancient Japanese thought.

Conversely, Western countries during the same period saw a dominance of the mind–body dichotomy and the paternal principle. Childcare was considered to be a woman's role exclusively, whereas men were responsible for children's education after infancy. Illich references the preface to a fifteenth-century grammar book stating that "Women feed and nurture infants who cannot speak, and men teach and guide children who can speak (Illich, 1981 p. 113)." Furthermore, Lacan, the French psychoanalyst who developed Freud's theories, noted that it is the father who cuts off the unification between mother and child shaped during the fetal period and infancy and socializes the child (Izcovich, 2006).

It is crucial to examine how the maternal principle and "*mi*" perspective motivated fathers to be involved in childcare, with reference to the discourse of Edo-period diaries and childrearing books.

In particular, fathers' involvement in childcare can be understood through the diaries of certain warriors in the Edo period, such as the Kuwana and Kashiwazaki diaries.[8] The author of the former diary was Heidayu Watanabe, a low-ranking warrior in Kuwana (now in Mie Prefecture), and of the latter diary was Heidayu's adopted son, Katsunosuke.

Two sets of family diaries were written because Katsunosuke was transferred to Kashiwazaki (now in Niigata Prefecture) for his work, forcing the family to live apart. At Kuwana, Heidayu and his wife lived with their first grandson, who was a 2-year-old; while at Kashiwazaki, Katsunosuke and his wife lived with their 2-month-old daughter. Thereafter, Katsunosuke had two more children.[9] Under these circumstances, Heidayu and Katsunosuke kept daily diaries for 9 years to inform each other of their daily lives, including childrearing practices. They occasionally exchanged their diaries.

We explore the noteworthy points from these diaries that highlight how the authors cared for their children. First, both Heidayu and Katsunosuke slept holding the naked babies in their arms at night. Katsunosuke wrote in his diary, "The baby sleeps well when I hold her in my arms . . . While the baby was sleeping in my arms, she woke up crying when she urinated." Moreover, they wrote that they changed diapers and often took their children to the toilet to urinate or defecate, as toilet training, before they started walking.

Next, as the children grew older, they induced them to naturally learn habits like brushing their teeth by watching and imitating adults. The grand/parents gave them baths, told them old tales before going to sleep, and slept with them. When the children were sick, they nursed them to health.

The diaries also describe how they took the children outside. They carried them on their back or held them in their arms when they were toddlers. They played together outdoors after they could walk. Moreover, they taught the children how to read and write at home.

[8] These writings began in 1839. Watanabe and Watanabe (1971, pp. 501–765) contains the original text of both diaries.

[9] Katsunosuke also had a second daughter and a second son in Kashiwazaki.

Each diary clearly depicts what the father and grandfather did for the children and the extent to which they devoted themselves to childcare. The diaries also indicate how intimate the relationships between the men and the children were. Moreover, the mothers and grandmothers nurtured and cared for the children, yet, no difference between the men and the women can be observed in the commitment to the children.

Additionally, these diaries often described their own perceptions and observations of children's behaviors. Katsunosuke's note that "The baby woke up crying when she urinated" is such an example. Such detailed descriptions based on their personal observations and bodily sensations served to inform each other about the children's development. As the Japanese term *mi* shows, what they felt through their "body" included plenty of information.

Through these diaries, we also find that since birth, the parents spent considerable time in the same space with their children. In this period, the perspective of mind–body unity considered face-to-face interaction a key teaching method.

Soko Yamaga's *Yamaga Gorui*, which was mainly written for fathers in the warrior class, reads as follows:

> A newborn baby is small and weak. Over time, however, the baby becomes lively and energetic. It is just like when a plant first sprouts out of the ground. The plant does not suddenly grow, no matter how much you plow or fertilize the field.... However, educate your baby from birth, otherwise you will stray from the path of duty. Humans' morality develops from the time they are born. If you cultivate the baby's mind well, you will follow the path of duty (Nakae, 2003, p.67).

Thus, Yamaga emphasized that caregivers had to pay attention to the mental realm of babies, who would need physical care from adults. This shows that there was no clear distinction between the moral and the physical.

Yamaga compared a baby to "a plant bud." This analogy is often found in many other childrearing books. Nature was quite familiar to people during this age, as reflected in the Confucian belief that humans are an integral part of nature.

Wet nurses played an important role in cases where the mother's breast milk was insufficient. Most childrearing books in the Edo period provided instructions on how to choose a wet nurse. For instance, *Wazoku Doji kun* (*Precepts for Education*), written by Ekiken Kaibara in 1710, noted the following hiring conditions:

> The wet nurse should be calm, with no wickedness, modest, and not talkative. Those who tell lies, are smooth talkers, and have a twisted mind, are unsuitable as wet nurses (Nakae, 2007, pp.32–33).

Gozan Kazuki, a doctor, gave similar advice on the wet nurses' hiring in the 1703 *Shoni hitsuyo sodategusa* (*Handbook on the Essentials in Childcare*). Thus, the books emphasized the nurses' personality and character rather than their physical health or physical conditions, as it was believed that babies' physical and moral development could be largely affected by the wet nurses' moral behavior.

The importance of fetal education (*Taikyo* in Japanese) was also stressed in many childrearing books. Tekisai Nakamura's book for women, *Himekagami (A Mirror of Young Women),* written in 1661, presented fetal education as follows:

> Fetal education should be taken in the first place. Child education begins from the time when a fetus is inside your womb (Nakae, 1983, p.15).

Fetal education was considered important not only for health precautions during pregnancy but also for the moral and physical development of the unborn fetus. Moreover, it encompassed parents' preparation for childcare and child education (Nakae, 1983, p. 15). Similar to conditions for hiring a wet nurse, the moral practices of a pregnant woman were highlighted in fetal education (Koizumi, 2020, p. 45). Such practices were believed to affect a fetus through the mother's womb; according to the mind-body unity perspective, the moral state of wet nurses and mothers was imprinted into the "body" of the fetus or baby, through the women's "bodies."

Finally, the *Tokukyo* method for teaching children was noted, making children emulate their parents' behavior through the natural process of observation.

For instance, in *Okina Mondo* (*Dialogue with an Old Man*), Toju Nakae wrote as follows:

> Parents today believe that teaching is preaching to children with words, but true teaching is that the children learn spontaneously, being influenced by a parent's conduct (Ohta, 2017, pp. 203–204)

Childcare and childrearing focused on physical communication and children's spontaneous learning rather than linguistic communication. The key pedagogical technique was the imitation of the whole "body." Fathers attempted to go about their conduct with a moral sense to inspire their children to do the same. It was important for fathers to transmit all their inherited knowledge, skills, and virtues, to their children using their "body." Likewise, children were required to learn through their body, and embody what they learned. Education only through academic knowledge was insufficient in this period, which placed great value on family continuity.

In other words, as fathers provided their children with a holistic educational approach, they also constructed a close father–child bond. The earlier the face-to-face interaction between a father and children began, the more effective it became for the purpose of teaching and learning.

Thus, during the Edo period, fathers were earnestly involved in childcare, right from the birth of the child.

11.3 Sociocultural Factors Influencing the Loss of Fathers' Childcare Behaviors During the Meiji Period

In 1853, the United States pressured Japan to open up to the world. As a result, the Edo shogunate, which had severely restricted foreign contact and trade, ended the national isolation. Subsequently, Yoshinobu Tokugawa, the fifteenth shogun, returned the ruling power to the emperor in 1867, marking the end of the Edo period. At the beginning of the Meiji period, the government adopted Western ideas and methods. Soon thereafter, modernization and industrialization were introduced in Japan.

In 1898, the government established the family system with male primogeniture succession. At the end of the Edo period, such a patrilineal family inheritance system based on Confucian norms was no longer restricted to the warrior class. It was already widespread among almost all classes. In other words, the Meiji government institutionalized the family system in the Civil Code, using the Edo period family as a model. The position and authority of the family head were enshrined in the Civil Code. The emperor was placed at the top of all families to unite the modern nation–state.

As explained, one of the reasons why Edo period fathers were deeply involved in childcare was the continuation of the family. Indeed, the father's involvement in childcare was the first step in training successors. Why, then, did fathers in the Meiji period cease to act as caregivers? There are two major sociocultural reasons for this.

The first reason is the collapse of the social class structure and the establishment of the modern educational system. In 1871, the Meiji government abolished the strict social class structure based on the feudal system. It was now possible for people to freely choose their occupations. Following its abolition, the modern educational system was established in 1872. Equal access to education was ensured for every child aged 6 years and above, regardless of gender.

In the former class-stratified society, the purpose of education was to meet the practical demands of daily life and acquire specific knowledge and skills for the succession of the family's affairs. Thus, the fathers got involved with their children as early as possible. The collapse of the social class system and the establishment of new schools providing Western knowledge and skills to children took.

The second was the ideology prevalent from the mid-Meiji period, embodying "good wife, wise mother" and "childcare as a maternal vocation." These notions positioned children's care and education as the woman's role. The notion of a "good wife, wise mother" is defined as the view that women should fulfill the roles of being a wife and a mother as defined by the gendered division of labor, where men work outside, and women stay at home (Koyama, 1994, p. 46). This was first emerged in discourses among members of the enlightened movement, who were influenced by Western ideas around the beginning of the Meiji period (Takó, 2021). In 1902, Minister of Education Dairoku Kikuchi officially stated that the objective of female education was to ensure girls became "good wives and wise mothers" (Smith, 1987, p.8). Thus, this ideology was officialized as an ideal of women's education and prevailed among the people.

The notion of "wise mother" implies that women should be equipped with a good education in order to fulfill their responsibility as caregivers and educators. As mentioned previously, in the Edo period, childcare and education were primarily the fathers' responsibility, with mothers being deemed unsuitable for this responsibility because of their inferiority (Niwa, 1993, p. 72). Hence, in the Meiji period, female education became necessary to make women wise mothers.

Thereafter, the notion of "wise mother" was further developed, linking it to nationalism and Confucianism (Niwa, 1993, p.75). Eventually, children's care and education came to be considered a maternal "vocation." According to Confucian thought, each individual has an ordained vocation according to their social class;

this idea, in effect, was a foundation of the rigid social class structure during the Edo period (Sekiguchi, 2010).

Women first contributed to the nation–state indirectly by serving their families as good wives and wise mothers. However, with the promotion of the overseas expansion policy by the Meiji government, they were expected to use their abilities for the nation–state more directly (Koyama, 2012). In other words, women were expected to reproduce potential soldiers, and smarter and healthier citizens for the nation–state.

While being a good wife could hardly be regarded as a "vocation," childbearing and childrearing was easily accepted as such (Kinoshita, 1982). Women's role in children's care and education was valorized, and this change led to books and magazines on childrearing being targeted specifically at women after the mid-Meiji period (Kaneko, 1986, p. 362).

Thus, modernization and the change of social systems and ideologies led to the gendered division of labor: women were situated in the private sphere, while men were positioned in the public. This was similar to the phenomenon observed in Western countries from the nineteenth to twentieth centuries. The Meiji government aimed to establish a nation–state comparable to the Western countries under the political slogan of "rich country and strong army." For this reason, the family system, which placed the emperor at the top, was changed into a more patriarchal and hierarchal structure compared to that in the Edo period. It can be said that the separate spheres between men and women in this period was more or less influenced by Western ideologies with paternal principles.

However, in Western countries, women were seen as being closer to nature than men because of their natural functions surrounding reproduction (Ortner, 1972). Hence, it was thought that women should fulfill children's care and education as a natural act, while men should be associated with cultural or intellectual activities. This view is based on the dichotomy of culture-nature, originally from the mind–body dualism perspective (Oliver, 2000). Simultaneously, nature and culture were viewed as opposing yet complementary concepts. In this sense, men ultimately dominated both spheres.

In Japan, on the other hand, childcare and child education were regarded as women's "vocation", that is to say, women's duty. As a result, men and women were given an equal weight of duty in each sphere. The view of childcare as women's vocation was influenced in part by Confucianism with the perspective of body-mind unity. However, the egalitarian division of men's and women's duty did not arise from Confucianism which requires women to be obedient to men, but from complex historical and social contexts in Japan, where the maternal principle has been dominant. Thus, mothers exclusively occupied the private sphere and this tendency persisted even later.

This is the main reason why fathers were absent from their homes for a long time in Japan. Japanese fathers came to indirectly contribute to childcare as the economic provider.

11.4 Conclusion

This paper discussed the sociocultural factors facilitating fathers' involvement in childcare in the Edo period, and inhibiting it in the Meiji period.

Fathers' behavior as caregivers was found to be affected by the combination of various factors with varying impacts. Hence, their commitment to childcare had large variations depending on each historical context. It could be either facilitated or inhibited, even when society had a common system or thought patterns across different periods.

From the mid-Edo period, almost all fathers spent their energy and time caring for their children to ensure the continuity of the family. From the-mid Meiji period, however, their connection with the children was lost, even as their familial concerns remained similar under the family system. Moreover, although the mind–body unity perspective and the maternal principle were dominant throughout both periods, their impacts differed considerably: in the Edo period, these ideas prompted fathers to become more involved in childcare, while in the Meiji period, they had the opposite effect.

Next, a historical examination found the following to be important sociocultural factors motivating fathers toward childcare involvement:

First is the social environment that fathers pay attention to for the survival and maturity of their children. The Edo-period fathers' duty was to bring up their heirs for the continuity of their family.

Second is the ideological belief that embraces everyone as co-parents, regardless of gender. Although Edo society had a patriarchal family system, the dichotomous male-female roles were ambiguous.

Third, the Edo period's prevalent method of childrearing emphasized the importance of fathers' physical interaction with their children from an early age.

Fourth is the existence of means to share norms and values surrounding fathers' childcare. In the Edo period, childrearing books for fathers were widespread. Fathers widely shared information on their roles through such books, and were inspired to be involved in childcare.

It is conceivable that these sociocultural factors helped arouse fathers' potential to care for their children as acquired in the evolutionary process. As stated before, the parental investment theory highlights co-parenting and fathers' concerns about the survival and maturity of their offspring. The conditions assumed by the parental investment theory came into existence in the social environment and the ideological belief of the Edo period.

Moreover, fathers in the Edo period often had physical contact with their children, which is currently considered a key element provoking the transition to fatherhood. The closeness of father-child relationship seen in this period appears to be due to fathers' physical contact with their children from an early age.

In summary, the Edo period was a unique age when the paternal behavior was widespread in almost all fathers. It is conceivable that during this period, a combination of factors led men to be involved in childcare, with an interaction between sociocultural factors motivating fathers' childcare, and biological factors evoking fathers' behavior.

Ultimately, the findings obtained from this historical perspective seem to suggest that fathers' childcare is "nature," "nurture," and their dynamic reciprocity.

Acknowledgements This work was supported by JSPS KAKENHI Grant Numbers JP18K02017 and JP21K12517

References

Arendell, T. (2000). Conceiving and investigating motherhood: The decade's scholarship. *Journal of Marriage and Family, 62*(4), 11921207. https://doi.org/10.1111/j.1741-3737.2000.01192.x

Arichi, T. (1986). *Nihon no oyako nihyakunen [Two hundred years of Japanese parent and child]*. Sinchosha.

Ariga, K. (1954). The family in Japan. *Marriage and Family Living, 16*(4), 362–368. https://www.jstor.org/stable/348444

Boyette, A. H., & Gettler, L. T. (2019). Evolution and human fatherhood. In L. M. M. Welling & T. K. Shackenlford (Eds.), *The Oxford handbook of evolutionary psychology and behavioral endocrinology* (pp. 179–199). Oxford University Press.

DeMarco, M. (2015). *Medieval warrior cultures of Europe and Japan: Body, mind, sword*. Via Media.

Dore, R. (1976). *The diploma disease. Education, qualification and development*. George Allen & Unwin.

Dore, R. (2010). *Education in Tokugawa Japan*. Routledge (First edition was published in 1965).

Feldman, R., Gordon, I., Schneiderman, I., Weisman, O., & Zagoory-Sharon, O. (2010). Natural variations in maternal and paternal care are associated with systematic changes in oxytocin following parent–infant contact. *Psychoneuroendocrinology, 35*(8), 1133–1141. https://doi.org/10.1016/j.psyneuen.2010.01.013

Fuess, H. (1997). A Golden age of fatherhood? Parent-child relations in Japanese historiography. *Monumenta Nipponica, 52*(3), 381–397. https://doi.org/10.2307/2385634

Geary, D. C. (2007). Evolution of fatherhood. In C. A. Salmon & T. K. Shackelford (Eds.), *Family relationships: An evolutionary perspective* (pp. 115–144). Oxford University Press. https://doi.org/10.1093/acprof:oso/9780195320510.003.0006

Gordon, I., Zagoory-Sharon, O., Leckman, J. F., & Feldman, R. (2010). Prolactin, oxytocin, and the development of paternal behavior across the first six months of fatherhood. *Hormones and Behavior, 58*(3), 513–518. https://doi.org/10.1016/j.yhbeh.2010.04.007

Gregory, A., & Milner, S. (2011). What is "new" about fatherhood? The social construction of fatherhood in France and the UK. *Men and Masculinities, 14*(5), 588–606.

Hirai, A. (2016). Kinsei koki ni okeru ie no kakuritsu :Tohoku nouson to seinan kaison no jirei [The formation of ie in the late modern period: A farming village in the northern eastern region and a village by the sea in southern western region]. In A. Kato et al. (Eds.), *Kazoku Kenkyu no Saizensen, Ie to Kyoudousei [The frontline of family studies Ie and cooperativeness]* (pp. 93–113). Nihonkeizaihyoronshya.

Hirayama, A. (2002). Bosei syakairon no datsu kochiku [Deconstruction of "Japan as maternal society" doctrine]. *Nihon Kenkyu: Journal of the International Research Center for Japanese Studies, 24*, 125–145. https://doi.org/10.15055/00000689

Hrdy, S. B. (2007). Evolutionary context of human development: The cooperative breeding model. In C. A. Salmon & T. K. Shackelford (Eds.), *Family relationships: An evolutionary perspective* (pp. 39–68). Oxford University Press.

Hrdy, S. B. (2009). *Mothers and others: The evolutionary origins of mutual understanding*. Harvard University Press.

Ichikawa, H. (1993). *Mi no Ko-zo-: Shintairon wo Koete [Structure of Mi: Overcoming the theory of the body]*. Kodansha Gakujutsu Bunko.

Illich, I. (1981). *Shadow work*. Marion Boyars.

Iriguchi, A. (2019). Edo Jidai no Yojo-sho Shuppan to Sono Fukyu [Publication and diffusion of Yojo-sho in the Edo period]. *Japanese Journal of Health and Human Ecology, 85*(1), 14–17. https://doi.org/10.3861/kenko.85.1_14

Izcovich, L. (2006). Du Nom-du-Père au père qui nomme. *Champ lacanien, 1*, 23–31. https://www.cairn.info/revue-champ-lacanien-2006-1-page-23.htm

Kalland, A., & Pedersen, J. (1984). Famine and population in Fukuoka domain during the Tokugawa period. *Journal of Japanese Studies, 10*(1), 31–72. https://www.jstor.org/stable/132181

Kaneko, S. (1986). Jyunyuron ni arawareta hahaoyakan no hensen [Transformations of perspectives on mothers as seen in the theories on breastfeeding]. *Bulletin of the Faculty of Education Ehime University, 32*, 361–378.

Kaplan, H. S., & Lancaster, J. B. (2003). An evolutionary and ecological analysis of human fertility, mating patterns, and parental investment. *Offspring: Human Fertility Behavior in Biodemographic Perspective, 1*, 170–223.

Kawai, H. (1997). *Boseishakai Nihon no Byori [Pathology of Japan as a society of maternity]*. Kodansya (First edition is published in 1976).

Kinoshita, H. (1982). Meijiki ni okeru ikujitenshokuron to jyoshikyoiku, [Maternal vocation and female education in the Meiji period]. *The Japanese Journal of Education Research, 49*(3), 255–264.

Koizumi, Y. (2020). *Shinkyo wo Motte Sho to Nasu: Edo ni Manabu Ningenkyoiku no Chie [The wisdom of the people in Edo-Period on "Human education"-Leaning spontaneously through observation]*. Keibunsha.

Kojima, H. (1986). Japanese concepts of child development from the mid-17th to mid-19th century. *International Journal of Behavioral Development, 9*(3), 315–329. https://doi.org/10.1177/016502548600900304

Koyama, S. (1994). The "good wife and wise mother" Ideology in post-World War I Japan. *US-Japan Women's Journal. English Supplement, 7*, 31–52.

Koyama, S. (2002). *Kodomotachi no kindai: Gakko kyoiku to katei kyoiku [Children modernization: School and home education]* (Vol. 143). Yoshikawa Koubunkan.

Koyama, S. (2012). *Ryōsai Kenbo: The educational ideal of good wife, wise mother in modern Japan*. Brill.

LaRossa, R. (1988). Fatherhood and social change. *Family Relations, 37*(4), 451–457. https://doi.org/10.2307/584119

Lewis, C., & Lamb, M. E. (2007). *Understanding fatherhood*, Joseph Rowntree Foundation. Retrieved March 14, 2023, from https://policycommons.net/artifacts/1799362/understanding-fatherhood/2531006/. CID: 20.500.12592/x14wgd.

Matsuyama, I. (2020). Nihonjin ni okeu bakumatsu kara meiji ni kaketeno kodomo to kosodate ni Taisuru mikata [Japanese recognition for children and the childcare from late Tokugawa period to Meiji period]. *Journal of the Faculty of Culture and Education, Saga-University, 4*(1), 173–180.

Moore, R. A. (1970). Adoption and samurai mobility in Tokugawa Japan. *The Journal of Asian Studies, 29*(3), 617–632. https://www.jstor.org/stable/2943247

Nakae, K. (1983). Taikyo shiso no rekisiteki kento [Fetal education from historical examination]. *The Japanese Journal of Education Research, 50*(4), 343–352.

Nakae, K. (2003). *Edo no kosodate [Childrearing in the Edo period]*. Bungeishunju.

Nakae, K. (2007). *Edo no shitsuke to kosodate [Discipline and childrearing in the Edo period]*. Shōdensha.

Niwa, A. (1993). The formation of the myth of motherhood in Japan. *U.S.-Japan Women's Journal English Supplement, 4*, 70–82. http://www.jstor.org/stable/42772053

Ohta, M. (2017). *Edo no oyako: Chichioya ga kodomo wo sodateta jidai [Childrearing in the Edo period: The age when fathers raised children]*. Yoshikawa koubunkan (First edition is published in 1994).

Oliver, K. (2000). Conflicted love. *Hypatia, 15*(3), 1–18. https://doi.org/10.1111/j.1527-2001.2000.tb00328.x

Ortner, S. B. (1972). Is female to male as nature is to culture? *Feminist Studies, 1*(2), 5–31. https://www.jstor.org/stable/3177638

Roopnarine, J. L., & Yildirim, E. D. (2019). *Fathering in cultural contexts: Developmental and clinical issues*. Routledge.

Saito, O., & Takashima, M. (2017). Jinkou to toshika idou to shugyo [Population, urbanization, mobility, and job]. In M. Nakamura et al. (Eds.), *Iwanami koza nihon keizai no rekishi kinsei 16seiki kouhan kara 19seiki zenhan [History of Japanese economy: From the latter half of 16th century to the first half of 19th century]* (pp. 61–104). Iwanami shoten.

Sekiguchi, S. (2010). Confucian morals and the making of a "Good wife and wise mother": From "Between husband and wife there is distinction" to "As husbands and wives be harmonious". *Social Science Japan Journal, 13*(1), 95–113. http://www.jstor.org/stable/40961243

Smith, R. J. (1987). Gender inequality in contemporary japan. *The Journal of Japanese Studies, 13*(1), 1–25. https://doi.org/10.2307/132584

Takó, F. (2021). Education and "civilization": Westernization through centralization and the concept of women's education in late 19th Century Japan. *Távol-keleti Tanulmányok, 13*(1), 255–268.

Watanabe, H., & Watanabe, K. (1971). Kuwana nikki, Kashiwazaki nikki. In K. Tanigawa (Ed.), *Nihon shomin seikatsu shiryo shusei [Compilation of historical materials on Japanese commoners' lives]* (Vol. 15, Toshi fūzoku [Urban customs], pp. 501–765). Sanichi Shobo.

Tsujimoto, M. (2000). Maturing of a literate society-literacy and education in the Edo period (17th-19th century). *Journal of Japanese Trade and Industry, March/April*, 45–48.

Chapter 12
Culture and the Microbiome-Gut-Brain

Gabe Ignatow

Abstract There is a longstanding heterodox tradition within sociology that has addressed the nature-nurture debate by integrating concepts and findings from the life sciences into sociological theory and research methods. In this chapter I suggest that recent trends in the life sciences away from genetic determinism and molecular reductionism and toward systems-based theorizing complement this sociological tradition, but that trends in sociology and in the organization of science are not favorable to heterodox interdisciplinary theorizing. I review recent life science research on the microbiome-gut-brain axis that I suggest has obvious relevance for social science research practice, and further suggest that social scientists who pursue research on pathways between the microbiome-gut-brain, society and culture are likely to make substantial contributions to knowledge even if their contributions remain on the margins of mainstream social science for some time.

12.1 Nature-Nurture and Life Science-Social Science Interdisciplinarity

Sociology has addressed the nature-nuture debate since the discipline's formation in the nineteenth century, and there is a longstanding stream of theoretical speculation and empirical research within sociology that has contributed to this debate. Sociological research within this now-heterodox tradition has integrated concepts and findings from the life sciences into social science rather than insisting on the superiority of social/nurture over biological/nature explanation. Contributions within this sociological tradition include Spencer's Social Darwinism and modern evolutionary sociology, the French tradition of psychologically informed sociology (Ignatow, 2012, 2019), and modern cognitive sociology and neurosociology (Brekhus & Ignatow, 2019). In this chapter I suggest that contemporary trends

G. Ignatow (✉)
Department of Spanish, University of North Texas, Denton, TX, USA
e-mail: gabe.ignatow@unt.edu

J. Pérez-Jara, Í. Ongay (eds.), *Beyond Nature and Nurture*, Synthese Library 497,
https://doi.org/10.1007/978-3-031-87715-5_12

within the life sciences in the direction of epigenetics, systems biology, and research on the microbiome-gut-brain axis make the life sciences remarkably compatible with multi-level sociological explanation. Intellectually, these trends portend a radically new era of social and behavioral research. Institutionally, however, long-term trends in sociology and in the organization of basic science are not favorable to heterodox interdisciplinary inquiry. These trends in sociology include disciplinary fragmentation, the end of grand theory, and a turn away from basic research and toward advocacy. Trends in the organization of science include hyper-specialization and shifting incentives for scientists pursuing basic research. Still, despite an unfavorable institutional landscape, social and life scientists who pursue research on pathways between biological factors, society, and culture are likely to make substantial contributions to knowledge even if their contributions remain on the margins of mainstream social science for some time to come.

12.2 Epigenetics and Systems Biology

In two recent papers (Ignatow, 2021, 2022) I reviewed recent trends in the life sciences away from genetic determinism and molecular reductionism and toward systems-based theorizing that I argued complement the sociological tradition. To recap, in the wake of the widely acknowledged failure of the Human Genome Project conducted from 1990 to 2003, several new life sciences research paradigms emerged that placed a greater emphasis on the higher-level emergent dynamics of complex biological systems rather than on lower-level basic cellular or molecular mechanisms. The post-Human Genome Project life sciences thus moved away from a deterministic research paradigm that had supported a "biological Weltanschauung that accords priority to DNA as a causal agent within the cell" (Powell & Dupré, 2009: 56). Today the life sciences are more oriented toward demonstrating how social and physical environments influence genes and their expression, and the functional interdependencies between the human microbiome, gut, and brain. Biologists have turned toward paradigms such as "systems biology" and "relational biology" (Meloni, 2014) that are based on the idea of a "reactive genome" (Gilbert, 2003) that changes in response to changes in the environment. Contemporary relational approaches in life science include epigenetics, ecological development biology, and systems biology. Rather than analyzing genetic sequences or biological structures one at a time, researchers working in these areas focus on relations between organisms and environments, the "biology of the in between" (Van de Vijver, 2009).

The so-called microbiome-gut-brain (MGB) axis is a centrally important theoretical construct within modern relational biology. The MGB refers to the electrochemical communication network connecting the human microbiome, gastrointestinal tract, and brain. Signaling in this network occurs via immune and neural cellular and molecular circuits involving neurochemicals such as catecholamines, dopamine, GABA, glutamate, and serotonin. These neurochemicals and their related receptors are widely dispersed throughout nature, even in microorganisms and plants, allowing for communication between phylogenetically distinct organisms.

The human microbiome is the aggregate of all microorganisms that reside on or within human tissues and biofluids along with their corresponding anatomical sites. Types of human microorganisms include bacteria, fungi, and viruses. The population size of one person's microbiome is estimated to be 100 trillion microorganisms; by most estimates a person's microbial cells outnumber their human cells. While at this time the role of most microorganisms in the microbiome is not well understood, there is a growing body of evidence showing how gut microorganisms produce neuroendocrine hormones that modulate cognition, mood, and behavior. Researchers have discovered associations between gastrointestinal pathology and psychiatric conditions such as autism and schizophrenia, and that fundamental neurological processes such as myelination (the formation of fatty sheathing that insulates nerve fibres) can be influenced by gut microorganisms. There is evidence that microbiome imbalance and gut dysbiosis result in changes in social behavior, and there is a large body of MGB-related research findings that would appear to have obvious relevance for sociological theory and empirical research in several sociological subfields. For instance, in the area of crime, deviance, and social control, research on cholesterol and serotonin shows that very low plasma cholesterol levels are associated with increased mortality rates as a result of suicide, violence, and accidents. The causes of these associations are hypothesized to be low cholesterol levels leading to reduced brain serotonin because of the requirement of adequate cholesterol in nerve cell membranes to maintain the functional integrity of serotonin receptors (Diaz-Sastre et al., 2007; Engelberg, 1992; Kaplan et al., 1997). The neurotransmitter serotonin modulates appetite, mood, sleep, reward, learning, and memory, and is found primarily in the ENS in the gastrointestinal tract where it regulates intestinal movements. It is also synthesized in the CNS. Low plasma cholesterol (hypocholesterolemia) is a result of inflammatory diets high in ultra-processed foods (UPFs), and of cholesterol-lowering pharmacological treatments such as statin drugs. Evidence of the association of hypocholesterolemia (more specifically, low levels of high density lipoprotein [HDL]) with suicide and interpersonal violence is by now overwhelming (e.g. Eriksen et al., 2017; Tomson-Johanson & Harro, 2018).

In research on social and political attitudes the MGB concept facilitates theoretical integration of nutritional psychiatry research that has demonstrated associations between poor nutrition, oxidative stress, and psychological distress on the one side, and social psychology research that has demonstrated associations between psychological distress and political extremism on the other. Canetti-Nisim and her colleagues have published several studies examining how political left- and right-wing extremists are, psychologically, similar to one another and different from moderates (2013, 2019; also Hirsch-Hoefler et al., 2016; van Prooijen & Krouwel, 2019). Van Prooijen and Krouwel have suggested that "psychological distress stimulates adopting an extreme ideological outlook; extreme ideologies are characterized by a relatively simplistic, black-and-white perception of the social world; because of such mental simplicity, political extremists are overconfident in their judgments; and political extremists are less tolerant of different groups and opinions than political moderates" (2019: 159). Compared with moderates, political extremists express stronger anxiety about their economic future (van Prooijen et al., 2015). Experiments have shown that people psychologically compensate for feelings of uncertainty and fear through strong ideological convictions (McGregor et al., 2013), and that inducing a loss of significance increases extreme beliefs (Webber et al., 2018). For instance, Zmigrod et al. (2020) found that partisan extremity was related to lower levels of cognitive flexibility, regardless of political orientation.

Research on religion from anthropology, sociology, psychology and cognitive science has turned to the life sciences for insight into neurophysiological processes associated with intense religious experience. For example the sociologists Wellman et al. (2014) have used cognitive neuroscience research on oxytocin to study social bonding and emotional energy in religious rituals such as megachurch services, while the cognitive anthropologist Xygalatas and his colleagues have studied heart rate synchrony between participants and audiences in fire walking rituals (Konvalinka et al., 2011; Xygalatas et al., 2011). There are many other examples of social science research on religion borrowing concepts from the life sciences, but there has been very little work done to integrate these research streams. The MGB concept can help in this regard by specifying some of the electrochemical processes and pathways involved in religious experience, and by providing synthetic theoretical language that is not dependent on the theoretical vocabulary of any one academic discipline. Geertz (2010), claiming that the future of the study of religion lies at the intersection of cultural and neurobiological sciences, has asked for just this sort of cross-disciplinary synthesis.

12.3 Culture, Dual-Process Theory, and the MGB

Neuroscience research on interpersonal violence, social and political attitudes, and religious ritual suggests possible relationships between the MGB and culture. Sociological dual-process theories are the most prominent recent attempt to theorize these relationships, although I will argue that on balance these theories are not the best path forward. Dual-process theory was imported into sociology from cognitive neuroscience in the early 2000s; it has also been imported into many other areas of research (Melnikoff & Bargh, 2018: 1). Dual-process theory posits that there are two types of cognition: "Type 1" cognition that is unintentional, uncontrollable, unconscious, and efficient; and "Type 2" cognition that is intentional, controllable, conscious, and inefficient. In his influential 2009 article introducing dual-process theory into cultural sociology, Vaisey notes that in cognitive psychology the idea that there are two types of cognition "is now uncontroversial" (p. 1683) and maps neatly onto sociological practice theory's dualism of practical versus discursive consciousness (Bourdieu, 1984; Giddens, 1984). Vaisey and his colleagues (Vaisey, 2009; Vaisey & Frye, 2019), Lizardo et al. (2016), and other sociologists working with dual-process theory (e.g. Miles et al., 2019) claim that dual-process theory is a consensus view in cognitive psychology, and that the amygdala (Leschziner, 2019: 171) is the key brain region generating "knee-jerk" (Tavory, 2011: 274) Type 1 cognition. However, cognitive psychologists have developed variations of dual-process theory that emphasize that Type 1 and Type 2 cognition occur in parallel, and that reasoning is always a hybrid process involving both types of cognition (De Neys, 2017). A number of sociologists have made similar arguments that cognition in vivo involves iterations between Type 1 and Type 2 cognition. In her study of olfactory meaning-making, Cerulo (2018) emphasizes how deciphering scents involves dynamic interactions between practical and discursive consciousness (p. 363; see also Leschziner & Green, 2013; McDonnell, 2014; Moore, 2017; Vila-Henninger, 2015). However, while sociologists have questioned the idea that Type 1 cognition is the default form of cognition that is only occasionally interrupted by Type 2 deliberation, they have not engaged with several more fundamental critiques of the dual-process framework from cognitive neuroscience.

The psychologists Melnikoff and Bargh (2018) review criticisms of dual-process theory based on the theory's lack of empirical support and weak internal coherence and its incompatibility with established findings. They argue that there is no evidence that human cognitive processing features cluster together into two distinct groups, and there is substantial evidence that these features do not cluster this way. Further, Melnikoff and Bargh claim that the theory has never been tested and perhaps cannot be tested, and that there is no statistical test in the dual-process literature pertaining to the degree of correspondence between processing features. While researchers have demonstrated that processing features do sometimes align as dual-process theory predicts, there are many examples of ordinary cognitive phenomena where features align differently. Melnikoff and Bargh list dozens of examples of misalignment between Type 1 and Type 2 cognition (pp. 285–288).

The sociologist Leschziner (2019) has provided a more detailed account of the neurobiological foundations of dual-process theory than has generally been the case in sociology. She suggests that "two neurological systems" have been found to be associated with "reflexive and reflective cognitive processes, respectively" (p. 171). Based on research from social cognitive neuroscience (Lieberman, 2003; Lieberman et al., 2004), she claims that Type 1 cognition is associated with "the amygdala, basal ganglia, and lateral temporal cortex," while Type 2 cognition is associated with brain modules including "the cingulate cortex, prefrontal cortex, and the medial-temporal lobe" (p. 171). Addressing this often made claim directly, Pessoa and Adolphs (2010) have shown that the brain regions involved in Type 1 and Type 2 cognition are characterized by broad neural connectivity to each other and to other brain regions. They summarize neuroscience research on the speed and coarseness of emotional evaluations, and also on brain anatomy and physiology, in order to question the validity of dual-process typologies that are premised on the existence of a "low road" of visual emotion processing from the visual cortex to the amygdala. Pessoa and Adolphs suggests that because the brain is characterized by broad connectivity between regions rather than hyper-modularity, a network model of emotion processing is better supported than a "locationist" one (Lindquist et al., 2012). They argue that rather than modeling emotion processing in terms of a low road and a high road, there are multiple roads from stimulus to behavior, and the stimulus-behavior link occurs in "multiple waves" of neural activation across brain regions, a "cortical sweep" of visual information processing occurring in parallel across brain regions in multiple waves or iterations. Multiple waves of visual processing occur within milliseconds. While data supportive of a multiple waves model are "well known to parts of the neuroscience community," the "lack of appreciation for these data by 'affective science' researchers could partly explain the widespread acceptance" of the dual-process framework (Pessoa & Adolphs, 2010: 3).

The iterative reprocessing model introduced by the psychologist Cunningham and his colleagues (Cunningham et al., 2007; Van Bavel et al., 2012) resembles Pessoa and Adolphs' "multiple waves" model of neural activation in response to visual stimuli. Cunningham et al. theorize that information processing does not occur via two forms of cognition, but rather via a course-to-fine updating strategy in which "quick and dirty" processing can be followed by processing that is more nuanced and possibly reflective. Several iterations of neural activation may occur within a few hundred milliseconds of perception (Cunningham et al., 2007: 744) and involve unconscious and conscious cognition in parallel. In contrast to the dual-process framework, the dynamics of the iterative reprocessing model are essentially Bayesian: the same attitude weights and patterns of activation generated in the first iterative cycle are used in subsequent evaluative processing (p. 747), but these weights can gradually be updated over time. Another iteration begins updating the evaluation within a few hundred milliseconds of initial processing, but in each subsequent iteration additional processes, such as physiological responses from the body, can influence the evaluation. Multiple iterations may occur before one is consciously aware of the current evaluation (Cunningham et al., 2007).

Sociologists have relied on social cognitive neuroscience, via cognitive psychology, for support for the sociological dual-process model's claim that distinct and independent brain regions are associated with practical and discursive consciousness. For instance, Leschziner (2019) cites an influential paper by the psychologist Evans, who in turn cites social cognitive neuroscience research by Lieberman (2003; Lieberman et al., 2004) that, although "in its early stages" (Evans, 2007, 2008: 270) at the time of the publication of Evans' paper in 2008, provided "perhaps the strongest basis in the literature for maintaining some form of dual-system distinction" (Evans, 2007, 2008: 270). Sociological reliance on research from dual-process psychology, which in turn relies on Lieberman's dual-process social cognitive neuroscience, has become problematic because dual-process social cognitive neuroscience cannot plausibly be claimed to be "now uncontroversial" (Vaisey, 2009: 1683), in either psychology or neuroscience, as dual-process social cognitive neuroscience is not currently the consensus approach in either field. Rather, while dual-process models are recognized to "provide generative frameworks for understanding a wide range of psychological phenomena," recent developments in cognitive neuroscience "suggest that the human evaluative system, like most of cognition, is widely distributed and highly dynamic" (Van Bavel et al., 2012: 438). The "notion of a strict dissociation between implicit and explicit processes/representations is unlikely to be accurate" (Cunningham et al., 2007: 750).

For sociologists, dual-process models of specific empirical phenomena are theoretically defensible, and they are compatible with a broader microbiome-gut-brain theoretical framework. In an MGB framework, cognition is not limited to the brain (Clark, 2008; Milkowski, 2013) but is rather distributed across the microbiome, gut, and brain (Ignatow, 2021, 2022). In order to understand the interactions between culture on the one side and social interactions and social structures on the other, it is necessary to understand the fundamental biological mechanisms that connect the microbiome, gut, and brain. This framework has several implications for sociology, pointing away from a variety of theoretical dualisms (Keren & Schul, 2009; Kruglanski et al., 2006; Melnikoff & Bargh, 2018). It also unlocks the "black box" (Boudon, 1998: 175) of the sociological concept of *habitus*, and the related concepts *habitus clivé* and *hysteresis*, by specifying electrochemical mechanisms through which social interactions and social structures influence learning and behavior.

12.4 Biology of the Habitus

An MGB theoretical framework can generate new ways of specifying and operationalizing core cultural concepts in sociology. For instance, it can specify Bourdieu's influential but perhaps somewhat abstract *habitus* concept, a concept that was first introduced by Aristotle and subsequently developed by Aquinas, Hegel, Weber, Durkheim, Mauss, and Husserl (Wacquant, 2006). Bourdieu revived *habitus* and the closely related concept of *hexis* in the late 1960s, and refined the concepts in several of his major works. The opacity with which Bourdieu discussed the *habitus* has led to charges that it operates as a "black box" able to do all sorts of conceptual and theoretical work (Boudon, 1998: 175). Arguably the microbiome-gut-brain axis *is*, for all intents and purposes, the *habitus/hexis*, and accordingly processes of oxidative stress and psychological distress are centrally implicated in Bourdieu's concepts of *habitus clivé* (divided *habitus*) and *hysteresis* (the experience of a mismatch between *habitus* and field). Further, neuroplasticity research from cognitive neuroscience and demographic research on the relationships between dietary and lifestyle factors and neuroplasticity appear directly relevant to theoretical debates over the flexibility and reflexivity of the habitus (Adams, 2006; Decoteau, 2016). Thus sociologists who are so inclined now have theoretical concepts and methodological tools that can potentially settle these debates (Changeux, 2006; Downey, 2014; Wacquant, 2014: 129).

12.5 Heterodox Interdisciplinary Theory and the Science of Science

While there are today many areas of overlap between the social and life sciences, in terms of theoretical, epistemological and ontological compatibility as well as substantive areas. And yet trends in the organization of science are not favorable to heterodox interdisciplinary theorizing. First, within sociology there is a division between sociology as basic social science on the one side, and critical public sociology on the other (Turner & Turner, 1990). Within sociology as basic social science there is not an agreed upon theoretical core; rather sociology is a multiple-paradigm, fragmented field in which research is conducted within highly specialized subfields and there is little communication across these subfields. The sociological tradition of integrative "grand theory" all but ended in the 1990s (Turner, 2006; cf. Turner & Boyns, 2001). Accordingly, interdisciplinary theorizing is marginal within contemporary sociological theory, with few exceptions.

Within this context, interdisciplinary theoretical projects such as Walsh's *Biosociology* (2017), evolutionary sociology (Maryanski et al., 2015), cognitive sociology (Brekhus & Ignatow, 2019), and the French tradition of psychologically informed sociology (Ignatow, 2012) remain peripheral within sociology, with their proponents occasionally expressing frustration that sociology plows ahead as though for the past century the life sciences have discovered nothing of value about human nature. The exception might be sociological dual process theory, which has achieved a degree of mainstream acceptance despite some limitations discussed above.

While contemporary sociology tolerates the heterodox tradition of integrative social science-life science theory and empirical research, disciplinary fragmentation and related trends are not conducive to theory-driven empirical research. And the heterodox tradition faces even more obstacles beyond the boundaries of sociology. Research from the Science of Science (SciSci) has "revealed choices and trade-offs that scientists face as they advance both their own careers and the scientific horizon," and that scientists tend to be risk-averse, "preferring to study topics related to their current expertise, which constrains the potential of future discoveries" (Fortunato et al., 2018). Einstein (1949) claimed that individual scientists inevitably become narrower in their expertise as scientific knowledge expands. Electronic publication may be accelerating this trend as online searching encourages scientists to anchor their findings in prevailing opinion within their subfield rather than in past scholarship (Evans, 2007, 2008). In a 2016 Australian study of the "paradox of interdisciplinarity"—the paradox being that interdisciplinary research is often encouraged at the policy level but poorly rewarded—Bromham, Dinnage and Hua used a novel interdisciplinary distance metric to discover that research with a higher degree of interdisciplinarity had a lower probability of being funded. Their data included almost 19,000 Australian proposals in the arts, humanities and sciences.

Given the current landscape of science, perhaps the best interdisciplinary theorists can do is to accept our heterodox status secure in the knowledge that as discoveries continue to emerge from multiple disciplines the value of theory for synthesizing new discoveries and directing future research is likely to increase. For those so inclined, life science research on the microbiome-gut-brain provides a solid foundation for the next phase of development of the heterodox tradition.

References

Adams, M. (2006). Hybridizing habitus and reflexivity: Towards an understanding of contemporary identity? *Sociology, 40*(3), 511–528.

Boudon, R. (1998). Social mechanisms without black boxes. In P. Hedstrom (Ed.), *Social mechanisms: An analytical approach to social theory* (pp. 172–203). Cambridge University Press.

Bourdieu, P. (1984). *Distinction: A social critique of the judgement of taste*. Harvard University Press.

Brekhus, W. H., & Ignatow, G. (Eds.). (2019). *The Oxford handbook of cognitive sociology*. Oxford University Press.

Bromham, L., Dinnage, R., & Hua, X. (2016). Interdisciplinary research has consistently lower funding success. *Nature, 534*(7609), 684–687.

Canetti, D., Cavari, A., Rapaport, C., Shalev, H., & Hobfoll, S. (2019). Individual exposure to terror and political attitudes: A physiologically-based model of militancy. *Terrorism and Political Violence, 33*(5), 1055–1070.

Canetti, D., Hall, B., Rapaport, C., & Wayne, C. (2013). Exposure to political violence and political extremism. *European Psychologist, 18*, 263–272.

Cerulo, K. (2018). Scents and sensibility: Olfaction, sense-making, and meaning attribution. *American Sociological Review, 83*(2), 361–389.

Changeux, J.-P. (2006). *Les Bases Neurales de L'habitus. Croyance, Raison et Déraison* (pp. 143–158). Odile Jacob.

Clark, A. (2008). *Supersizing the mind: Embodiment, action, and cognitive extension*. Oxford University Press.

Cunningham, W., Zelazo, P., Packer, D. J., & Van Bavel, J. J. (2007). The iterative reprocessing model: A multilevel framework for attitudes and evaluation. *Social Cognition, 25*(5), 736–760.

Decoteau, C. (2016). The reflexive habitus: Critical realist and Bourdieusian social action. *European Journal of Social Theory, 19*(3), 303–321.

De Neys, W. (Ed.). (2017). *Dual process theory 2.0*. Routledge.

Diaz-Sastre, C., Baca-Garcia, E., Perez-Rodriguez, M. M., Garcia-Resa, E., Ceverino, A., Saiz-Ruiz, J., & De Leon, J. (2007). Low plasma cholesterol levels in suicidal males: A gender-and body mass index-matched case-control study of suicide attempters and nonattempters. *Progress in Neuro-Psychopharmacology and Biological Psychiatry, 31*(4), 901–905.

Downey, G. (2014). 'Habitus in extremis': From embodied culture to bio-cultural development. *Body & Society, 20*(2), 113–117.

Einstein, A. (1949). *The world as i see it* (A. Harris, Trans.). Covici Friede, 1934, 242.

Engelberg, H. (1992). Low serum cholesterol and suicide. *The Lancet, 339*(8795), 727–729.

Eriksen, B., Bjørkly, S., Lockertsen, Ø., Færden, A., & Roaldset, J. (2017). Low cholesterol level as a risk marker of inpatient and post-discharge violence in acute psychiatry—A prospective study with a focus on gender differences. *Psychiatry Research, 255*, 1–7.

Evans, J. (2007). On the resolution of conflict in dual-process theories of reasoning. *Thinking and Reasoning, 13*, 321–329.

Evans, J. (2008). Dual-processing accounts of reasoning, judgment, and social cognition. *Annual Review of Psychology, 59*, 255–78.

Fortunato, S., Bergstrom, C. T., Börner, K., Evans, J. A., Helbing, D., Milojević, S., Petersen, A. M., Radicchi, F., Sinatra, R., Uzzi, B., Vespignani, A., Waltman, L., Wang, D., & Barabási, A. L. (2018). Science of science. *Science, 359*(6379), eaao0185.

Geertz, A. (2010). Brain, body and culture: A biocultural theory of religion. *Method & Theory in the Study of Religion, 22*(4), 304–321.

Giddens, A. (1984). *The constitution of society: Outline of the theory of structuration*. University of California Press.

Gilbert, S. (2003). The reactive genome. In G. B. Müller & S. A. Newman (Eds.), *Origination of organismal form: Beyond the gene in developmental and evolutionary biology* (pp. 87–101). MIT Press.

Hirsch-Hoefler, S., Canetti, D., Rapaport, C., & Hobfoll, S. (2016). Conflict will harden your heart: Exposure to violence, psychological distress, and peace barriers in Israel and Palestine. *British Journal of Political Science, 46*(4), 845–859.

Ignatow, G. (2012). Mauss's lectures to psychologists: A case for holistic sociology. *Journal of Classical Sociology, 12*(1), 3–21.

Ignatow, G. (2019). Cognitive sociology and French psychological sociology. In W. H. Brekhus & G. Ignatow (Eds.), *The Oxford handbook of cognitive sociology* (pp. 137–152). Oxford Academic.

Ignatow, G. (2021, December). Cognitive sociology after relational biology. *Sociological Forum, 36*, 1253–1270.

Ignatow, G. (2022). The microbiome-gut-brain and social behavior. *Journal for the Theory of Social Behaviour, 52*(1), 164–182.

Kaplan, J., Klein, K., & Manuck, S. (1997). Cholesterol meets Darwin: Public health and evolutionary implications of the cholesterol-serotonin hypothesis. *Evolutionary Anthropology: Issues, News, and Reviews, 6*(1), 28–37.

Keren, G., & Schul, Y. (2009). Two is not always better than one: A critical evaluation of two-system theories. *Perspectives on Psychological Science, 4*(6), 533–550.

Konvalinka, I., Xygalatas, D., Bulbulia, J., Schjødt, U., Jegindø, E. M., Wallot, S., & Roepstorff, A. (2011). Synchronized arousal between performers and related spectators in a fire-walking ritual. *Proceedings of the National Academy of Sciences, 10820*, 8514–8519.

Kruglanski, A., Erbs, H.-P., Pierro, A., Mannetti, L., & Chun, W. Y. (2006). On parametric continuities in the world of binary either ors. *Psychological Inquiry, 17*(3), 153–165.

Leschziner, V. (2019). Dual-process models in sociology. In W. H. Brekhus & G. Ignatow (Eds.), *The Oxford handbook of cognitive sociology* (p. 169). Oxford Academic.

Leschziner, V., & Green, A. (2013). Thinking about food and sex: Deliberate cognition in the routine practices of a field. *Sociological Theory, 31*(2), 116–144.

Lieberman, M. D. (2003). Reflective and reflexive judgment processes: A social cognitive neuroscience approach. In J. P. Forgas, K. D. Williams, & W. Von Hippel (Eds.), *Social judgments: implicit and explicit processes* (pp. 44–67). Cambridge University Press.

Lieberman, M. D., Jarcho, J. M., & AJay B. Satpute. (2004). Evidence-Based and Intuition-Based Self-knowledge: An fMRI Study. *Journal of Personality and Social Psychology, 87*, 421–435.

Lindquist, K. A., Wager, T. D., Kober, H., Bliss-Moreau, E., & Barrett, L. F. (2012). The brain basis of emotion: A meta-analytic review. *Behavioral and Brain Sciences, 35*(3), 121–143.

Lizardo, O., Mowry, R., Sepulvado, B., Stoltz, D. S., Taylor, M. A., Van Ness, J., & Wood, M. (2016). What are dual process models? Implications for cultural analysis in sociology. *Sociological Theory, 34*(4), 287–310.

Maryanski, A., Machalek, R., & Turner, J. H. (2015). *Handbook on evolution and society: Toward an evolutionary social science*. Routledge.

McDonnell, T. (2014). Drawing out culture: Productive methods to measure cognition and resonance. *Theory and Society, 43*(3–4), 247–274.

McGregor, I., Prentice, M., & Nash, K. (2013). Anxious uncertainty and reactive approach motivation (RAM) for religious, idealistic, and lifestyle extremes. *Journal of Social Issues, 69*(3), 537–563.

Melnikoff, D., & Bargh, J. (2018). The mythical number two. *Trends in Cognitive Sciences, 22*(4), 280–293.

Meloni, M. (2014). How biology became social, and what it means for social theory. *The Sociological Review, 62*(3), 593–614.

Miles, A., Charron-Chénier, R., & Schleifer, C. (2019). Measuring automatic cognition: Advancing dual-process research in sociology. *American Sociological Review, 84*(2), 308–333.

Milkowski, M. (2013). *Explaining the computational mind.* MIT Press.

Moore, R. (2017). Fast or slow: Sociological implications of measuring dual-process cognition. *Sociological Science, 4*, 196–223.

Pessoa, L., & Adolphs, R. (2010). Emotion processing and the amygdala: From a 'low road' to 'many roads' of evaluating biological significance. *Nature Reviews Neuroscience, 11*(11), 773–782.

Powell, A., & Dupré, J. (2009). From molecules to systems: The importance of looking both ways. *Studies in History and Philosophy of Science Part C: Studies in History and Philosophy of Biological and Biomedical Sciences, 40*(1), 54–64.

Tavory, I. (2011). The question of moral action: A formalist position. *Sociological Theory, 29*(4), 272–293.

Tomson-Johanson, K., & Harro, J. (2018). Low cholesterol, impulsivity and violence revisited. *Current Opinion in Endocrinology & Diabetes and Obesity, 25*(2), 103–107.

Turner, J., & Turner, S. (1990). *The impossible science: An institutional analysis* (p. 224). Sage.

Turner, J. H. (2006). American sociology in chaos: Differentiation without integration. *The American Sociologist, 37*(2), 15–29.

Turner, J. H., & Boyns, D. E. (2001). The return of grand theory. In J. H. Turner (Ed.), *Handbook of sociological theory* (pp. 353–378). Springer.

Vaisey, S. (2009). Motivation and justification: A dual-process model of culture in action. *American Journal of Sociology, 114*(6), 1675–1715.

Vaisey, S., & Frye, M. (2019). The old one-two. In W. H. Brekhus & G. Ignatow (Eds.), *The Oxford handbook of cognitive sociology* (p. 101). Oxford Academic.

Van Bavel, J. J., Jenny Xiao, Y., & Cunningham, W. A. (2012). Evaluation is a dynamic process: moving beyond dual system models. *Social and Personality Psychology Compass, 6*(6), 438–454.

Van de Vijver, G. (2009). No genetics without epigenetics? No biology without systems biology? On the meaning of a relational viewpoint in a complex account of living systems. *Annals of the New York Academy of Sciences, 1178*(1), 305–317.

van Prooijen, J. W., & Krouwel, A. P. (2019). Psychological features of extreme political ideologies. *Current Directions in Psychological Science, 28*(2), 159–163.

van Prooijen, J. W., Krouwel, A. P., Boiten, M., & Eendebak, L. (2015). Fear among the extremes: How political ideology predicts negative emotions and outgroup derogation. *Personality and Social Psychology Bulletin, 41*(4), 485–497.

Vila-Henninger, L. (2015). Toward defining the causal role of consciousness: Using models of memory and moral judgment from cognitive neuroscience to expand the sociological dual-process model. *Journal for the Theory of Social Behaviour, 45*(2), 238–260.

Wacquant, L. (2006). Pierre Bourdieu. In R. Stones (Ed.), *Key contemporary thinkers*. Macmillan.

Wacquant, L. (2014). Putting habitus in its place: Rejoinder to the symposium. *Body & Society, 20*(2), 118–139.

Walsh, A. (2017). *Biosociology: Bridging the biology-sociology divide*. Routledge.

Webber, D., Babush, M., Schori-Eyal, N., Vazeou-Nieuwenhuis, A., Hettiarachchi, M., Bélanger, J. J., Moyano, M., Trujillo, H. M., Gunaratna, R., Kruglanski, A., & Gelfand, M. J. (2018). The road to extremism: Field and experimental evidence that significance loss-induced need for closure fosters radicalization. *Journal of Personality and Social Psychology, 114*(2), 270.

Wellman, J., Corcoran, K., & Stockly-Meyerdirk, K. (2014). 'God is like a drug . . . ': Explaining interaction ritual chains in American megachurches. *Sociological Forum, 29*(3), 650–672.

Xygalatas, D., Konvalinka, I., Bulbulia, J., & Roepstorff, A. (2011). Quantifying collective effervescence: Heart-rate dynamics at a fire-walking ritual. *Communicative & Integrative Biology, 4*(6), 735–738.
Zmigrod, L., Rentfrow, P., & Robbins, T. (2020). The partisan mind: Is extreme political partisanship related to cognitive inflexibility? *Journal of Experimental Psychology: General, 149*(3), 407.

Gabe Ignatow is Professor of Sociology and Digital Social Science Program Coordinator at the University of North Texas. His recent papers on social science-life science interdisciplinarity have been published in Sociological Forum and the Journal for the Theory of Social Behaviour.

Chapter 13
From Territorial Behavior to Imagined Territories and Back: Exploring Spatiality in International Relations

Jun J. Nohara

Abstract Why are people willing to fight for territories they have never seen? What are territories and why do they matter in international relations? This chapter seeks to address some of these questions by confronting several perspectives on the subject. In particular, the recent introduction of evolutionary theories in the field of international relations (IR) has provoked a debate on the role of hardwired cognitive mechanisms behind international dynamics. These theories appear to directly challenge some of the tenets of well-established research programs in IR, notably constructivism and its emphasis on social constructs. This chapter argues that while these evolutionary theories may redefine our conception of territories, their current shortcomings—regarding their transposition to the territories of complex political units such as modern states, among other things—undermine their argument and highlight the continued need for cultural interpretations of territories in IR. In order to do so, this chapter reviews the way various research programs in IR have conceptualized the territory into their theoretical framework, how evolutionary theories have progressively come to interact with IR theories—notably through the issue of war and territoriality—and how evolutionary theories and seemingly opposite propositions such as constructivism IR theories may offer conflicting yet complementary approaches to our understanding of territories in international relations.

Keywords International relations · Evolutionary theories · Territory · States · Human territoriality · Conflict

In a discipline concerned with relations between various entities across defined space and boundaries, the notion of territory—or more generally, spatiality—ought to figure prominently in studies and investigations. After all, the delimitation of territories, just like the identification of one's own kin and social group, appears

J. J. Nohara (✉)
Faculty of International Relations, Kyoto Sangyo University, Kyoto, Japan
e-mail: j.j.nohara@cc.kyoto-su.ac.jp

J. Pérez-Jara, Í. Ongay (eds.), *Beyond Nature and Nurture*, Synthese Library 497,
https://doi.org/10.1007/978-3-031-87715-5_13

to reinforce the distinction between 'us' and 'them' and to act as a powerful understructure that supports the very existence of international relations (Neuman, 1996; Schmitt, 2007).[1] It is no wonder that many etiological myths of ancient polities place a great emphasis on the sacred bonding acts between man and the land; by plowing a square furrow around the Palatine Hill to demarcate the location of the walls of Rome, or using very thin strips of an ox hide to encircle the hill of Byrsa and mark the emplacement of Carthage, Romulus and Dido gave birth to a tangible expression of the political body. Eventually, the rise of the modern international system enshrined the territory—alongside population and government—within the holy trinity of the sovereign state, which, in the eyes of modern IR theory, marks the starting point of modern international relations (Herz, 1957; Philpott, 1995; Ruggie, 1993).

Yet, in the field of IR, the territory does not quite get the attention it deserves. For long, the term eluded serious scrutiny (Jacobson, 1997). Worse, the way many scholars define territories often preserves and feeds on their own narrow conceptualization of the state and on how they understand the nature of its relations with other political units. Nevertheless, a small but active group of scholars who reflect on the implications of evolutionary theories in international relations have recently presented fresh—but controversial—ideas that could connect with the way we understand and value territory. This presents IR scholars with an opportunity to think anew on a better theoretical integration of spatiality in current frameworks. The first section will introduce the way various research programs in IR understand the significance of territories. The second section will review the overlapping research interests in evolutionary theories and international relations, and how the two fields have come to engage with one another. The third and final section will discuss the recent attempts made by evolutionary theorists to employ genetic variables in explaining territorial disputes.

13.1 Territory and International Relations

In conceptualizing territory, none remain more minimalist—or reductionist—than Neorealism, who sees states as "black boxes" or "billiard balls," hermetic and impenetrable objects with nothing to distinguish them but their given attributes of power. The territory becomes at once an embodiment of the international structure and one of the many other materialistic manifestations of state power—the territory shelters a population, provides resources, etc. (Finel, 2001; Fordham & Asal, 2007; Kapstein 1995; Waltz, 1959). Other brands of Realism such as Classical or Neoclassical theories converge on their interpretation of the national territory as an

[1] In this text, the words 'international relations' (lowercase) designate international relations as phenomena and a research object, whereas the capitalized 'International Relations' or 'IR' refer to the academic discipline itself.

essential characteristic of the state, at once an unequivocal condition to its survival and the manifestation of its vitality. So obvious has the notion become that, by force of habit, Realists tend to gloss over its importance. To Realism, the territory is a given; if a state were to lose its territory, it would cease to exist. This is the greatest calamity that can befall on a state. As with the Jews chased out of Judaea, or Aeneas and the Trojans escaping the burning of their city, diasporas may survive but the state would wither, until new territories are found or conquered. Defending territorial integrity therefore becomes one of the most important missions of the state, and any form of territorial encroachment should be met with no small dose of alarm and determination (Walt, 1991; Waltz, 1979).

The national territory is also source of power (Morgenthau, 1967). From it, the state extracts resources; on it, food is grown, raw materials processed, and goods manufactured. It helps define—although never entirely determines—the nation's demography and the strength of its population. As such, Realists recognize that territories are sometimes coveted by envious neighbors and that the perspective of enjoying the spoils of conquest invites international conflicts (Huth, 1996; Liberman, 1993).

Aside from the territory, Realist theories tend to approach the more general question of spatiality through two other angles—geography and location—that, while retaining some distinctive characteristics, remain fundamentally intertwined with their view of territory. Realists often see geography, defined as the natural arrangement of the earth's physical features, as an additional variable in the calculation of the balance of power. Some, and particularly when discussed as part of a state's territory, view it as an integral part of a state's power; whether the state may benefit from favorable positions or be at a strategic disadvantage depends on the territory's geographic parameters (Morgenthau, 1967). Others look at the more general properties of these physical features, and the influence they may or may not have on international relations. For instance, Mearsheimer observes how large bodies of water such as seas and oceans have a "stopping power," which hinders power projection and prevents regional powers from becoming world hegemons (Mearsheimer, 2001). In deciding whether or not offensive capabilities are generally superior to defensive capabilities at any given point in time, Jervis also takes into account topographic features such as rivers or mountains (Jervis, 1978). Finally, Realists acknowledge that location—the place where the state *is* in relation to other states—also plays an important role. Close proximity generates more interaction, and also correlates well with the probability that a conflict will occur; the closer states are, the more likely they are to go to war against each other (Walt, 1985).

The idea that proximity begets hostility is widely shared outside Realism. Constructivists such as Buzan and Waever, who coined the term Regional Security Complex, argue that security issues should be analyzed through a regional perspective because geographical proximity generates more security interaction (Buzan & Waever, 2003). On the other hand, the concept of International Society advocated by the English School or the idea of Security Community developed by Karl Deutsche stress that states being near one another could also be a factor of cohesion that fosters common values and a greater sense of cooperation (Bull, 1977; Deutsche et al., 1957).

Because liberals ground their premises in the acceptance of a wider range of meaningful actors on the international scene, their interpretation of the territory is far more organic—as opposed to the Realists' monolithic and mechanical conception (Agnew, 1994; Elden, 2013). To begin with, territories never really ceased to be shifting spatial constructions over which multiple political units exert their influence, in one form or another (Badie, 2014). The dominion of supranational entities, from the feudal Holy Roman Empire to the confederation-like European Union sometimes complements, sometimes competes with the sovereignty of nation-states (Herz, 1957; Ruggie, 1993). In turn, the flow of people, ideas, and money on intranational and transnational levels redefine territories according to their image or their needs. State sovereignty may also be undermined by political units of a different nature, denying the state control over its territory. Regionalized independence movements, transnational criminality, or terrorist groups fall into this category (Elden, 2009). At times, even international institutions and agreements may have a role to play in the management of territories, as in the resolve of territorial disputes, the protection of ecosystems, or the exploitation of resources (Cullis-Suzuki & Pauly, 2010; Simmons, 2002). And by promoting legal norms that prohibit the use of force for "national aggrandizement or territorial gain," international law prevents law-abiding states from initiating wars out of pure power politics (Kocs, 1994).

Additionally, the Realist position that boundaries are impervious lines delimitating strictly one political unit from another is received by Liberals with skepticism at best. They tend to downplay the significance of frontiers by showing that the movement of people and things across borders directly contribute to complex interdependences—those many connections that tie countries to one another in ways that undermine traditional power relations and territorial logic (Nye & Keohane, 1977). While Liberals share some assumptions regarding states and space, the fact that global relations are transnational as much as they are international means that their conception of space is far less rigid, where territories can become a place of interaction between state actors and non-state actors alike and not simply 'containers' of societies (Agnew, 1994).

The continued phenomenon of globalization, the fall of the Soviet Union, and the progress made in the construction of the European Union seemingly supported Liberal claims while also briefly renewing scholarly interest for the territory (Elden, 2013; Painter, 2010; Postel-Vinay, 1998). In spite of this, IR continuously falls short of offering a vigorous debate on the matter, as, once outside the familiar and comforting realm of 'territorial disputes,' most contributions still hail from other fields such as geography, anthropology, sociology, or philosophy (Elden, 2009). This state of affairs would not be particularly distressing if it were not lowering the visibility of—and the curiosity for—a concept that acts as a cornerstone of our understanding of international relations.

In parallel, in another remote corner of IR, there has been a small but noticeable effort to include perspectives drawn from natural sciences—and notably evolutionary theories—into the field. Drawing bridges between evolutionary studies and international phenomena, this new set of theories sought to buttress, modify, or disprove prior claims. Initially concerned with the biological drives of human nature and its implication for state behavior on the international scene, it largely ignored the relationship between the evolutionary perspectives on the 'territory' and international dynamics at first. In other social sciences, the controversial sociobiology had already begun problematizing the issue of territoriality since the 1960s, albeit from a social and political—rather than truly IR—viewpoint (Jacobson, 1997). But scholars have recently introduced evolutionary psychology in order to explain international territorial disputes, which may lead to a much-needed reassessment of our understanding of territory in IR in future research.

13.2 International Relations and Evolutionary Theory—Where Are we at?

IR is the quintessential multi-disciplinarian social science, as evidenced by the discipline's hundred years' history (Schmidt, 1994). On top of addressing an incredibly diverse variety of research objects, grafting theories from other fields has often been the norm rather than the exception. International law played a key role in early liberalism, and neorealists heavily drew from economic theories (Kirshner, 2015). Constructivism builds upon on the works of prominent sociologists, while the English School is defined by its methodological eclectism (Copeland, 2003; Makinda, 2000). IR scholars have also brought notions from more distant fields such as psychology (Jervis, 1976). Each of these contributions has helped IR grow, and its ability to internalize these debates continues to define the discipline today.

Although scholarly interest for biological determinants of state behavior on the international scene has remained marginal, it is certainly not new. In fact, a certain number of scholars and public intellectuals attempted to explain the bellicose dynamics of international relations using biological variables 50 years *before* the discipline of IR was even formally established.[2] This was especially true in Germany, where the social applications of Darwin's natural selection theory were readily embraced. Soon after the translation of *On the Origin of Species*, many German scientists enthusiastically endorsed militarism, imperialism, and racial competition on evolutionary grounds (Weikart, 1993). Offering their own synthetic interpretations of Darwinian natural selection and Malthusian ideas of population pressure, prominent figures such as Ernst Haeckel, paleontologist Friedrich Rolle, or biologist Ziegler explained how "according to Darwin, war has constantly been of the greatest importance for the general progress of the human race." Whether or not these individuals, later lumped together under the label of Social

[2] The creation of a chair of International Relations at the Aberystwyth University (formerly University of Wales) in 1919 is usually remembered as the founding act of international relations as an academic discipline. It goes without saying that the topic elicited a large number of intellectual enquiries long before.

Darwinists, effectively influenced political leaders of that time or merely colored their rhetoric is debatable. But the idea that war was driven by biological—and therefore irresistible—necessities, and that the origin of great power competition rested in the principles of natural selection and the resulting 'survival of the fittest' clearly permeated the atmosphere of pre-WWI German society (Crook, 1994; Hawkins, 1997; Williams, 2000).[3] In order to accommodate population pressure and the survival of the German 'race,' territorial expansion became eminently salient, leading to fantasized reinterpretations of the *Ostsiedlung* through the slogan *Drang Nach Osten*, and ultimately culminating with the concept of *Lebensraum*.

Already weakened by mounting criticisms from within the academic society, its alleged association with Nazi and nationalist ideas proved nearly fatal to Social Darwinism (Claeys, 2000). In IR, early classical realist scholars of the post-war era such as Reinhold Niebuhr or Hans Morgenthau prudently distanced themselves from it, preferring to ground the urge of men to dominate one another—the *animus dominandi*—in theological or metaphysical factors rather than biological ones. This is not to say that the lexicon of Social Darwinism completely disappeared however, and some of its remnants may still be found in mainstream research programs. Indeed, one cannot fail but notice the persistence of the state's 'survival' in both classical realist and neorealist theories (Morgenthau, 1962; Waltz, 1979; Wight, 1995). Considering how structural the concept has become in every of the discipline's paradigm, it is surprising that the concept has long eluded any serious scrutiny from IR scholars (Paul, 1999; Wendt, 1992).

Aside from this epistemologically dubious tendency to anthropomorphize states, few other bridges were built between the two disciplines, and fewer yet durably influenced IR. E.O. Wilson's popularization of 'sociobiology'—the new face of Social Darwinism, according to its critics—in the late 1970s did not significantly alter this situation (Laitinen & Maude, 1986). In *Sociobiology: The New Synthesis* published in 1975, Wilson sought to apply principles from biology and evolutionary theory to social organization, sparking an intense controversy but also renewing scholarly interests for biology in other disciplines (Wilson, 1975). In the case of IR however, and with rare exceptions, such cross-disciplinarian investigations remained peripheral. These mainly focused on the biological factors behind aggression and cooperation applied to international relations. Of the two topics, aggression and war initially proved the more prolific. In 1953, Raymond Dart hypothesized that interpersonal aggression, violence, and war had driven human evolution, and that these instincts remained hardwired in humans (Dart, 1953; Milam, 2019). Playwright and anthropologist Robert Ardrey built on Dart's works in *African Genesis* (Ardrey, 1961), cementing the image of the congenital violence of the 'killer ape.' A few years later, Ardrey also developed the idea that man is biologically driven to expand and defend his territory (Ardrey, 1966). In the same period, ethologist

[3] The expression 'survival of the fittest' is not from Darwin himself, but from Herbert Spencer. See Schuurman, P. (2016). Herbert Spencer and the Paradox of War. *Intellectual History Review, 26*(4), 519–535.

Konrad Lorenz investigated the biological reasons behind aggression, paving the way for an evolutionary approach to war that only followed years later (Fry, 2013; Haas & Piscitelli, 2013; Lorenz, 1966).

IR scholars knew of this literature, which is hardly surprising since some of these works were widely distributed in circles beyond academia (Morgenthau, 1970). It is then puzzling that, in spite of the centrality of war in their own discipline, these scholars, and especially Classical Realists, failed to engage on any significant level with these findings. In the decades that followed, debates on war and evolution periodically resurfaced but always outside the scope of IR. For instance, in the late 1980s, a series of works revisited the biological drives behind war. In a 1987 issue of the *International Studies Quarterly*, R. Paul Shaw and Yuwa Wong coauthored two articles in which they developed a model of ethnic grouping with a propensity for war based on kin selection and inclusive fitness (Shaw & Wong, 1987a; Shaw & Wong, 1987b). The two authors later expanded on their initial arguments in a book entitled *Genetic Seeds of Warfare: Evolution, Nationalism, and Patriotism* (Shaw & Wong, 1989). In response to Shaw and Wong's initial articles, Joshua Goldstein and Philip Kitcher argued that the ideas developed by the authors brought nothing new to the debate and that they equally failed to address the criticism formulated against the existing evolutionary explanations of war, namely a propensity to interpret violence as an innate feature of human nature, and a tendency to fall for confirmation biases, among other things (Goldstein, 1987; Kitcher, 1987). The following year, a collective of researchers produced a new series of essays on the evolutionary explanations behind human nature and aggression (Klama, 1989).

On the other end of the spectrum, debates on the place of cooperation in natural selection had begun in earnest following the publication of Darwin's works. The theories of Peter Kropotkin on the pragmatic benefits of cooperation in evolution, which the Russian naturalist had developed at the end of the nineteenth century, provided the starting point for future research programs (Kropotkin, 2009). In the 1930s, Haldane and Fisher helped advance understanding on puzzling traits such as altruistic behaviors that benefited groups at the expense of individuals (Fisher, 1930; Haldane, 1932). William Hamilton's seminal work on 'inclusive fitness' and 'kin selection' scientifically demonstrated Haldane and Fisher's earlier observation (Hamilton, 1964). Later, Hamilton worked with political scientist Robert Axelrod on *Evolution of Cooperation*, which helped confirm the hypothesis of Kropotkin (Axelrod & Hamilton, 1981). Axelrod's application of evolutionary principles and popular game theories to the problem of cooperation was met with considerable success but failed to convincingly engage IR scholars, some of whom describing the findings of limited use to real-world international relations (Lebow, 2013). In the years that followed, work on cooperation, evolution, and international relations continued intermittently, although most did not receive the same degree of appraisal as Axelrod's (Maxwell, 1990).

Since the turn of the millennia, scholarly attempts to bring evolutionary theories into the realm of IR have become more numerous, more structured, but ultimately not necessarily more convincing than preceding ones. For instance, in an article published in 2000, Bradley A. Thayer proposed a reevaluation and a consolidation of classical realism by refocusing the level of analysis on the first image—the individual political leaders (Thayer, 2000). Evolutionary psychology, he contended, could explain international relations because it could uncover the natural instincts hardwired in the brain of decision-makers. He later expanded the scope of his research in a book (Thayer, 2004), in which he argues that "humans wage war for reasons predicted by evolutionary theory—to gain and protect vital resources but also for the physically and emotionally stimulating effects of combat." A few years later, Lopez, McDermott, and Petersen continued to explore the possibilities offered by advances in evolutionary psychology in order to explain 'coalitional behavior'—especially how within-group and between-group dynamics coordinate, cooperate, or compete in a given environment or faced with a particular challenge (Lopez et al., 2011). In parallel, Dominic Johnson wrote extensively on the various ways evolutionary psychology affects conflict and cooperation in international relations, ranging from the role of religion and ancient fears of supernatural punishment as a cement for cooperation to that of psychological biases and overconfidence in the probability of conflict (Johnson, 2016; Johnson & Tierney 2011; Lopez & Johnson, 2017).

The attempts are not a concerted effort to breach the walls of the field however, and their approach to IR sometimes differ widely—but not always contradict each other—in how they deal with some of the most fundamental aspects of the discipline. The level of analysis is a case in point. In IR, scholars often have to determine the level from which they will investigate their research object—the individual, the state, or the system (Waltz, 1959).[4] This distinction is not unlike the multilevel selection theory (MLS) proposed by David Sloan Wilson and Elliott Sober, which distinguishes various levels at which natural selection operates, ranging from genes to cells, organism, and group (Wilson & Sober, 1994). Likewise, some of the scholars applying evolutionary concepts to IR prioritized psychological mechanics on the level of the 'organism'—which would correspond to Kenneth Waltz' 'first image,' or the individual level of analysis—while others drew parallels between group selection and the system level of analysis, and a few even tried to juggle with both (Thayer, 2000; Thayer, 2004; Waltz, 1959).

But each of these attempts was met with vigorous pushback. Duncan Bell and Paul MacDonald noted that the methodology of sociobiology remained scientifically controversial, also adding that, even if it were to overcome its limits, a revised Classical Realist theory based on sociobiological principled would still fail to inform us on whether humans are altruistic or selfish (Bell & MacDonald, 2001). Duncan Bell later expanded on his initial argument (Bell, 2006), concluding that evolutionary psychology was "politically indeterminate." Looking at Thayer's focus on the individual level, Richard Lebow also noted some important inconsistencies,

[4] Those three levels are also sometimes referred to as the first, second, and third images. The choice of a level of analysis often reveals the epistemic preferences of the researcher for one or the other research programs that are Realism, Liberalism, etc.

criticizing how natural selection was hardly applicable to political leaders since "political skill is not inherited" from one leader to the next (Lebow, 2013).

Employing evolutionary principles to inform us on the causes of war and other forms of organized social violence—a central theme in IR—has also exposed some of the most important methodological flaws found in evolutionary theories. For instance, Lopez argued that war has represented a relatively constant ancestral challenge to man, and that therefore, one could reasonably assume that human brains have developed "information-process systems" that hardwired psychological response to war, or the perspective of war (Lopez et al., 2011). One of the most pressing issues that immediately arises is that of the temporality of evolution. In other words, what exactly is the pace of evolution? And how far back, or how recently do we need to look at in order to find traces of psychological traits that have evolved out of necessity? Donald Symons, one of the founders of evolutionary psychology (Symons, 1992), argued that "the majority of human psychological mechanisms are adapted to reproductive problems frequently encountered in Pleistocene environments." The Pleistocene, a geological epoch of a few million years, ended around 11,000 years ago. This would correspond to the oldest evidence to date of mass human violence of a group towards another, uncovered by archeologists at Jebel Sahaba, in the Nile valley (Kelly, 2000). Unsurprisingly, the evidence does not square well with the claims found in the arguments of Lopez or Thayer since it would suggest that most of our psychological mechanisms predate the appearance of systematic intergroup violence.

While it is impossible to assert that intergroup violence did not frequently take place before this date, anthropologist Raymond Kelly confirms that nomadic lifestyle and the low population density of early hunter-gatherer societies meant that the risks of conflict remained low during the Paleolithic and well after the appearance of *Homo sapiens* (Kelly, 2000). In fact, it was the rise of sedentism, the Neolithic Revolution, and the consequent upsurge in population density from around 5000 BC that provoked the systemic occurrence of war (Lee, 2015). Therefore, when contrasting the 'recent' rise of systemic warfare with the fact that *Homo sapiens* is thought to have emerged around 300,000 years ago, one can wonder whether war is truly 'ancestral,' and whether we have been doing war long enough for it to have influenced our psychological mechanisms to such profound degrees (Lopez, 2016). By the same token, it is equally dubious that our relatively recent practice of statehood—large and complex social groupings—has led to any form of psychological evolution in such a short time span.

Indubitably, a relevant and scientifically sound use of evolutionary theory in international relations remains marred by epistemic issues. More than a century ago, Herbert Spencer himself had already recognized that any attempt to apply biological principles to complex social dynamics faced considerable obstacles, and these obstacles continue to challenge proponents of sociobiology in IR. Although some sociobiologists contend that the methodological flaws are fundamentally not too different from those found in other social sciences, many other scholars are disgruntled by what they see as a travesty, a theory that attempts to mobilize the legitimacy of natural sciences when its epistemic foundations are not more—but also perhaps no less—solid than other social and human sciences (Gannon, 2002).

13.3 Territory and Evolutionary Theory

In comparison to IR, evolutionary theories embraced the integration of spatial considerations more fully and decisively. Above all else, the environment plays a critical role in the evolution of species. The biophysical environment refers to the sum of biotic and abiotic factors present in the surrounding of an organism or a population that, taken together, make up a more or less complex ensemble called an ecosystem. These factors significantly influence the survival of organisms living in its midst. For instance, *Homo sapiens* developed most of its cognitive mechanisms in response to reproductive challenges it encountered in Pleistocene environments (Symons, 1992).

Understandably, the notion of environment figures prominently in the evolutionary psychology literature, and it has also attracted the attention on neighboring terms that are more specifically—but never exclusively—focused on spatiality, such as 'habitat,' 'milieu,' or 'territory.' Although some tenuous similarities may exist between the concept of ecosystem and that of the international system, neither the environment nor the habitat—whether or not taken in the context of evolution—were found to have an especially suitable use in IR, which is perhaps the result of their complex and multilayered nature. On the other hand, the notion of territory—and some of its derivatives, including 'territoriality' and 'territorial behavior'—has garnered attention from scholars applying evolutionary principles to IR. This may come as no surprise. After all, the term benefits from some semantic proximity across various disciplines, from political science and IR to anthropology and ethology. Here again, sociobiologists helped create bridges between those various fields, as they sought to explain the evolutionary drives behind displays of territorial behavior in modern societies.

But where most sociobiologists limited their analysis of the role of territory and territorial behavior to the level of individuals or small groups of individuals, some scholars attempted to replicate the results of these findings to the interaction between units that are among one of the highest levels of social groupings in existence. The appeal is understandable, as territorial disputes remain the single most powerful variable explaining wars (Huth, 1996).

In a study on territorial conflicts, Dominic Johnson and Monica Toft argued that the cognitive mechanism known as territorial behavior—or territoriality—is "soft-wired" in *Homo sapiens* and helps explain why, under certain prevailing conditions, territorial aggression takes place. The territory, they contend, is a "proxy through which organisms secure access to key resources and protect them from competitors." Based on the premise that territoriality is subject to convergent evolution, the authors look at some of the territorial behaviors displayed by other territorial animals such as birds, to which they superimpose models of "ecological rationality" and evolutionary game theory. While variations in population, resource density on said territory, and a great number of other factors also lead to variations in territorial behaviors, Johnson and Toft show that some common patterns are found across territorial species. Once transposed to international relations, they identify three mechanisms that help explain why violent aggression may occur in territorial disputes. One is the difference in raw military power between contestants. A second mechanism is the "value asymmetry" that states attach to a given territory, and the third is the "economic defensibility"—a simple cost-benefit calculation—of said territory (Johnson & Toft, 2013).

Their research elicited a response from Raymond Kuo, who criticized the transposition of behavioral evolutions across species, and their problematic use of levels of analysis, among other things (Kuo et al., 2015). While Johnson and Toft offered a point-by-point response to the critiques formulated by Kuo, their remark on the issue of levels of analysis appears to leave the problem unresolved. Indeed, if we accept that individual beings matter in explaining politics—and therefore the actions of the state—then it would be reasonable to assert that whatever influences those individuals—i.e., genes—matters too and would also influence politics. However, in the case of large and complex social groupings such as states, knowing precisely what genes or evolutionary mechanism may be at play in explaining certain types of behavior remains extremely difficult, and the case of human territoriality and international conflict is a case in point.

When it comes to the defense of a territory, previous research has shown that single individuals display territorial behavior and may promptly react to the invasive presence of a stranger in their personal spheres or spaces to which they attach a sense of ownership.[5] Evolutionary psychologists would also argue that kin selection explains why such reactions might extend to similar occurrences that affect other individuals belonging to the same social groups such as families and small communities—even if it comes at great personal expense. So far, the relatively limited scope of human territoriality should come as no surprise. After all, early human tribes rarely exceeded 150–200 individuals, which supposedly corresponds to our brains' capacity to compute and handle relationships (Dunbar, 1998). Among small-sized groups, it is also likely that most group members have firsthand knowledge of the territory on which they claim some sort of ownership. In addition, some anthropologists imply that location as a context to shared social interactions within a group plays a key role in defining non-genetic kinship, which in turn sheds a new light on the determinants of cooperative and altruistic behavior (Holland, 2012).

But that this same cognitive mechanism explains why modern societies of millions of individuals possess the faculty to mobilize citizens across the entire country and make them fight for a given territory that most have often never been to is not readily demonstrated by Johnson and Toft. Admittedly, there is ample empirical evidence suggesting that individuals display signs of territorial behavior regarding territories of which they have no firsthand knowledge and/or that are of concern to non-kin members. For instance, a recent governmental survey conducted in Japan found that 63.6% of respondents remain 'interested' in the territorial

[5] Animals use a variety of strategies in the defense of their territory, including withdrawal or confrontation.

dispute opposing South Korea and Japan over a group of small islets (Government of Japan, 2023). Taken out of its social context, the high percentage is rather stunning, considering the fact that the islets are barely habitable, the surrounding waters hold few valuable natural resources, and they have no obvious strategic interest. On top of these telltale signs of irredentism—a type of territoriality that appears to be peculiar to humans—very few Japanese have actually had the opportunity to visit the islets in person, since they have been under Korean control for over 60 years. To the overwhelming majority of the Japanese population, this territory ceases to be a tangible reality and falls into the category of the imagined—much like Anderson's 'imagined communities' and Said's 'imagined geographies.'

Simply put, such conflict has very little chances of recreating to the same degree the cognitive conditions conducive to the type of reactions defined by territorial behavior in most individual—whether political leader or ordinary citizen—because it has so little to do with the premises under which territoriality operates, *unless* the dispute was socially construed, socially accepted, and politically sustained as such. And while social contextualization arguably occurs on the individual and kinship level as well, the sheer amplitude of aggregated social behavior involved in international relations may obfuscate genetic variables to the point that they truly become indeterminate without a great deal of social and cultural contextualization, and lose much of their explanatory value (Bell, 2006). In any case, the territory transcends its simplistic definition as a "proxy through which organisms secure access to key resources and protect them from competitors," and becomes a powerful social construct, capable of conjuring up striking imageries that stimulate the national psyche and provoking immense emotional responses. Evolutionary theory would rather do well to focus on the genetic determinants behind leadership, obedience, coalitional psychology, cooperation, or altruistic behavior extended to non-kin to elucidate the behavior of political units in international territorial disputes.

Last, Johnson and Toft take the genetic origin of territorial behavior in humans for granted. This is problematic because the nature of human territoriality remains controversial. Although the subject is well-documented (Malmberg, 1980), scholars continue to disagree on whether human territoriality is 'homologous' to animal territoriality—that it is driven by 'instinctual' imperatives—or merely 'analogous'—similar in patterns but fundamentally different in nature (Gold, 1982; Sack, 1983). Evolutionary theorists point to the near-universal occurrence of human territoriality to support their position, while those who argue that it is 'nurtured' contend that this same frequent occurrence simply shows that it is a particularly effective yet culturally derived strategy applied to problems faced by humans. Particularly virulent in the 1970s and the 1980s, the debate died down and remains inconclusive, with instinct theorists and their opponents still at loggerheads (Brighenti, 2010).

The point is crucial to their argument, yet the authors fail to discuss the matter entirely. Scholars may understandably rely on controversial assumptions for the sake of laying out their hypotheses, provided that they do not construe these assumptions as scientifically valid and unequivocal.

13.4 Conclusion

The practice of applying evolutionary theories to social sciences has come a long way. They have generated a considerable amount of debate in many disciplines, forcing scholars to revisit long-established claims. And while they were often met with strong pushbacks, they also contributed to the emergence of new ideas or fresh takes on old issues.

The case of human territoriality and international disputes presented in this work is one example. Often overlooked by IR scholars in spite of its conceptual significance, some of the thought-provoking contributions of evolutionary theories have the potential to stimulate research programs on the territory. A better integration of the territory in current IR frameworks would be a welcomed development, even if it were to conclusively disprove the explanatory significance of genetic determinants in the process.

As things currently stand however, points of contact between IR and evolutionary theories remain too few to create more meaningful debates. In order to do so, some efforts will be needed on both sides of the bridge. Should contributions from evolutionary theories grow to levels on par with other paradigms, IR scholars will be asked to familiarize themselves with the basic principles of biology—and no doubt that much needs to be done there. At the same time, evolutionary theorists will need to refine the methodological groundworks behind some of their more problematic assumptions, and get a better grasp of the existing diversity in IR. For while current evolutionary theorists have had the tendency to side with—not to say parrot—Realist thinkers (Thayer, 2004; Lopez, 2016; Johnson & Toft, 2013), a dialogue with paradigms that place a greater emphasis on patterns and outcomes of social interactions such as Liberalism or Constructivism could prove more fruitful in the long run. This is because ultimately, the reductionist approach proposed by some evolutionary theories will likely become unattainable when applied to IR. The recognition that some social behaviors might be 'soft-wired' (Johnson & Toft, 2013) carries the implication that instincts can be situational, overridden, or complemented, and that a balance between genetic determinants and cultural variables will have to be struck.

References

Agnew, J. (1994). The territorial trap: The geographical assumptions of international relations theory. *Review of International Political Economy, 1*(1), 53–80.

Ardrey, R. (1961). *African genesis*. Atheneum.

Ardrey, R. (1966). *The territorial imperative: A personal inquiry into the animal origins of property and nations*. Atheneum.

Axelrod, R., & Hamilton, W. (1981). Evolution of cooperation. *Science, 211*(4489), 1390–1396.

Badie, B. (2014). *La fin des territories*. CNRS éditions.

Bell, D. S. A. (2006). Beware of false prophets: Biology, human nature, and the future of international relations theory. *International Affairs, 82*(3), 493–510.

Bell, D. S. A., & MacDonald, P. (2001). Correspondence: Start the evolution without us. *International Security, 26*(1), 187–194.
Brighenti, A. M. (2010). On Territorology: Towards a general science of territory. *Theory, Culture & Society, 27*(1), 52–72.
Bull, H. (1977). *The anarchical society: A study of order in world politics*. Macmillan.
Buzan, B., & Wæver, O. (2003). *Regions and powers: The structure of international society*. The Press Syndicate of the University of Cambridge.
Claeys, G. (2000). The "survival of the fittest" and the origins of social Darwinism. *Journal of the History of Ideas, 61*(2), 223–240.
Copeland, D. C. (2003). A realist critique of the English school. *Review of International Studies, 29*(3), 427–441.
Crook, P. (1994). *Darwinism, war and history. The debate over the biology of war from the "origin of species" to the first world war*. Cambridge University Press.
Cullis-Suzuki, S., & Pauly, D. (2010). Failing the high seas: A global evaluation of regional fisheries management organizations. *Marine Policy, 34*(5), 1036–1042.
Dart, R. (1953). The predatory transition from ape to man. *International Anthropological and Linguistic Review, 1*(4), 201–217.
Deutsch, K. W., et al. (1957). *Political community and the North Atlantic area: International Organization in the Light of historical experience*. Princeton University Press.
Dunbar, R. I. M. (1998). The social brain hypothesis. *Evolutionary Anthropology, 6*(5), 178–190.
Elden, S. (2009). *Terror and territory: The spatial extent of sovereignty*. University of Minnesota Press.
Elden, S. (2013). *The birth of territory*. The University of Chicago Press.
Finel, B. I. (2001). Black box or Pandora's box: State level variables and progressivity in realist research programs. *Security Studies, 11*(2), 187–227.
Fisher, R. A. (1930). *The Genetical theory of natural selection*. Clarendon Press.
Fordham, B. O., & Asal, V. (2007). Billiard balls or snowflakes? Major power prestige and the international diffusion of institutions and practices. *International Studies Quarterly, 51*(1), 31–52.
Fry, D. P. (Ed.). (2013). *War, peace, and human nature: The convergence of evolutionary and cultural views*. Oxford University Press.
Gannon, L. (2002). A critique of evolutionary psychology. *Psychology, Evolution & Gender, 4*(2), 173–218.
Gold, J. R. (1982). Territoriality and human spatial behavior. *Progress in Human Geography, 6*(1), 44–67.
Goldstein, J. S. (1987). The Emperor's new genes: Sociobiology and war. *International Studies Quarterly, 31*(1), 33–43.
Government of Japan, Cabinet Office. (2023). Opinion Poll on Takeshima 2013. https://survey.gov-online.go.jp/hutai/r04/r04-takeshima/index.html
Haas, J., & Piscitelli, M. (2013). The prehistory of warfare: Misled by ethnography. In D. P. Fry (Ed.), *War, peace, and human nature: The convergence of evolutionary and cultural views* (pp. 168–190). Oxford University Press.
Haldane, J. B. S. (1932). *The causes of evolution*. Longmans, Green & Co.
Hamilton, W. D. (1964). The Genetical evolution of social behaviour. *Journal of Theoretical Biology, 7*(1), 1–16.
Hawkins, M. (1997). Social Darwinism in European and American thought, 1860–1945
Herz, J. (1957). The rise and demise of the territorial state. *World Politics, 9*(4), 473–493.
Holland, M. (2012). *Social bonding and nurture kinship: Compatibility between cultural and biological approaches*. Createspace Press.
Huth, P. K. (1996). *Standing your ground: Territorial disputes and international conflict*. University of Michigan Press.
Jacobson, D. (1997). New Frontiers: Territory, social spaces, and the state. *Sociological Forum, 12*(1), 121–133.

Jervis, R. (1976). *Perception and misperception in international politics*. Princeton University Press.

Jervis, R. (1978). Cooperation under the security dilemma. *World Politics, 30*(2), 167–214.

Johnson, D. D. P. (2016). *God is watching you: How the fear of god makes us human*. Oxford University Press.

Johnson, D. D. P., & Tierney, D. (2011). The Rubicon theory of war: How the path to conflict reaches the point of no return. *International Security, 36*(1), 7–40.

Johnson, D. D. P., & Toft, M. D. (2013). Grounds for war: The evolution of territorial conflict. *International Security, 38*(3), 7–38.

Kapstein, E. B. (1995). Is realism dead? The domestic sources of international politics. *International Organization, 49*(4), 751–774.

Kelly, R. C. (2000). *Warless societies and the origin of war*. University of Michigan Press.

Kirshner, J. (2015). The economic sins of modern IR theory and the classical realist alternative. *World Politics, 67*(1), 155–183.

Kitcher, P. (1987). On the Crest of 'La Nouvelle Vague. *International Studies Quarterly, 31*(1), 45–52.

Klama, J. (1989). *The myth of the beast within*. John Wiley & Sons.

Kocs, S. A. (1994). Explaining the strategic behavior of states: International law as system structure. *International Studies Quarterly, 38*(4), 547–548.

Kropotkin, P. (2009). *Mutual aid: A factor of evolution, (reprinted)*. Freedom Press. (Original work published in 1902).

Kuo, R., Johnson, D. D. P., & Toft, M. D. (2015). Correspondence: Evolution and territorial conflict. *International Security, 39*(3), 190–201.

Laitinen, A., & Maude, G. (1986). Biologism, politics and international politics. *Acta Sociologica, 29*(2), 113–128.

Lebow, R. (2013). You can't keep a bad idea down: Evolutionary biology and international relations. *International Politics Review, 1*, 2–10.

Lee, W. E. (2015). *Waging war: Conflict, culture, and innovation in world history*. Oxford University Press.

Liberman, P. (1993). The spoils of conquest. *International Security, 18*(2), 125–153.

Lopez, A. C. (2016). The evolution of war: Theory and controversy. *International Theory, 8*(1), 97–139.

Lopez, A. C., & Johnson, D. D. P. (2017). The determinants of war in international relations. *Journal of Economic Behavior & Organization, 178*, 983–997.

Lopez, A. C., McDermott, R., & Petersen, M. B. (2011). States in mind: Evolution, coalitional psychology, and international politics. *International Security, 36*(2), 48–83.

Lorenz, K. (1966). *On aggression*. Methuen.

Makinda, S. M. (2000). International society and eclecticism in international relations theory. *Cooperation and Conflict, 35*(2), 205–216.

Malmberg, T. (1980). *Human territoriality: Survey on the Behavioural territories in man with preliminary analysis and discussion of meaning*. De Gruyter Mouton.

Maxwell, M. (1990). *Morality among nations: An evolutionary view*. State University of New York Press.

Mearsheimer, J. (2001). *The tragedy of great power politics*. Norton & Company.

Milam, E. L. (2019). *Creatures of Cain: The hunt for human nature in cold war America*. Princeton University Press.

Morgenthau, H. (1962). The problem of the National Interest. In H. Morgenthau (Ed.), *Politics in the twentieth century, Vol. 1, the decline of democratic politics* (pp. 79–112). University of Chicago Press.

Morgenthau, H. (1967). *Politics among nations: The struggle for power and peace* (4th ed.). A.A. Knopf.

Morgenthau, H. (1970). Supplement: International Relations, 1965–1969. *The Annals of the American Academy of Political and Social Science, 390*, 114–119.

Neuman, I. B. (1996). Self and other in international relations. *European Journal of International Relations, 2*(2), 139–174.

Nye, J., & Keohane, R. (1977). *Power and interdependence: World politics in transition*. Little Brown and Company.

Painter, J. (2010). Rethinking territory. *Antipode, 42*(5), 1090–1118.

Paul, D. E. (1999). Sovereignty, survival and the Westphalian blind alley in international relations. *Review of International Studies, 25*(2), 217–231.

Philpott, D. (1995). Sovereignty: An introduction and brief history. *Journal of International Affairs, 48*(2), 353–368.

Postel-Vinay, K. (1998). La transformation spatiale des relations internationales. In M. Smouts (Ed.), *Les nouvelles relations internationales* (pp. 161–181). Presses de Sciences Po.

Ruggie, J. (1993). Territoriality and beyond: Problematizing modernity in international relations. *International Organizations, 47*(1), 139–174.

Sack, R. D. (1983). Human territoriality: A theory. *Annals of the Association of American Geographers, 73*(1), 55–74.

Schmidt, B. C. (1994). The historiography of academic international relations. *Review of International Studies, 20*(4), 349–367.

Schmitt, C. (2007). *The concept of the political, (1932; reis)*. Chicago University Press.

Shaw, R. P., & Wong, Y. (1987a). Ethnic mobilization and the seeds of warfare: An evolutionary perspective. *International Studies Quarterly, 31*(1), 5–31.

Shaw, R. P., & Wong, Y. (1987b). Inclusive fitness and central tendencies in warfare propensities. *International Studies Quarterly, 31*(1), 53–63.

Shaw, R. P., & Wong, Y. (1989). *Genetic seeds of warfare: Evolution, nationalism, and patriotism*. Unwin Hyman.

Simmons, B. A. (2002). Capacity, commitment, and compliance: International institutions and territorial disputes. *The Journal of Conflict Resolution, 46*(6), 829–856.

Symons, D. (1992). On the use and misuse of Darwinism in the study of human behavior. In J. H. Barkow, L. Cosmides, & J. Tooby (Eds.), *The adapted mind: Evolutionary psychology and the generation of culture* (pp. 137–159). Oxford University Press.

Thayer, B. A. (2000). Bringing in Darwin: Evolutionary theory, realism, and international politics. *International Security, 25*(2), 124–151.

Thayer, B. A. (2004). *Darwin and international relations: On the evolutionary origins of war and ethnic conflict*. University Press of Kentucky.

Walt, S. M. (1985). Alliance formation and the balance of world power. *International Security, 9*(4), 3–43.

Walt, S. M. (1991). The renaissance of security studies. *International Studies Quarterly, 35*(2), 211–239.

Waltz, K. (1959). Man, the state, and war: A theoretical analysis.

Waltz, K. (1979). *Theory of international politics*. Addison-Wesley.

Weikart, R. (1993). The origins of social Darwinism in Germany, 1859–1895. *Journal of the History of Ideas, 54*(3), 469–488.

Wendt, A. (1992). Anarchy is what states make of it: The social construction of power politics. *International Organization, 46*(2), 391–425.

Wight, M. (1995). Why is there no international theory? (1966). In J. Der Derian (Ed.), *International theory* (pp. 15–35). Palgrave Macmillan.

Williams, R. (2000). Social Darwinism. In J. Offer (Ed.), *Herbert Spencer: Critical assessment* (pp. 186–199). Routledge.

Wilson, E. O. (1975). *Sociobiology: The new synthesis*. Belknap Press of Harvard University Press.

Wilson, D. S., & Sober, E. (1994). Reintroducing group selection to the human behavioral sciences. *Behavioral and Brain Sciences, 17*(4), 585–654.

Chapter 14
The Constitutive Component of Sexes, Genders, and Sexualities: The Case of Thailand

Porranee Singpliam

Abstract To what extent are our bodies natural? This is the question that begs us to consider the state of our bodies as to whether bodies are mere natural biological constructs or are canvases to be governed and regulated by our culture and society? This chapter gives an overview of the case of Thailand and its structural relations to its citizens' bodies by inspecting closely how society theorizes and understands the bodies to be and, in the later section, how bodies are mediated by giving them or depriving them of the legal statuses. The viability or legality of the bodies, as this chapter will present, will be closely related to the contingency of socio-cultural and historical factors regarding the biological deterministic discourses on the sexes, as well as genders and sexualities. Why are our bodies designated into the either/or female or male categories? This question incites us to inquire how our cultural practices can shape the biological nature of our bodies (based on reproductive systems) into fixed and (non)viable entities. This chapter discusses issues of marriages and uniforms, for instance. Ultimately, this chapter wishes to make a case that the bodies are sites for political control and struggle, making them a central point for nature, culture, and society to collide. The chapter ends by providing a proposal and a case example on how the governing of the bodies or the way society and culture can interact with our biological natures could be most inclusive and can potentially start eliminating the way natural dimorphism affects our lives.

Keywords Biologism · Sexed bodies · Legal section

P. Singpliam (✉)
Faculty of Arts, Chulalongkorn University, Bangkok, Thailand
e-mail: porranee.s@chula.ac.th

J. Pérez-Jara, Í. Ongay (eds.), *Beyond Nature and Nurture*, Synthese Library 497,
https://doi.org/10.1007/978-3-031-87715-5_14

14.1 Introduction

On October 29, 2020, Thailand saw saddening news involving a transgender woman government official and her male partner, which showed how the couple could not utilize the government official's spousal benefits as they cannot be lawfully wedded. Her partner, having advanced stage of liver cancer, had to spend their own savings in order to better their health. Unlike other heterosexual couples who can enjoy health care consent, spousal tax benefits, and government pensions, to name only but a few, the (biological) same-sex couple's basic rights that they could receive by conjugal union are deprived from them. In June 2024, Thailand's Senate passed the marriage equality law, which came into effect in January 2025 after royal endorsement. Thailand became the third country in Asia to recognize this law. Same-sex couples will be able to legally wedded at the age of 18 years and older, they will be recognized as spouses rather than wife and husband (Tangsathaporn, 2024). Though, the issues concerning whether or not individuals can change their name titles according to their self-identified gender identity, and their apparent gendered bodily appearance, are still under debate in the Committee.

The natural bodies play an important role in many domains of Thai society. The bodies are being maneuvered by the state for them to achieve certain purposes. This chapter examines the root of why it took so long for Thai state to legalize same-sex marriage and give it an official status the way citizens ought to receive. The chapter investigates first, how the body goes through the process of being sexed under the biologistic discourse. The second part scrutinizes how the sexed bodies are given legal status, one of which conserves the natural dimorphism or binaristic structure that encompasses only females and males, and discursive construction of normative familial institution. Ultimately, the chapter wishes to suggest that the legislators' conceptualization is trapped under the discursive biologism, effectively preserving the sexual difference making the bodies no longer a natural entity, but entities where socio-cultural norms are being mediated.

14.2 Sexed Bodies: Theoretical Approach to How the Bodies Are Sexed

Our bodies bear great significance. The fact that lawmakers and representatives vote for or against the legalization of same-sex marriage means that bodies and sexual orientations (or sexualities) are regulated. While our bodies, our desires, and what we decide to do with them should be privatized and individualistic, this has proven to be otherwise in the case of Thailand. This section will discuss the importance of the bodies and how they can act as a double edged sword given the unique characteristics of the mainstream ideology on genders/sexualities in Thai society. It will further make a case that the bodies in Thai society are inherently regulated by biologistic discourse where the "natural dimorphism" (Ammaturo, 2017) between those of females and males are considered factic.

The debate that emerged after the Senate's approval of marriage quality begs the question as to how natural our bodies are. To what extent are our bodies natural? Why are our bodies regulated, disciplined, and penalized? Can we express our desires the way we wish through our bodily expressions? These are central questions that many scholars in Cultural Studies and Body Studies have put forth on the "nature" of the bodies. Some scholars (Ammaturo, 2017; Evans, 2002; Sheldon, 2002) have questioned the naturalness of the body by underscoring the "natural dimorphism" (Ammaturo, 2017) and its restrictions on the individuals in the society whilst acting as a key mechanism which can prolong the binary categories that only include females and males.

The bodies are always interpreted and shaped by our society and culture. Sexed bodies, therefore, are not in any way "natural," but they go through the process of historicism and can be said to be a historical product (Sheldon, 2002). Susan Bordo writes in "Feminism, Foucault and the Politics of the Body" on Foucauldian historicism that "the intellectual orientation which insists that ideas neither descend from a timeless heaven nor are grounded in the necessities of "nature," but develop out of the imaginations and intellects of historical human beings" (1994). According to Bordo, this deconstructionist notion leads the author and many like her to critique the way in which bodies, particularly that of women, become materialized and politicized in such a way that women's bodies may become "docile," "domesticated," and "monitored" (1994) by the regime of the homogenization of images. This is not to reiterate the way in which female bodies are perceived to be "speaking" something "provoking," "dangerous," and "inviting" just because her bodies are females (Bordo, 2003). The bodies thus shoulder the meanings differently contingent upon temporalities and historicities.

What is the body, then? Bodies are both natural and cultural constructs. More importantly, our bodies are sites where they both free us while, at the same time, domineering over our genuine wishes and desires. In the former instance, the bodies in Thai society are where individuals choose to communicate and convey their genders/sexualities daily whether through make up, gender affirmation surgery, or hormone taking (Evans, 2002). The bodies are always in close interaction with society's technology and what it has to offer. In the latter instance, however, the very same bodies can indeed oppress an individual's free reign in terms of expression, especially when the bodies deviate away from what society deems as normative. In Evans' words, "the body is thus our means of freedom and the location of our enslavement" (2002). This is specific in the case of how the bodies are sexed. We ought to question why is it that the bodies are designated into categories of females and males on official documentation? (It must be noted that in factuality, there are intersex bodies as well.) What sorts of consequences and expectations are understood in society after the procedure of being sexed is carried out?

Additionally, it seems that the "natural" state of the bodies should be reconsidered because first, the anatomical components that make up women's bodies obscure other facts there are to be noted and learned about the bodies. This view, as Blackman terms "biologism", is rigid and pre-determines what we could know and learn about the body as it is fixating and reductionist. In other words, the reductionist view "reduce[s] the complexity of human psychological and social life to the biological make-up of individual and groups" (2008). This first view of the "naturalistic body" is seen by Blackman as a form of biological determinism. But, as time evolves and the challenges in paradigms of thoughts arise, Blackman observes that the way the bodies are determined by their biological make-up is not the only factor that regulates one's lives and, more intimately, one's bodies.

Following Emile Durkheim's sociological view of the body, the bodies bear the effects of social structure's cultural inscription. This process emphasizes the social structure and its exertion of power on the individual's body to study how and in what ways society could achieve social unity. In the case of a same-sex couple mentioned in the Introduction; this view on cultural inscription would highlight the interrelations between the body, ideology, and power, but in such a way that it underscores "an examination of the constraint and imposition of social structures on the formation of human subjects … [to] achieve a uniformity" (Blackman, 2008). This very same idea of cultural inscription could extend to cover and influence ideas, practices, or beliefs regarding sexed bodies in society, particularly how women's bodies and their cognitive capacity are put aside, and that the society only asks that because their bodies are female, they are expected to be heterosexual, compliant to the societal demands, which only goes on to maintain and sustain the overarching ideology that is sexism and androcentrism. Blackman, seeing that the bodies are both understood as restrictive by the social structure (sociological view of the body) and predetermined by the biological criteria (essentialism), states "the move to this body as an alternative to the naturalistic is a move from one form of determinism, *essentialism*, to another: *discourse* or *social determinism*, which is seen by many to be more sophisticated form of essentialism" (2008). What is for certain about the views on the bodies are the way in which the bodies are regulated and meanings, demands, and restrictions are imposed upon by the society making the understanding of the bodies in no way, "natural" constructs.

The fact that there exist only two sexed bodies, at least in the bureaucratic official documentation, means that the binary sexes have the potentiality to sediment the "determining constraints" (Evans, 2002) onto the people in the society. We are born with certain anatomical structures that make us idiosyncratically different between females, males, and intersex. Our bodies are thus genuinely natural. The hard-boiled chromosomes and genetic algorithm are innately inbred with us since birth. How our bodies function in society is in no way near natural. The way in which we take for granted the biological categories is what Smart terms as biologism (1990). In fact, Evans states in *Real Bodies* that "the body is a social construction" (2002). The reciprocity between nature, culture, and society makes up the bodies rather than either one of these. Sheldon (2002), too, probes into how "pure" the body could be by referencing Judith Butler in "the masculine body" that, "there is no reference to a pure body which is not at the same time a further formation of that body". By stating that a woman's body ought to bear children and a man's body ought to provide or protect others (namely, the controversial move from the opposition party to abolish mandatory military conscription, deemed a duty of Thai males) is none other than the process of sedimenting, materializing, and concretizing the constitutive binary system. In Thailand, males with Thai nationality aged twenty years and above with abled-bodied are eligible up to two-year military service. One way to avoid the service is for Thai males to enroll in Reserve Officer Training Corps (ROTC) for three years during high school. The 1954 Military Service Act materializes constitutive binaristic sexed system and keeps biologism intact. This law makes matters more complicated for trans individuals who must undergo the recruitment process where additional relevant medical certificates are required from them in order to be exempted. How then, is it just that bodies must undergo such a physically rigorous regimen merely because one is born male?

This section informs the illusive, yet potent way sexed bodies are situated within social discourse, how the bodies ought to be perceived as both "natural" entities and a product of a "discursive social construct" all at once. The way in which sexed bodies come to bear potency are first and foremost illusory because they cannot exist on their own. Sexed bodies go through the modality of performativity (Butler, 1988, 1999, 2011) where repetition and materialization are key. The bodies are overloaded with presumptions or punitive actions in the case where the bodies are read as non-normative. The Thai Court's previous recognition of only opposite-sex couples as constitutional or the rights to adopt that though, granted, but complicated by the essentialism of the binary sexed bodies, speak of the leverage the discourse of sexed bodies and its knowledge has in society. Bodies that are sexed either as females or males in the mainstream Thai discourse are thus an ideal manifestation of "the illusion of dichotomous sexes [that] is (re)formed at the site of the body (re)constituting men and women in context" (Browne, 2004).

Gender performativity and the attempt to cement normative heterosexuality is in no way a simplistic scheme. From the news mentioned in the Introduction, that various groups have called for the legalization of same-sex marriages and the rights that should be consigned to them as humans shows that the performative acts of materializing the biologistic discourse is resisted by the society. This resistance is particularly done by the actors, the people in society themselves who refuse to be sexed in the bifurcated opposites. When such a case happens and takes on a large scale, it can upend the "factic" qualities and beliefs that come with the binary opposition. In Rae's analysis of Merleau-Ponty's notion of the body, existence, and sexuality, the author underlines how sexuality is always part and parcel of our bodies, and that it is also a historical lineage. It is in this sense that when people start questioning the restrictions that are deployed in the society just as they are exploring other alternatives and possibilities of sexual orientations and preferences, it shows that sexed bodies and their emphasis on females and/or males do not make up their realities as they experience through the "living body" (Rae, 2020). It is in this sense that Rae's words highlight how sexed bodies must go through the modality of performativity, one of which is not entirely natural. In the author's words, there is "the "cultural" additions that have brought the body to its factic determinations" and that there exist "bodily schemas that are the condition of and manifested through specific factic sexual determinations" (Rae, 2020).

Furthermore, the procedure where the bodies become sexed speaks of its potency. This process, the deployment of sexing an individual's body reveals its greatness when it is maneuvered by certain individuals in a society, a particular belief, or a distinct practice within a certain historicity. Effectively, it does hold its force so powerfully that same-sex couples or individuals who are transgenders/transsexuals or those who prefer homosexual inclinations are deprived of their basic rights, to form a family as mentioned above, one that should be bestowed to them as they are the most basic. The force of the "biological dichotomisation of sex" (Browne, 2004) is in no way stable. It needs, much like its sedimentation process through repetition and reification that is mentioned earlier, institutional practices to carry out the "presumed naturalness" and make it known within society. Practices of wearing "gendered" civil servants' ensembles, the military enlistment of Thai male individuals, or the pre-school young children's school uniforms are all but a repetition of the way the bodies are sexed.

The process of sexing the bodies can be said to be restrictive more than freeing the bodies of the individuals. The discourse of biologism can only display its force so long as the materialized bodies are reproduced and re-deployed within the various facets of society. As it stands in Thailand today, biologistic discourse still regulates and monitors the bodies of the citizens, but things are changing as the ruling of same-sex marriage shows. Much is clear that there can never be a pure body existing outside of social discourse (Ammaturo, 2017; Browne, 2004; Evans, 2002; Rae, 2020; Sheldon, 2002). The sexed bodies and their process that we have looked at in this section indeed shows the "impossibility of their purified categories" (Browne, 2004). As the indignation grows through the visibility of movements and various discontented voices in the parliament and in forms of online public discussion following the marriage that is *almost* thoroughly equal (Way Magazine, 2024), it is crucial that we ask the question about how our understandings of the bodies, sexualities, and our desires are mediated. What purpose(s) does biologism serve? How can our bodies that are overlaid with presumptions and assumptions become a politicized matter and politically affect the individuals within society?

14.3 The Legality of the Sexes: Mediated Cultural Bodies

The bodies are also sexed by laws and regulations. It is of significance that one understands how marriage, its regulation, and officialization are a major part in legitimating the biologistic discourse. Numerous authors highlight how marriage is prime evidence of a social institution that shapes the subjects into sexed individuals. Flynn's work titled "Sex and (sexed by) the state," and Prasad's work on "Potential and perils of same-sex marriage" show that marriage institution is one founded on the heterosexist discourse. Prasad's work underlines one potent religious ideology, Christianity, that serves to reify marriage between different sexed persons to be seen as valid. In this article, Prasad writes "how marriage is customarily premised on the essentialist foundation of ontological sexual difference. Any discussion on marriage should, however, commence with consideration of the realm in which the institution is most commonly situated, that is, of course, religion" (Prasad, 2008). The case where the above mentioned scholars analyze is Westocentric. However, a few things can be aptly read from the case of Thailand as well, namely, essentialist perception on biologism, which the institution puts emphasis on.

Orthodox marriage in Thailand under Thai Family Laws is founded upon the heterosexist construct. In the Conditions of Marriage, Section 1448 writes that "A marriage can occur only when the man and woman have finished their seventeenth year of age. But the Court may, in case of having appropriate reason, permit them to marry before attaining such age" (Thailand Law Forum, 2023). Section 1457 further states that "Marriage under this Code shall be effected only on registration being submitted" (Thailand Law Forum, 2023). With regards to the Relationship of Husband and Wife, Section 1461, the Law writes that "Husband and Wife shall cohabit as husband and wife. Husband and wife shall provide for and support each other according to his or her ability and condition in life" (Thailand Law Forum, 2023). In the case that there be ailment or sickness to one of the spouses, Section 1463 affirms that "If one of the spouses is adjudged incompetent or quasi-incompetent, the other becomes guardian or curator by operation of law. But on application of any interested individual or Public Prosecutor, the Court may on substantial grounds, appoint another person as guardian or curator" (Thailand Law Forum, 2023).

In this way, the ontological sexual difference can be seen as the statist mechanism, a *sine qua non* where one's wedlock is lawfully recognized (Flynn, 2004). The statist mechanisms or "organs" in Jackson (2004)'s term include those that extend beyond that of the religion of Buddhism, which include the bureaucracy, the military, and the police force. In the case of Thailand, the officialization of the conjugal union done by the statist personnel in this case concretizes the normative monogamous heterosexual relations as either "viable" and "non-viable" spouses (Ammaturo, 2017). Apart from the bureaucratic procedure, for example, adoption regulation mentioned prior, it should be noted that there had always been challenges made towards the people who hold the power within the society by the public at large in the forms of parades, movements, and parliamentary oppositions. *The Bangkok Post* reports that:

> In 2012, a same-sex couple was denied the right to register marriage because the Civil and Commercial Code does not recognise same-sex union. But the military coup in 2014, led by Gen Prayut Chan-o-cha, stalled the civil partnership proposal. After the process resumed, the cabinet approved the government-sponsored draft bill in July 2020, but it underwent further review. Meanwhile, the opposition Move Forward Party has sought an amendment to the Civil and Commercial Code. After parliament received the draft bill in June 2020, it remained in queue for deliberation for some time until lawmakers in the House of Representatives voted in favour of it in February, but it would remain under review for 60 days. Finally, parliament passed the marriage equality draft bill and three others, including government-sponsored civil partnership, in the middle of June....
>
> The fight for marriage equality has faced opposition from some religious and legal institutions. An MP from the Prachachart Party told a House meeting in February that he rejects the draft bill on behalf of other Muslims for flouting their belief which assigns the

> definition of husband and wife to man and woman. An expert from the Sheikhul Islam Office expressed concern that if the bill passed into law, young followers would disown their tenets and marry those of the same sex. Last November, the Constitutional Court delivered a landmark ruling on Section 1448 of the Civil and Commercial Code, which allows only a heterosexual couple to tie the knot, after a same-sex couple denied the right to marry in 2020 filed a complaint. A 12-page document said it is not unconstitutional because, by tradition, marriage is the union of a man and a woman for reproduction. A separate law should be enacted for LGBTI people. (Boonlert, 2022)

Marriage, a conservative union between a female and a male, mainly serves the interests of the state. As it appears, marriage endorses opposite-sex unions as "heterosexism par excellence" (Prasad, 2008) and that, as an official institution, it binds the state of being humans so closely to that of one's sexualities. One's rights and dignities can thus be taken from one's fundamental prerogatives as a human all because of one's sexual orientation. This case, unlike other already developed nations where progressive thinking towards sexualities and genders are upheld, detains one's agency to be, to exist as a human. As Rae's (2020) work on "Merleau-Ponty on the sexed body" asserts that sexuality is part and parcel of the body and that sexuality is very much historical, the author also points out that Merleau-Ponty's work on the bodies delineates how one's body and one's preference towards certain sexualities are seen to stand in for oneself, to represent oneself as a human, and a member of a society. This "lived body" is a constituted product, one in which cultural process gives it a factic determination. Sexuality, then, being perceived as one and the same as the state of being, is inescapably political.

> Sexuality is one of the fundamental ways in which each individual becomes what he is. Rather than determining existence, sexuality is coterminous with it, shaping, expressing, and revealing the way an individual exists. Instead of being merely a part of a larger bodily existence, sexuality "symbolize[s] a whole attitude" to existence. There is no separation from one's sexuality; one's sexual being reveals, without determining, what one is. As a consequence of the intimate bond between sexuality and (individual) existence, Merleau-Ponty maintains that "the question is not so much whether human life does or does not rest on sexuality, [but] what is to be understood by sexuality." (Rae, 2020)

In Thailand, it is evident that one's basic human rights are to be given to those who do not deviate from the state's acclaimed "natural dimorphism" (Ammaturo, 2017). But, is this just? Are lawmakers and the proclaimed "representatives" of the people such as those who are members of the parliament or the ruling regime seeing the fluidity that is sexualities? The degree of worthiness and viability regarding one's humane existence is being founded upon one's sex, specifically one's sexed body. As a result, a social stratum is established. In Cossman's (2002) work titled "Sexing citizenship, privatizing sex," the author defines sanctioned sexual citizenship as "sexual subjects [which] are privatized, de-eroticized and depoliticized". This effectively creates a sense of belonging or being Othered. In Cossman's words, "sexual outlaws" are engendered and considered products of non-normativity. Exclusion and marginalization are further carried out towards those whose sexualities do not fit in the "phallocentric culture" (Smart, 1990) where legal makers arbitrarily construct sexed subjects.

Bodies, in this case, are the sites for power struggles to unfold and take place. There is no moment in time where the bodies are pure or unchained to the cultural inscriptions or impositions. Bodies are mediated cultural products and bear historical graveness depending on who is studying, talking, or embodying the bodies. This was why marriage equality's legalization took so long. Marriage, as a social institution, was founded upon binary constructs, biologism, resulted in a calamity mentioned in the Introduction. Lives were lost and legal rights were deprived from Thais as their sexual bodies were treated as "sexually charged bodies, these sexual outlaws, remained outside of the claim to citizenship of power. These bodies were too sexed, too sexually marked as difference, to fall within the borders of belonging" (Cossman, 2002). While marriage equality finally became effective in January of 2025, many facets of life are still sexed. This leads to the legality of the body being coerced for some and penalized or shamed for others. In some domains, our sexes become, "the ultimate truth of a person, it [becomes] each individual's most secret yet telling property" (Smart, 1990).

Thailand's case on the politics of the sexes and sexualities is intriguing in the sense that the expected binaristic bodies are being challenged and negotiated by many groups and individuals within the society, one of which is the religion. A renowned example from a religious figure, an abbot in Thailand, namely Venerable Shine Waradhammo (hereafter Phra Shine) sees that there are no constraints made towards homosexualities from the teachings of Buddhism. The "conservative Buddhist-majority kingdom" (AFP, 2019) that is Thai society in fact has different ideas about what a Buddhist society ought to be. In 2020, the people with power's words are disparate from that of the religious figure itself as the ontological sexual difference seen as the law of the wedding that can only be held between opposite sexes is a view that is not shared by every institutional subject. Phra Shine Waradhammo provides his thought on the teachings of Buddha in relation to sexualities and bodies that:

> "treating LGBT people badly goes against the Buddha's teachings," emphasized Phra Waradhammo, who was ordained at the age of 21 and now defines himself as a neo-Buddhist. "LGBT people are also humans, they are also Buddhists, and as a monk I support and accept all Buddhist people, and aim to reduce their suffering." (Thomson Reuters Foundation)
>
> "Buddhism [in Thailand] was created only for heterosexual people. Male and female are the only genders that are recognized by the religion," Phra Waradhammo said, observing that people whose identities don't fall into either of those two groups "have no place to be." He continued: "It's already difficult for *Katoey* [a Thai word widely used to describe both gay and transgender people] to integrate into society, but religion makes it worse." (*Bangkok Post*)
>
> "As a monk, I believe that the Buddhist teachings should be interpreted in accordance with social realities. If they are not interpreted in this way, they are useless, said Phra Waradhammo. "Between people and religion, people should come first. We should try to free people from their suffering. If we want to protect these teachings, we have to find a way to interpret them according to the life and customs of the people." (Lewis, 2020)

The resistance towards the way in which the bodies are seen as the objects of control and monitoring by the State and its institutions is evident. The public outcries, especially through Popular Press and online platform probe the Parliament to truly and thoroughly contemplate the gaps that exist even with the Senate passing same sex marriage. Certain rights such as adoption, cultural biases, and mistreatments that may occur to children of the same sex parents remain in the cultural environment led by normative binary sexes. These incidents call out the way in which the discourse of biologism no longer serves the people's lived realities. Rather, the legality of the sexes through official documentation such as that of national ID card, passport, or even one's driver's license only serves to strengthen the constitutive nature between the bodies and biological sexes, effectively enhancing the power of the "naturalness" when, in reality, this preservation by legal means only serves certain ideological needs, dominant and normative ones. The preservation only sustains patriarchy, women and gender diverse bodies' subordination, and ultimately sexism. Queer marriages will, thus, dismantle the altar that is sexual binarism.

The presumed *a priori* connection between that of the binary sexed bodies to heterosexism deemed as normative and compulsory can be broken through queer marriages. Prasad (2008) writes clearly that "the queer marriage similarly functions to emasculate patriarchy, the autonomous system of women's subordination in society". A similar case is already taking place in Thailand. The statist regulation that is the Marriage Law is further being probed into by not only the public, but the opposition parties themselves such as figure like an openly identified LGBT Member of Parliament (MP) Tunyawaj Kamolwongwat, who is pressing the state to recognize titles' change accordingly to self-determined genders, rather than sexes. When citizenship, rights, and benefits are taken away from gender diverse individuals and those whose sexual orientations cannot be traditionally defined, demands and contestations emerge in various forms. This is where we can see that the bodies no longer act as "static slaves to their biology" (Prasad, 2008), but that the bodies are historical products, being shaped, imagined, and sanctioned or penalized all according to the working and manipulating of the natural biological make up of one's body. Bodies, seen in this way, are being used as tools and effective mechanisms by the lived people themselves who may embody genders and sexualities seen outside of the realm where "natural dimorphism" can only sanction the "conservative egalitarianism" (Ammaturo, 2017).

Bodies, as they are lived, appear to be both natural and cultural at the same time. Depending on who is living it and making use of it in order to sustain certain ideologies and power relations, the body then should be thought of as a historical construct. Contemporary Thailand has witnessed many campaigns as well as lawsuits filed against the Constitutional rule that is the Marriage Law. This speaks of where bodies are seen to be "disrupting" the dominant gender/sexuality within the society (Cossman, 2002). Bodies seen in this way are all connected to what our natural givens in terms of sexes are assigned at birth; bodies go through the process of being sexed to fit and serve the traditional institutionalized practices, and bodies, too, are close to that of the political issues. There is no way that bodies can be one of these facets, but exist altogether at once, liberating one's agency, and simultaneously acting as their constraints. In the words of Cossman, the author affirms that there is a "political nature of sex and sexuality; an assertion of public sphere in which sex and sexuality is negotiated and performed" (2002). This negotiation is performed through institutional figures, corporeal movements on the streets, and within the parliament as fundamental as advocating for legal gender recognition bill. As it appears, the most foundational relations that sustain society, the symbolic that is linguistic usage, is under challenge.

14.4 Conclusion

This chapter provides a conceptual proposal to the constitutive components of the sexed bodies in Thai society. It discusses specifically the case of same-sex marriage, which was recently legalized in Thailand, basic human rights deprived of same sex couples before the legalization, and the conscription law. In order to understand why this is in contrast to what the country is reputed for, a queer hub (Jackson, 2011), there is a need to grasp the hegemonic gender/sexuality ideology which is embodied by those in power. Biologistic discourse is upheld by the state, the bureaucracy, and those who work for the statist organs. This discursive constitution of the bodies only adheres to and reveres the way in which the bodies are sexed in a binary way, imposing cultural expectations onto the female and male sexed bodies. Such expectations are in parallel with heterosexism, exerting cultural renditions onto the bodies. The bodies, seen in this way, are being maneuvered by the state to sustain certain orthodox gender/sexual norms, as well as serving politico-economic purposes. The legitimization of opposite-sex bodies is, however, being challenged by numbers of parties within the society because the way the bodies undergo the process of being sexed does not cohere with the phenomenological way the bodies are lived by the individuals and the citizens.

What would be the best solution for this issue? To progressively eliminate binaristic dimorphism is a difficult task because binary thinking does seep into every microunit of our society, starting from the day we are born until death as we are still classified and documented as either/or. The challenge, as I view it, needs to be participated by major institutional players within the society, namely, education, bureaucracy, religion, and legislation. These major institutions are key players that could implement and enforce certain cultural changes, specifically, how the society tends to naturalize socio-historical views about the bodies. These institutions can start by producing and disseminating certain discourses and practices, beginning with the process of educational change by highlighting how the bodies are both natural and cultural construct. Bodies are our own and the way we are attracted to others are discrete and should be privatized. Nevertheless, they are also a part of the society which should be treated based on the principle of human rights, equality, and equity. How could this be done and how could this be achieved, one may challenge? I would say that to successfully eliminate binaristic dimorphism would need to be done from the roots up, i.e., from our individualistic lens. To regulate and to monitor how bodies are discursively sexed could derive from those who are minorities, non-normative, and at times, oppressed, but who have a position to have a pronounced say and effects at the structural level. We are starting to see progress as the newly elected MP who self identifies as LGBT and an active member within the community pushes for the elimination of sex-based student and civil servant's uniforms, and to recognize titles based on an individual's self-identification. This move by MP Tunyawaj Kamolwongwat shows that by having inclusivity within and amongst those with power, legislatively, could be a potent start for change regarding queering policies and legislation.

Ultimately, the way our bodies are regulated or governed ought to be conceptualized structurally. History affects our bodies, our expressions, and how we live or the benefits we may gain in our lives. All of these that are our acts and our desires are no longer self constructed, no longer mere biological nature, but are heavily shaped by historically contingent reasonings dictating the way our bodies are to be and exist within the society. Changes are indeed difficult, but impending. Conclusively, sexualities here should thus be seen as "socially constructed, socially located, embodied, and both personally and politically meaningful" (Dunk-West & Brook, 2015). Biologism that is widely upheld by those in power must be contended with as there is no way that our bodies can be deemed as pure natural essence, but rather a culturally mediated corporeality. As a historical product and political entity, one can expect many possibilities stemming from the bodies, ones that are not confined to the binaristic dimorphism.

References

AFP. (2019). 'Not here for decoration': Transgender MPs make history in parliament. The Phuket News, June 6.

Ammaturo, F. R. (2017). *European sexual citizenship: Human rights, bodies and identities*. Palgrave Macmillan.

Blackman, L. (2008). *The bodies: The key concepts*. Berg.

Boonlert, T. (2022, September 13). On the road to equality: Years of campaigning has raised hopes the same sex marriage bill will be passed into law next year. *Bangkok Post*

Bordo, S. (1994). Feminism, Foucault and the politics of the body. In R. Miguel-Alfonso & S. Caporale-Bizzini (Eds.), *Reconstructing Foucault: Essays in the wake of the 80s* (pp. 219–243). Rodopi.

Bordo, S. (2003). *Unbearable weight: Feminism, Western culture, and the body*. University of California Press.

Browne, K. (2004). Genderism and the bathroom problem: (re)materializing sexed sites, (re)creating sexed bodies. *Gender, Place & Culture, 11*, 331–346.

Butler, J. (1988). Performative acts and gender constitution: An essay in phenomenology and feminist theory. *Theatre Journal, 40*, 519–531.

Butler, J. (1999). *Gender trouble: Feminism and the subversion of identity*. Routledge.

Butler, J. (2011). *Bodies that matter: On the discursive limits of "sex"*. Routledge.

Conditions of marriage. (2023). Thailand Law Forum. Accessed February 15, 2023, from http://thailawforum.com/database1/marriage-law-thailand-2.html

Cossman, B. (2002). Sexing citizenship, privatizing sex. *Citizenship Studies, 6*, 483–506.

Dunk-West, P., & Brook, H. (2015). Sexuality. In V. Robinson & D. Richardson (Eds.), *Introducing gender and women's studies* (pp. 150–165). Palgrave Macmillan.

Evans, M. (2002). Real bodies: An introduction. In M. Evans & E. Lee (Eds.), *Real bodies: A sociological introduction* (pp. 1–13). Palgrave Macmillan.
Flynn, T. (2004). Sex and (sexed by) the state. *Women's Rights Law Reporter, 25*, 217–218.
Jackson, P. A. (2004). The Thai regime of images. *Sojourn, 19*, 118–218.
Jackson, P. A. (2011). *Queer Bangkok: Twenty-first-century markets, media, and rights*. Hong Kong University Press.
Lewis, C. C. (2020). Buddhist monk seeks equality, acceptance for Thailand's LGBT communities. *BDG*, August 24.
Marriage (that is almost) equality: Restricted rights to adopt. (2024). Way Magazine. Accessed August 2, 2024, from https://waymagazine.org/lgbtq-and-child-adoption/
Prasad, A. (2008). On the potential and perils of same-sex marriage: A perspective from queer theory. *Journal of Bisexuality, 7*, 191–215.
Rae, G. (2020). Merleau-Ponty on the sexed body. *Journal of Phenomenological Psychology, 51*, 162–183.
Sheldon, S. (2002). The masculine body. In M. Evans & E. Lee (Eds.), *Real bodies: A sociological introduction* (pp. 14–28). Palgrave Macmillan.
Smart, C. (1990). Law's power, the sexed body, and feminist discourse. *Journal of Law and Society, 17*, 194–210.
Tangsathaporn, P. (2024). Senate passes same-sex marriage law. Bangkok Post, June 19.

Chapter 15
Naturalizing the "Unnatural" Beauty: Making Sense of the Iconic Performance of the Body in the Social Media Platform of China

Renxue Wan

Abstract Drawing from the online fitness instruction of fitness influencers on "Bilibili.com," the article examines how bodily performance on Chinese social media platforms shapes and influences contemporary body aesthetics. The developments in media visual technology and virtual interaction triggered by the proliferation of social media platforms have fundamentally changed how the body is gazed at and the way individuals imagine the shape of the body. Through the analysis of textual and visual data from the top 9 rated fitness influencers from February 2020 to May 2021, the empirical aim of the paper is to examine how fitness influencers "naturalize" the body ideal, once considered "unnatural" in the Chinese context, by re-aligning two opposing discourses on the body and beauty that coexist in Chinese society and creating the "performative scripts" that legitimize the objectification of the body. Drawing on iconic theory, which takes celebrities as typical aesthetic forms that encapsulate profound meanings and purposes, the article brings in the aesthetic dimension of the body and identifies twofold bodily performative scripts used by fitness influencers, the glamorizing script and the moralizing script. The article jettisons the critical-constructivist approach in the literature that describes the commodification and objectification of the body as a process of de-meaning and exploitation. Instead, the paper advocates a cultural sociological perspective that takes the marketing and commodification of the body as a meaning-making process in which the discourse of body aesthetics is purposefully reconstructed.

Keywords Social media platform · Fitness influencer · The performativity of the body · Iconic theory

---

R. Wan (✉)
School of Sociology, Central China Normal University, Wuhan, China

J. Pérez-Jara, Í. Ongay (eds.), *Beyond Nature and Nurture*, Synthese Library 497,
https://doi.org/10.1007/978-3-031-87715-5_15

15.1 Introduction

"Get moving! This is to help you exercise your gluteus maximus and outer thigh muscles!" During a fitness live-streaming event on a popular social media platform, a highly influential fitness trainer and his spouse led the audience in an aerobics workout. The video was enhanced with special effects, creating a porcelain doll-like glow on their skin and a slender and proportionate physique. Real-time comments displayed at the bottom of the screen allowed viewers to engage with the hosts, with many expressing admiration, such as "You guys are so hot!" This energizing fitness session lasted almost an hour and attracted nearly 10 million viewers.

Although online fitness streaming is not a new phenomenon, the scale of participation in China during the "zero-COVID" policy period was unprecedented, which was mainly due to restrictions on physical mobility resulting from pandemic-related policies. Previously an ordinary activity, online fitness training became a massive carnival-like event. Beneath the cult-like following of online fitness lies an embrace of novel fitness culture in Chinese society, centered around naturalizing what was deemed an "unnatural" way of body management.

The term "unnatural" beauty refers to beauty ideals based on virtual aesthetics, where digitally altered and photoshopped bodies are considered the norm. The theoretical tension of this chapter arises when the ideal of "unnatural" beauty created on the Internet is transformed into a "natural" way of perceiving the body under the massive proliferation of social media platforms (SMPs). This tension echoes the debate in previous body work literature centered around judgments and social implications of natural vs. manufactured beauty (Peiss, 1998/2011; Wegenstein, 2006, 2012; Holliday & Elfving-Hwang, 2012).

Previous discussions, however, are dominatingly based on the critical-constructivist model, from which two arguments are drawn towards the transition of body work from natural to artificial. The first holds that people's belief in manufactured or artificial beauty, promoted and practiced, especially by individuals who are employed in the beauty industry, is a process of the commodification of the body, which will lead to self-alienation under the capitalist mode of production (Wolf, 1991/2002; Baudrillard, 1970/1998; Wegenstein, 2006, 2012). In this process, the body disconnects from the mind and soul, thus damaging the integrity of the subject. The second argument deals with the natural/cultural dichotomy, underscoring the cultural struggles behind the beauty, e.g., Western-non-Western, men-women (Bordo, 1993; Dworkin & Wachs, 2004; Xu & Feiner, 2007; Elias et al., 2017; Dippner, 2018). The above two arguments show that the critical-constructivist perspective describes the commodification and objectification of the body as a process of de-meaning and exploitation.

The arguments proposed in this chapter are the opposite. First, building on the discussions of Miller (1987) and Alexander (2020) regarding the "meaning" of material goods and commodities, this article takes the bodily presentation performed by fitness influencers on social media platforms as a meaning-making process. In his book *Material Culture and Mass Consumption*, Miller (1987, p. 86) advocates the retainment of the positive connotations of Hegel's view between objectification and subjectification. He points out that "objectification refers to a process of externalization and sublation essential to the development of a given subject." Alexander (2020) furthers the perspective by analyzing the "performativity" of objects, taking the process of objectification as the combination of aesthetic surface and discursive depth. Building on the "anti-commodification" model, I take fitness influencers' presentation and marketing of the body as a process of defining the meaning of fitness, beauty, and health by transforming and utilizing their corporeal attributes.

Second, the "meaning" attached to beauty practices such as fitness, cosmetics are also culturally specific. Instead of following the "cultural-hegemonic" model, which takes cultural representations and the power relations behind them as one-sided and static,[1] the chapter adopts a pragmatic model (Alexander, 2004) by showing how influencers creatively and tactfully combine different body aesthetics with the moral discourses that better suit the cultural background of their target audience, while simultaneously constructing a self-idiosyncrasy.

In China, promoting fitness and beauty signifies a challenge to deeply ingrained moral ideals which worships natural beauty, balance, and harmony between the body and mind. Any attempt to modify the body through external or artificial forces was considered harmful and unethical (Yang, 1996; Wu & Tsiang, 2005). Less than half a century ago in China, emphasizing physical appearance and an aesthetic of the glamorous body was still considered a cultural taboo (Brownell, 1995); and just 15 years ago, there was no such profession as "fitness trainer" in China, let alone full-time fitness influencers (Wan, 2022). It suggests that we need to situate the bodily performance fitness influencers in a specific cultural-historical context and make a more nuanced examination of the cultural meanings they present.

Previous studies have also focused on the symbolic dimension of bodily performance without focusing on the precise corporeal attributes that make it possible. The aesthetic dimension — of sensory experiences made of flesh and blood — is downplayed (Featherstone, 2010). Drawing on iconic theory (Alexander, 2010a, 2020), the chapter argues that the "meaning" demarcated by class, gender, and ethnic concepts can only be manifested and mediated by aesthetic experiences elicited by the shape, makeup, and demeanor of the actor's body. In everyday life, these bodily representations are aesthetic negotiations between actors in the midst of complex discourses.

Using Chinese fitness influencers as a case, the objective of this chapter is to explain how fitness influencers "naturalize" body management that is once considered "unnatural" by realigning two seemingly opposing discourses regarding the body and beauty exist in Chinese society and creating the "performative scripts" that legitimize the objectification of the body. Drawing on the celebrity-icon analysis of the iconic theory, which considers celebrities as typical aesthetic

[1] See Holliday and Elfving-Hwang's (2012) critique on the Western dominance model in the study of beauty culture in Asian academic societies.

forms encapsulating profound meanings and purposes (Alexander, 2010a, 2020), this chapter brings in the aesthetic dimension of the body and identifies twofold bodily performative scripts used by fitness influencers: the glamorizing and the moralizing scripts.

Theoretically, I jettison the materialistic and the critical-constructivist approach in body and fitness studies, presenting a potential street for understanding the objectification of the body by taking the exercising body as an icon composed of aesthetic surface and discursive depth (Alexander, 2010a, 2010b, 2020). The bodily performance of fitness influencers on SMPs is thus considered a meaning-making process during which the discourse of body aesthetics is purposefully re-constructed.

15.2 Natural or Self-Made: Discourses of Body Aesthetics in China

The ubiquity of fitness influencers with perfectly proportioned and sexualized figures on social media has marked a fundamental shift in body aesthetics from centuries ago. In traditional Chinese society, the display of a curvy, athletic, and seductive body is unimaginable. Before introducing how fitness influencers design their "performative scripts," this section examines two body aesthetic discourses that coexist in Chinese society, namely the "natural" body aesthetic discourse that advocates the unadorned and homogeneous body, and the "self-made" body aesthetic discourse that welcomes the well-managed and trained, modified body.

15.2.1 The "Natural" Body Aesthetics Discourse

The aesthetic discourse of the natural body in China advocates an unadorned, restrained, and disciplined body. Researchers studying the iconicity of the body in this period agree that the body in traditional Chinese visual culture has been stripped of its inherent physical glory compared to its Western counterparts (Wu & Tsiang, 2005). As Mark Elvin notes:

> Chinese pictures of the human body, clothed or semi-clothed (in furtive pornography), are – to Western eyes – meager, schematic, and inadequate. (Elvin, 1989, p. 267)

The restrained and introverted body aesthetic is closely related to Confucianism's "xíng-qì-xīn" (shape-energy-mind) theory of oneness. This theory holds that the human body serves as a medium to connect with the universe. By maintaining the body, humans coexist with nature, achieve balance, and nourish the soul, and this connection is achieved through a mysterious energy called "qì" (Yang, 2011). In the concept of body-nature connection, people must maintain simplicity in appearance and emotional stability to achieve harmony. It means that people must achieve health through introspection. In the field of fitness and sports, China's traditional martial arts also emphasize "naturalness." The purpose of strengthening muscles through exercise is not to show off one's physique, but rather the opposite.

As a form of physical training, martial arts practice is depicted as painful, arduous, and requiring strong willpower. Individuals are able to activate their muscles and connect them with nature as they are cultivated to have a mindful, forbearing, and tranquil mentality (Allen, 2015). By creating bodily pain, the body can become more restrained, allowing personal character development. These ideas are clearly illustrated in depictions of martial arts legends such as Bruce Lee (Bowman, 2013), which have had a powerful influence on fitness concepts throughout the broader East Asian cultural sphere.

The traditional discourse of bodily aesthetics is also highly gendered. While men's bodies are expected to be strong, women are expected to present themselves as elegant yet restrained. Women's beauty practices usually aim to maintain fair skin and a delicate physique, reflecting their moral ethos as obedient and pure ((Spielvogel, 2003). Foot binding was one of the most extreme examples of this emphasis on feminine aesthetics of the time. The bound feet in the nineteenth century were reckoned as part of the decoration of the body, which decorated women as dainty and refined (Ko, 1997; Mann, 2011).

15.2.2 The "Self-made" Body Aesthetics Discourse

In the year 1986, the fourth National BodyBuilding Competition was held under the support of the General Administration of Sports of China with the slogan "Bringing Chinese Bodybuilding to the International Stage," attracting thousands of people to visit. It was also the first time that women appeared in a national sports competition wearing a bikini. At that time, "Shen Pao" (Shanghai News) reported this event as "liberating the human instinct for pursuing beauty for the very first time."

The self-made aesthetic discourse defines the beauty of the body as workable, athletic, and seductive. Like the famous slogan in the Chinese cosmetics industry indicates, "There are no ugly women, only lazy women," the maintenance of health and beauty is seen as the responsibility of the individual and is achieved through continual changes to the body (Glassner, 1989). Studies have shown that "change" is often portrayed as self-investment or self-empowerment in consumerist ideologies (Maguire, 2002; Featherstone, 2010).

Fitness also implies that beauty requires individual effort. Its secret lies in tireless body management through multiple technologies. Based on modern sports medicine and health science, this management views the body as a material entity composed of limbs, muscles, and organs that rely on various nutrients to maintain their functions (Zhao, 2015). "Fitness" is an activity that breaks down the entire body into individual muscles and trains them according to prescribed steps. Each muscle has its required shape, and the body can only be considered "beautiful" when all parts conform to specific shapes and proportions.

Since the mid-2000s, the craze for bodybuilding, or what scholars have characterized as "neoliberal" body ideals, has been further fueled by the rampant development of the influencer economy and China's "Internet +" agenda (Dippner, 2018; Craig et al., 2021). The General Administration of Sport of China initiated the "Internet + Fitness" agenda, which is seen as coincidentally prescient after the outbreak of the COVID-19 pandemic in late 2019, to promote "streaming home exercise" supported by SMPs (The State Council, 2015).

The advent of SMPs has further changed the way people perceive and manage their bodies. Instead of accepting their actual bodies, people are increasingly concerned with whether their bodies are photogenic. Leading SMPs in China, such as Tik-Tok, Bilibili, and Weibo, provide users with various Photoshop functions, ranging from adjusting the brightness, contrast, and angle to modifying the lines, shades, and symmetry of the body. In addition, fitness influencers on SMPs also try to portray physical exercise as a relaxing and even enjoyable pastime rather than a strenuous activity.

This section introduces two body aesthetic discourses that exist in Chinese society, namely the "natural" and the "self-made" body aesthetic discourses. They define what is considered "beauty" and how to achieve it in fundamentally different ways. In the following research, I will show how fitness influencers create "performative scripts" that re-narrate the "self-made" body aesthetic as a way to enhance "natural" beauty through fitness training.

15.3 Research Methods

15.3.1 Conceptualizing the Iconic Performance in Cyberspace

To understand the performativity of the body in fitness exercise, I draw on useful concepts and insights from iconic theory, which sees objects as living things condensed with discursive depth (Alexander, 2010a, 2010b, 2020). In this case, the cultural meanings that animate the objects must and can only be revealed through the aesthetic surface. Following the thread, celebrities are seen as one of the most potent icons in everyday life, whose physical appearance, as the material surface of objects, commands attention and compels attachment (Alexander, 2010b: 324). I define fitness influencers' elaborate instruction and multiple ways of bodily presentation on SMPs as iconic performances. In the process, fitness influencers act as active agents who, based on their awareness of the iconic power of the body, consciously deploy iconic scripts (gendering, problematizing, and moralizing the body) to market and monetize the iconic body as an aesthetic ideal through multiple social media functions (i.e., photoshopping, video editing, e-brand marketing, and social commerce).

According to Alexander (2020), before an object is staged, it is ingeniously created and aesthetically designed by agents with scripts and access to the means

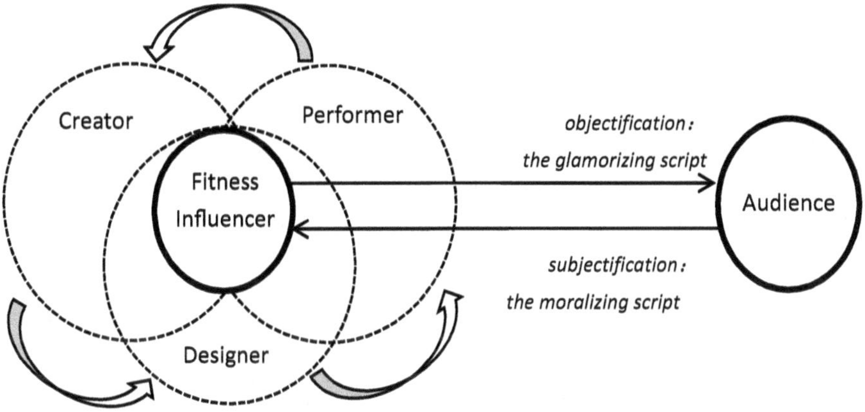

Fig. 15.1 Theoretical frame of the iconic performance of the body

of symbolic production. Once produced, the object enters the stage of circulation when it is packaged and sold by a group of professional promoters such as marketers, advertisers, and PR specialists. This analytical framework assumes that iconic production is a linear process in which creation, design, and promotion occur chronologically and independently. This is probably true in conventional organizations or industries where the production process is highly institutionalized.

However, in the realm of SMPs, the production process is decentralized because the content is massively generated by the users themselves, who have different interests and backgrounds (Buckley & Prashantham, 2016). Those who gain influence from the grassroots must act as creators, designers, directors, and performers throughout the iconic production process. Based on this, I propose to understand the performativity of the body in social media as an iterative rather than linear process in which creation, design, direction, and performance occur synchronously. Figure 15.1 above illustrates the theoretical framework of the iconic performance of the body, centering on the role of the fitness influencer, who simultaneously acts as a creator, designer, and performer. These roles are interconnected through dynamic feedback loops, shaping the influencer's bodily presentation.

The solid arrow from the fitness influencer to the audience represents the process of body objectification, wherein the influencer constructs an "aesthetic surface" by presenting an idealized and aestheticized physique. Conversely, the solid arrow from the audience to the fitness influencer signifies the moralizing script, through which influencers imbue their perfected bodies with moral virtues, fostering subjective identification among viewers.

15.3.2 The Collection and Analysis of Data

My research focuses on the social media platform Bilibili, chosen for its impressive popularity in China. As of June 30, 2020, Bilibili boasts over 172 million monthly active users, making it the most widely-used platform in the country. Unlike its competitors Weibo and TikTok, Bilibili offers a diverse range of content, featuring live streams and longer videos that span an array of topics, from entertainment to self-learning. Additionally, Bilibili was the first platform in China to introduce the innovative feature of "bullet messaging" (known as "danmaku" in Japanese video games). This feature allows viewers to leave rapid-fire comments in real-time while watching a recorded video, and has been praised for its ability to "reinvigorate the meaning of liveness while creating a sense of sharedness amongst audiences". (Zhao, 2016). Given its unique interactivity, Bilibili is an intriguing subject for my investigation into the interactivity of iconic performances.

I use Python 3.7 to access the API of the Bilibili platform and obtain 565 videos uploaded by the 9 most popular fitness influencers of the fitness channel from February 1, 2020 to May 31, 2021. The data consists of two parts. One is in the form of statistics regarding the total video views, the number of "likes," "comments," "danmaku," and the amount of "social tips." The other is the textual analysis regarding the titles and descriptions of the videos. I performed word frequency to categorize the words into organized themes. The statistical and textual data were primarily collected using data mining techniques.

To investigate the performativity of the body, I collected data by capturing and analyzing screenshots from the videos, as well as by observing and documenting the operational practices of fitness influencers on their channels. Specifically, I captured 56 screenshots at moments when danmaku appeared on the screen, signaling heightened audience engagement and the emergence of iconic consciousness. Given the non-textual nature of these screenshots, I employed a structural-hermeneutic approach to conduct a detailed analysis of how fitness influencers utilize various aesthetic elements—including body shape, facial expressions, postures, clothing, hairstyles, and the sensory effects they generate, such as video settings, background music, and other visual details—to construct and reinforce specific body ideals that align with the aesthetic preferences of their target audiences.

15.4 Negotiating the "Natural" Beauty

Interactive data illustrates the popularity and influence of the video, which indicates, though not necessarily, the success of the iconic performance of the body. Table 15.1 shows the interactive data of 9 fitness influencers from high to low according to the average number of video views, likes, social tipping, comments, and danmaku. In general, among the top 9 fitness influencers, the proportion of male and female is basically equal (female influencer accounts for 55.6%). As shown in the

Table 15.1 General interactive statistics of the top 9 fitness influencers of Bilibili.com

ID	Sex	Number of subscribers	Number of uploaded videos	Average video views	Average likes	Average social tipping	Average comments	Average danmaku
April Han	Female	2,415,702	33	3,511,318	71,009	31,152	5247	20,555
Zoey	Female	7,615,342	71	1,136,982	36,754	11,233	2077	6498
Pamela Reif	Female	6,533,058	65	2,155,025	34,197	11,653	1633	5318
Aurora Ouyang	Female	2,224,290	97	1,010,144	35,461	5291	1903	7699
Shuaisoserious	Male	1,997,520	30	650,100	41,068	58,163	4641	8657
Shugui	Male	2,011,783	70	762,048	47,785	23,878	1627	3273
Yan Shuaiqi	Male	1,643,456	97	676,752	35,744	21,659	1407	2626
Uncle Zhuo Zengzhong	Male	1,151,260	48	571,969	28,059	17,234	1187	2740
Shoubuliaomao	Female	1,241,568	54	399,662	12,263	4468	828	2462

data ranking, the top 4 fitness influencers are all female, followed by four male influencers.

The interactive data shows that each fitness influencer has more than a million followers on their channel, demonstrating their immense commercial value. Top-ranked influencers also benefit from their personal brands and commercial endorsements. In addition, the average video view for all 9 fitness influencers is over 100,000. Notably, the number of followers differs from the popularity of the video. For example, April Han's average video views reached 3 million, with three times fewer followers than Zoey, who has the highest number of followers. The same is true between Zoey and Pamela Reif.

In general, the statistics show the immense influence of the aesthetic surface. The crux of the matter is how fitness influencers can most quickly and maximally evoke the sensory resonance of the audience in front of the screen through the display of appearance. In this case, the influencers who gain exponential commercial benefit are those who set themselves up as image ideals come to life. By gendering and problematizing the body, they carve out the "iconic binaries" that distinguish appearance as hot or unattractive, form as symmetrical or deformed, and personal aura as appealing or repulsive.

15.4.1 Glamorizing the "Unnatural" Beauty

> For Babes who want to become beautiful, Attention! Here is the simplist fitness course regardless of your exercising experience: Swan-like neck, ruler-shaped shoulders, defined collarbone... After all, only those who meet all these standards can be called GODDESS! (Aurora Ouyang, 2020-04-01 17:50:08)

Above is the description of the video posted by a female influencer named Aurora Ouyang on April 1, 2020, titled "Get your ruler-like shoulder and angle-like back! Also, get rid of the hunched trapezius that makes you look slouchy! A must-watch 10-minute workout video by Aurora Ouyang." This video received over 3 million views within a week, giving Aurora more than a 30% increase in followers in 1 day. The video began with Aurora Ouyang adjusting the lights in her room. She said she used the "cold light effect" to light up the room and her skin to lift viewers' spirits. With a cat yawning on the back of the sofa, she also wanted to create a homey atmosphere to encourage those "lazy" girls to take advantage of every minute to burn off their calories.

The terms "swan-like" neck and "ruler-shaped" shoulders have gained popularity as trending hashtags originating from body challenges initiated by major social media platforms (SMPs) in China, such as Weibo and Bilibili. These challenges aim to encourage user participation in "streaming home exercises." The analogy between the human body and physical objects can be understood as what Lakoff and Johnson (2008) call the "ontological metaphor," which views physical phenomena as entities and substances bounded by a surface. This set of metaphorical constructs effectively "animates" the body, allowing viewers to cognitively envision and aesthetically appreciate the slenderness of the neck. As a result, fitness influencers frequently employ these metaphors, both textually and visually, to construct and reinforce feminine beauty ideals.

For instance, in a fitness video posted on April 1, 2020, Aurora Ouyang demonstrated a "visualized" criterion for achieving "ruler-shaped" shoulders. She placed a lipstick tube in the hollow of her collarbone while maintaining an upright posture, ensuring that the tube remained balanced. Dressed in an off-the-shoulder top, she addressed the camera, stating, "If your shoulders can maintain this position, then you have nothing to worry about for the summer—you can fully embrace stylish outfits."

In the "glamorizing script," fitness influencers present perfect bodies, which are actually far above everyday standards, to the audience as "normal" through the use of editing, camera angles, lighting, and in combination with their own physical condition. Meanwhile, they also rationalize the "unnatural" body aesthetics by treating those bodies that do not fit the "ontological metaphor" as problematic and ordinary body shapes in everyday life as abnormal. Just as Aurora Ouyang praised the beauty of a sexy collarbone and ruler-shaped shoulders, she simultaneously problematized the body in a highly materialized way. In the same video, Aurora Ouyang demonstrated how one would look "fat and pathetic" with rounded shoulders.

Beyond Aurora Ouyang, another influencer, April Han, employed precise measurement techniques to quantify body transformation. In her fitness video titled "10-minute efficient thigh slimming exercise - lose 3cm in a week! Beginners friendly! Practice lying in bed for longer legs!"(2020-03-06 19, 00:37), April Han showed herself measuring her thighs and emphasized how she could make her previously bulky and flabby thigh muscles firm with just 10 min of exercise. Meanwhile, the "danmaku" floating on the video comments, "her legs look whiter after a week" or "mark my first day of practice! Keep it up!" From the video's release date (March 16, 2020) to the recording date (October 21, 2020), the video was viewed 26,269,223 times and became the most viewed fitness video of the month, with a total of 247,220 social tips. In two other fitness videos that also received over 20 million views during this period, April Han used metaphors such as the "umbrella-shaped trapezius" or "fat-wrapped legs" to describe untrained or unmodified bodies as problematic.

Table 15.2 demonstrates the binary descriptions that fitness influencers used to describe the glamorous and problematic bodies using ontological metaphor.

Like female fitness influencers who design and perform according to mainstream feminine beauty ideals, the gendered script is also deployed by male fitness influencers who are under the spell of heterosexual masculinity. In analyzing everyday aesthetic essentialism, Alexander points out that the aesthetic surface of the celebrity icon is imbued with gender rhetoric:

> The surfaces of the male celebrity-icon are sublime and the female celebrity-icon beautiful, and each surface carries the mythical meanings of its (putatively) archetypically gendered forms. (Alexander, 2010b, p. 330)

Table 15.2 Word frequency of the binary descriptions of the body

Idealized Body	N = 110	Problemized Body	N = 93
Swan-like neck	32	Umbrella-shaped trapezius	24
Ruler-shaped shoulder	24	Rounded shoulder	21
Inverted triangle figure	22	Square-shaped face	18
Ant waist	11	Saggy breast	11
V-shaped face	9	Puppet face	7
Peach butt	7	Bucket waist	6
Pencil legs	5	Beer barel belly	6

In fact, the world of fitness is a historically masculinized space. Introduced by Zhao Zhuguang, known as the Chinese "Charles Atlas," the imagery of body shaping (i.e., toned muscles, broad shoulders) was supposedly associated with hetero-masculine appearances. Only by the end of 2018, the number of female consumers in fitness clubs outnumbered men, accounting for 53.2% of the total number of consumers. However, for most Chinese women, fitness is still a male-dominated activity. As Xiong (2021) note in their narrative analysis of Chinese women's fitness practices, compared to Western countries, the strength training domain was most often occupied by men who consciously projected fitness as a heterosexual prowess. In my case, the male fitness influencers all use this masculinized discourse in fitness to create and design the corresponding "aesthetic surfaces."

The most common script is the one that glamorizes the inverted triangle figure as a masculine ideal. For example, the famous fitness influencer named Uncle Zhuo, who showed in his most viewed video how he managed to evolve from a scrawny weakling to a powerful man with a rugged physique. The video began with him describing how being scrawny made him feel awkward growing up. "It's understandable that girls find it rude when you tell them outright that they're fat. As a guy, you can imagine how awkward it is to be called skinny. It's really offensive." (2021-03-16 18:05:00) The scene then went from him weighing less than 110 pounds to a finely groomed, V-shaped man in a suit. At that point, danmaku began to occupy the screen. He illustrated in the video how the toned muscles and broad shoulders helped him regain the masculinity he once lacked.

For men, a "problematic" body is one that appears feminine, which includes both being thin and being overweight. Male fitness influencers Shuaisoserious and Yan Shuaiqi introduced their audience to the medical term "gynecomastia," which in Chinese translates to a symptom of male obesity as feminization based on modern medicine. In his video "Save you From Gynecomastia: How to Get Rid of Breast Fat and Build the Perfect Breast Shape," (2021-01-18 12:13:24), Shuaisoserious described the enlargement of male breast tissue as a physical health problem that causes men embarrassment and lack of self-confidence. It signifies a hormonal imbalance that makes most men lack masculinity and appear unattractive. Therefore, to get rid of gynecomastia, he listed a number of diet and fitness strategies for the audience by collecting research from various medical journals. For example, he suggested avoiding foods that can increase estrogen levels, such as dairy or soy products, while exercising.

The introduction to the glamorizing scripts of fitness influencers typically reflects the "self-made" body aesthetic discourse, in which aesthetic experiences of beauty are defined by two opposing codes. On the one hand, the "natural" body, which has not been intentionally trained or modified from its initial state, is often seen as unhealthy, awkward, or inadequate. On the other hand, the "unnatural" body, which involves an ongoing process of artificial correction and adjustment, is praised as breathtaking, beautiful, and stylish.

15.4.2 Moralizing the "Unnatural" Beauty

Creating an attractive appearance through the glamorizing script is important for fitness influencers, but it is not enough for a performance to be "unforgettable" or "successful." According to Alexander (2020), an object's performativity is only considered successful when it can "fuse" with the audience, and this fusion occurs when profound meanings are encapsulated beneath its aesthetic surfaces. As a result, influencers must develop a "moralizing script" that attaches cultural and moral significance to fitness and makes audiences appreciate and empathize with what is under the skin (Wegenstein, 2006; Alexander, 2020).

Interestingly, as summarized in the previous article, although the glamorizing script of fitness influencers encourages body modification, it is common for them to borrow the discourse of "natural" body aesthetics to demonstrate moral superiority. Rather than presenting themselves as unattainable, perfect idols, fitness influencers present authenticity by "disenchanting" their audiences from beauty myths. In the video "Fake?! Revealing the Ugly Truth About Fitness Influencers! The Other Side of the Internet Celebrity You Don't Know." (2020-09-10 20:21:49), Zoey, the fitness influencer with the most followers on, showed her belly and the fat under her arms at the beginning of the video.

In the video, Zoey "revealed" the video shooting techniques of fitness influencers, demonstrating how the same body muscles, body size and posture can have completely different visual effects due to changes in camera position and viewing angle. At the end of the video, she urged her fans to use physical exercise as a channel to promote inner health. At the same time, Zoey, who is Malaysian-Chinese, used the phrase "xiū shēn yǎng xìng," a traditional Chinese idiom, to describe how a well-trained body helps nourish one's moral character. The bullet comments: "You are so real!" occupied the screen while the video was playing.

In addition to emphasizing the integration of fitness and introspection through the "demystify" approach, fitness influencers present different "aesthetic surfaces" for different audiences, which in turn reflects the moral ethos of "beauty" in different cultural systems. For example, Shoubuliaomao identified herself as a fitness influencer who specializes in "facial exercises" that can help maintain a youthful appearance. In her most popular video series, "Get Jawline Like Ishihara

Satomi, 30 Days to Improve Cheekbone Protrusion and Reduce Jaw Muscle." (2020-04-01 16:51:43), she used Japanese celebrity Ishihara Satomi as an example of how to achieve a natural "small V-shaped face" through facial muscle exercises, as opposed to plastic surgery. A small "V" face refers to a face shape with fuller cheeks and a smaller, more defined jawbone. According to Shoubuliaomao, a small and defined jawline can make you look younger and convey a sense of tenderness and vulnerability.

In contrast to Shoubuliaomao's strong advocacy of fairness, Zoey is more flexible in defining a multi-gender ethic that addresses both Eastern and Western aesthetic discourses. She presents herself in a tanned, glowing skin tone that she describes as indicative of health and vitality. As Spielvogel (2003) analyzes, the association of tan with a qualified fitness instructor demonstrates the influence of American fitness culture in Japan. At the same time, in contrast to Shoubuliaomao's or April Han's focused on how to be "thin," Zoey's videos asserted that it is problematic for Asian girls to equate thinness or skinny with beauty.

Instead, Zoey posted photos of herself being more "muscular" after years of working out and tells viewers that she "feels strong because she is strong." (2021-01-20 20:44:08). In the video in which she attempted to challenge the "guó fēng" makeup style (Chinese makeup style) (2020-07-04 18:30:46), she joked that even though her tiny eyes and dark skin did not fit the traditional definition of classical beauty, she still hoped to show China's alternative female beauty of chivalry, self-confidence, and autonomy by attempting to wear traditional Chinese costumes.

Conventionally, people believe that the worship of youth, fair skin, and slim figures has become a global rule under the influence of popular culture. However, the moral values behind these demands belong to different cultural systems. The popularity of aesthetic standards that refer to Japan and South Korea in China is not accidental; rather, it reflects the rise of the Asian idol industry, which emerged at the end of the twentieth century and swept the world (Galbraith & Karlin, 2012). This idol industry emphasizes a "kawaii" (cute) appearance style that emphasizes a childlike aesthetic experience. Typically, this appearance is characterized by round eyes, double eyelids, a small but prominent nose bridge, white skin, an oval face with soft contours, and slender limbs. This appearance embodies an aesthetic experience of innocence, purity, weakness, gentleness, and delicacy.

Through this experience, people can feel the female moral qualities prevalent in East Asian societies: compliance, naturalness, kindness, and devotion, which also imply purity (untainted) and sanctity, in contrast to the mature female image associated with sex appeal (tainted) (Yang, 2011; Hoang, 2015). Galbraith and Karlin (2012) suggest that the infantile and seemingly non-aggressive sensuality has a unique sexual appeal for countries that generally discourage individuals from overtly expressing their desires.

Male influencers also draw on different cultural representations of masculinity in their moralizing script. The following images show scenes from two influencers, Shuaisoserious and Uncle Zhuo, in their own videos in which they also cite Arnold Schwarzenegger as a spokesperson for male fitness. They cite this iconic figure of American popular fitness as their fitness mentor and spiritual leader, and share how Schwarzenegger's fitness journey helped them develop the moral qualities of perseverance and grit. In contrast, Yan shuaiqi explained in his video "What's the Secret to Bruce Lee's Physique? With a Real Bruce Lee Training Plan Attached!" (2020-04-10 11:42:39) that China also has its own influential fitness mentors worldwide. In his video, Yanshuaiqi introduced the philosophy behind Bruce Lee's Jeet Kune Do and the martial artist's thoughts —

> He is more like a Son Goku living in the real world, expressing the full potential of his body through continuous learning and integration of new knowledge. "Limitless" is the core of his Jeet Kune Do and philosophical ideas... There will always be some stagnation on the journey of life, but we must not stop here and continue to strive to go beyond (2020-04-10 11:42:39).

The discursive strategies of male fitness influencers reveal that multiple dominant masculinities are reflected not only in the dominant vs. subordinate/marginalized divide, but also within what is considered "hegemonic masculinity." Although both promote moralities such as perseverance, persistence, and self-transcendence, fitness influencers use different cultural references. As shown, male fitness influencers from different ethnic groups or regions construct different moral references for male masculinity. Yanshuaiqi's videos show how fitness influencers use culturally adaptive language to naturalize their perfect physique. Through an introduction to Bruce Lee's martial arts and fitness philosophy, Yanshuaiqi links robustness and resilience to traditional Chinese philosophical and moral concepts that are also compatible with the entrepreneurial ethics present in fitness.

15.5 Conclusions

In her book, *Beauty Myth*, Naomi Wolf (1991/2002) poignantly points out how beauty ideals presented in advertisements and on screens could bring about the disconnection between the body and the self, especially for women, and that individuals are passive agents follow the manipulation of political and economic forces, in the process of which the body is reduced to an (even hostile) "object" separated from the subject. In the twenty-first century, with the popularity of social media platforms, new developments have arisen in discussions of images and the alienation of the body (Wegenstein, 2006; Bailey, 2016). The integration of social media platforms and e-commerce has turned former "bloggers" into "influencers," creating a new profession relying on the marleting of looks. Research has argued that the influencers' reliance on the "attention economy" has made the exploitation of the body more insidious and at the same time caused reproduced social hierarchy (Abidin, 2016; Duffy, 2016).

However, it is hard to ignore the fact that actors on SMPs are by no means the puppets subject to structural forces. In fact, scholars have already recognized the need to move away from binary assumptions such as commodification-alienation and domination-subordination present in a critical-constructivist perspective to reveal the complex relationship between online presentations and discursive implications within SMPs[2] (see Smith, 2016). As to Kopytoff (1986/2013, p. 64), "from a cultural perspective, the production of commodities is a cultural and cognitive process. Commodities must not only be produced materially as things, but also culturally marked as being a certain kind of thing." The cultural perspective allows us to bridge the nature-culture dichotomy that exists in the body work literature (Gimlin, 2007), and fill the gap of current research on fitness in China.

In Chinese studies, fitness and beauty culture are seen as a product of modern consumerism and the influence of modern medicine (Tang & Xie, 2021; Xiong, 2021). It is implicitly acknowledged that modern body aesthetics are disconnected from traditional concepts of the body. This article suggests that in actual everyday body performances, due to differences in cultural origins and target audiences, fitness influencers appropriate traditional natural discourses to justify a particular type of body aesthetics. The performative scripts reflect a more complex relationship between natural body aesthetic discourse and self-made body aesthetic discourse, providing us with a space to understand the multiple possibilities of cultural translation. This is essential for the international academic community to understand fitness culture in China.

References

Abidin, C. (2016). Visibility labour: Engaging with influencers' fashion brands and# OOTD advertorial campaigns on Instagram. *Media International Australia, 161*(1), 86–100. https://doi.org/10.1177/1329878X16665177

Alexander, J. C. (2004). Cultural pragmatics: Social performance between ritual and strategy. *Sociological Theory, 22*(4), 527–573. https://doi.org/10.1111/j.0735-2751.2004.00233.x

Alexander, J. C. (2010a). Iconic consciousness: The material feeling of meaning. *Thesis Eleven, 103*, 10–25. https://doi.org/10.1177/0725513610381369

Alexander, J. C. (2010b). The celebrity-icon. *Cultural Sociology, 4*(3), 323–336. https://doi.org/10.1177/1749975510380316

Alexander, J. C. (2020). The performativity of objects. *Sociologisk Forskning, 57*(3–4), 381–409. https://doi.org/10.37062/sf.57.22321

Allen, B. (2015). *Striking beauty: A philosophical look at the Asian martial arts*. Columbia University Press.

Bailey, S. (2016). *Performance anxiety in media culture: The trauma of appearance and the Drama of disappearance*. Palgrave Macmillan.

Baudrillard, J. (1970/1998). *The consumer society: Myths and structures*. Sage.

Bordo, S. (1993). *Unbearable weight: Feminism, Western culture, and the body*. University of California Press.

Bowman, P. (2013). *Beyond Bruce lee: Chasing the dragon through film, philosophy, and popular culture*. Columbia University Press.

Brownell, S. (1995). *Training the body for China: Sports in the moral order of the People's republic*. University of Chicago Press.

[2] See Smith's (2016) advocation of perceiving celebrity culture as egalitarian and democratic.

Buckley, P. J., & Prashantham, S. (2016). Global interfirm networks: The division of entrepreneurial labor between MNEs and SMEs. *Academy of Management Perspectives, 30*(1), 50–70. https://doi.org/10.5465/amp.2013.0144
Craig, D., Lin, J., & Cunningham, S. (2021). *Wanghong as social media entertainment in China*. Springer Nature Switzerland AG.
Dippner, A. (2018). Social media celebrities and neoliberal body politics in China. *KFG Working Paper Series, 91*, 3–25.
Duffy, B. E. (2016). The romance of work: Gender and aspirational labour in the digital culture industries. *International Journal of Cultural Studies, 19*(4), 441–457. https://doi.org/10.1177/1367877915572186
Dworkin, S. L., & Wachs, F. L. (2004). "Getting your body back": Postindustrial fit motherhood in shape fit pregnancy magazine. *Gender & Society, 18*(5), 610–624. https://doi.org/10.1177/0891243204266817
Elias, A. S., Gill, R., & Scharf, C. (Eds.). (2017). *Aesthetic labour: Rethinking beauty politics in neoliberalism*. Palgrave Macmillan.
Elvin, M. (1989). Tales of Shen and Xin: Body-person and heart-mind in China during the last 150 years. In M. Feher, R. Naddaff, & N. Tazi (Eds.), *Fragments for a history of the human body part two* (pp. 267–349). Zone.
Featherstone, M. (2010). Body, image and affect in consumer culture. *Body & Society, 16*(1), 193–221. https://doi.org/10.1177/1357034X09354357
Galbraith, P. W., & Karlin, J. G. (Eds.). (2012). *Idols and celebrity in Japanese media culture*. Palgrave Macmillan.
Gimlin, D. (2007). What is "body work?" a review of the literature. *Sociology Compass, 1*(1), 353–370. https://doi.org/10.1111/j.1751-9020.2007.00015.x
Glassner, B. (1989). Fitness and the postmodern self. *Journal of Health and Social Behavior, 30*(2), 180–191. https://doi.org/10.2307/2137012
Hoang, K. K. (2015). *Dealing in desire: Asian ascendency, Western decline, and the hidden currencies of global sex work*. University of California Press.
Holliday, R., & Elfving-Hwang, J. (2012). Gender, globalization and aesthetic surgery in South Korea. *Body & Society, 18*(2), 58–81. https://doi.org/10.1177/1357034X12440828
Ko, D. (1997). The body as attire: The shifting meanings of footbinding in seventeenth-century China. *Journal of Women's History, 8*(4), 8–27. https://doi.org/10.1353/jowh.2010.0171
Kopytoff, I. (1986/2013). The cultural biography of things: Commoditization as process. In A. Appadurai (Ed.), *The social life of things: Commodities in cultural perspective* (11th ed., pp. 64–91). Cambridge University Press.
Lakoff, G., & Johnson, M. (2008). *Metaphors we live by*. University of Chicago Press.
Maguire, J. S. (2002). Body lessons: Fitness publishing and the cultural production of the fitness consumer. *International Review for the Sociology of Sport, 37*(3–4), 449–464.
Mann, S. L. (2011). *Gender and sexuality in modern Chinese history*. Cambridge University Press.
Miller, D. (1987). *Material culture and mass consumption*. Basil Blackwell.
Peiss, K. (1998/2011). *Hope in a jar: The making of America's beauty culture*. University of Pennsylvania Press.
Smith, D. R. (2016). "Imagining others more complexly": Celebrity and the ideology of fame among YouTube's "Nerdfighteria". *Celebrity Studies, 7*(3), 339–353. https://doi.org/10.1080/19392397.2015.1132174
Spielvogel, L. (2003). *Working out in Japan: Shaping the female body in Tokyo fitness clubs*. Duke University Press.
Tang, J., & Xie, Z. (2021). Discipline and distinction in the age of the internet: A sociological study of the fitness practice. *The Journal of Chinese Sociology, 8*(1), 1–25. https://doi.org/10.1186/s40711-021-00142-1
The State Council. (2015, September26). Instructions on constructing platforms to promote mass entrepreneurship. Retrieved September 11, 2020, from http://www.gov.cn/zhengce/content/2015-09/26/content_10183.htm

Wan, R. (2022). "Competing personas": Aesthetic labor in the Chinese fitness industry. *The Journal of Chinese Sociology, 9*(1), 1–18. https://doi.org/10.1186/s40711-022-00166-1
Wegenstein, B. (2006). *Getting under the skin: The body and media theory*. MIT Press.
Wegenstein, B. (2012). *The cosmetic gaze: Body modification and the construction of beauty*. MIT Press.
Wolf, N. (1991/2002). *The beauty myth: How images of beauty are used against women*. Harper Perennial.
Wu, H., & Tsiang, K. R. (Eds.). (2005). *Body and face in Chinese visual culture*. Harvard University Asia Center Publications Program.
Xiong, H. (Ed.). (2021). *The made of corporeal self*. Social Science Academic Press.
Xu, G., & Feiner, S. (2007). Meinü Jingji/China's beauty economy: Buying looks, shifting value, and changing place. *Feminist Economics, 13*(3–4), 307–323. https://doi.org/10.1080/13545700701439499
Yang, R. (1996). *"Ru Jia Shen Ti guan" (the Confucianist perspective on the body)*. Institute of Chinese Literature and Philosophy, Academia Sinica.
Yang, R. (2011). "*Liang Zhong Qi Xue, Liang Zhong Ru Xue*." (The dyadic theory of Qi and the Dyadic Confucianism behind it.), *185*(5), 143–148.
Zhao, B. (2015). Study of semantics of bodybuilding, body fitness and change, problem on the basis of logic. *Sichuan Sports Science, 04*, 23–27.
Zhao, E. J. (2016). Professionalization of amateur production in online screen entertainment in China: Hopes, frustrations and uncertainties. *International Journal of Communication, 10*, 5444–5462.

Chapter 16
Nature Loves Variety: On the "Backlash" Against Gender Equal Policies in Japan

Kimio Ito

Abstract At the end of the twentieth century, in response to the international movement for gender equality, various measures were created to eliminate gender discrimination in Japan: the Basic Law for Gender-Equal Society was enacted in 1999, and a variety of policies were developed. Since the beginning of the twenty-first century, however, a backlash against gender issues by conservative cultural figures, politicians, and the religious right has emerged. At issue was "gender-free" education. A campaign of lies, such as "they are forcing boys and girls to change clothes in the same room in the name of gender freedom," completely halted the movement toward gender equality in Japan. Conservative forces that criticized gender-free education began to target the abolition of the concept of gender itself. The target of criticism at that time was John Money, a sexologist who popularized the gender concept along with the New Guinea research of Margaret Mead, a cultural anthropologist who discovered that the temperaments of men and women change depending on culture. In criticizing Money, backlashers referred to the work of Milton Diamond, who also criticized Money, arguing that his perspective of gender as a socially constructed sex was false.

However, although Diamond was critical of Money's research methods, he clearly insisted on the perspective of gender as created by the social environment. For Diamond, the perspective of analyzing gender from the perspective of both gender as a social construct as well as biological sex differences (i.e., Nature and Nurture) was obvious. Nevertheless, the backlashers sought to distort Diamond's argument and deny the very concept of gender as a social construction.

Diamond's position could be summed up in his slogan, "Nature loves diversity." When we consider sex, we need to analyze the diversity of sexuality from both the perspective of sex, the biological sex difference, as well as the perspective of gender, which is deeply related to society and culture.

Keywords Backlash · Gender · Sex · Nature · Social environment

K. Ito (✉)
Kyoto University, Kyoto, Japan

Osaka University, Suita, Osaka, Japan

J. Pérez-Jara, Í. Ongay (eds.), *Beyond Nature and Nurture*, Synthese Library 497,
https://doi.org/10.1007/978-3-031-87715-5_16

From the mid-1970s onwards, the international community began to make serious efforts to address two major issues that had previously been 'invisible problems' and are regarded as common global problems that needed urgent solutions. The two issues here are the environment and human rights (particularly the protection of women's rights, defined by the UN as "the world's biggest human rights issue").

With regard to environmental issues, considerable progress has been made in corporate activities, as can be seen in the energy-saving policies adopted after the oil crisis. However, while progress has been made in environmental measures related to corporate production efficiency and profit growth, the reality in Japan is that environmental policies in areas related to human life, including safety management, have not been sufficient.

As to human rights, a comparison between the 1970 phase and the present clearly shows that there have been significant changes. Discrimination against people with disabilities, discrimination against foreigners (especially those from the former colonies who were called 'Zainichi'), discrimination against indigenous people such as the Ainu (a discriminatory law called the 'Law for the Protection of the Old Native People' existed in Japan until 1997), discrimination against the Buraku ostracized community)etc. are still problems in the country. It is true that there are still problems, such as discrimination against the Zainichi and the Buraku, but it is clear that the changes over the past 40 years have been significant (Ueda, 1999).

In a strange way, Japanese society has tackled the two universal challenges of the second half of the twentieth century in its own way. Curiously, however, Japanese society has made little substantive response to the issue of women's human rights, which has been described as the 'biggest' of all human rights. As early as 1990, it was already clear that an ageing society with declining birthrates would have to manifest itself in the most serious way (a reorganization of society around gender equality would have been the most strongly needed response to an ageing society with declining birthrates). The question of why Japanese society has been slower to respond to this issue of women's rights than other countries may, on reflection, be a very strange one (see Ito, 2009).

However, from around the mid-1990s, Japanese society began to respond to global trends in its own way, and the enactment of the Basic Law for a Gender-Equal Society in 1999 can be seen as the first step in this direction. (Almost 20 years after the rest of the world (including many developing countries as well as economically developed countries), Japanese society began to move towards gender equality in policy terms.

16.1 The Gender-Free Debate

Thus, at the beginning of the twenty-first century, when gender policies were finally beginning to appear in Japanese society, a strange movement, which could be called a backlash, began. This is the so-called 'uproar' over the term 'gender free'.

It was probably only in the 1990s that the term 'gender free' began to be used in Japanese society. In particular, it was often used by those who wanted to promote gender equality in the field of education. In fact, there is a reason for this. The Ministry of Education (now the Ministry of Education, Culture, Sports, Science and Technology = MECSST) positioned 'gender equality' education as insufficiently 'equal'. The Ministry of Education did not aim for full-fledged equality between men and women, but rather for 'men and women to recognize their differences and cooperate with each other' (the MECSST at the time of enactment of the revised Basic Education Law also referred to 'gender equality (Danjo-byodo)' in the revised Basic Education Law, which was a continuation of the old position of the former Ministry of Education). The Basic Law on Education was also revised, and the Minister of Education, Culture, Sport, Science and Technology at the time of its enactment gave an answer to the Diet that took over the position of the former Ministry of Education. Many people would feel that this is quite far from the image of 'gender equality' that is normally considered. Therefore, a term aiming for full-fledged 'gender equality', rather than the Ministry of Education's 'gender equality', became necessary, and the term 'gender-free' was developed (mainly from the education sector). However, this term has led to various misunderstandings and controversies (Ito, 2009).

The term 'gender-free' was often discussed as Japanese English. However, the term was already in use in the English-speaking world in the 1980s. A prime example is Barbara Houston's 1985 article entitled "Should public education be gender free? "(Huston, 1985, 1994).

At the beginning of this article, Houston classifies the implications of gender-free education into three categories: first, those that actively seek to disallow the existence of gender differences by assuming that there is no sexual gap; second, those that ignore gender or do not take gender into account; and third, those that seek to make public education gender-neutral (Gender-bias-free) by making it gender-conscious. Houston's position is naturally from the third perspective, and this freedom from gender bias is probably the dominant position in the practice of gender-free education in Japanese society.

Based on this categorization, let us summarize the gender-free debates of the 2000s.Those who argued against the gender-free debate often argued in this way. 'Those who argue for gender freedom are the ones who want to ignore gender differences between men and women, so they must be conducting physical examinations in the same room for men and women and changing clothes at PE time for boys and girls together. This is a problem." Those who make such claims are probably thinking of the gender-free concept in the first sense. In fact, education based on such misconceptions is an obstacle to achieving gender equality (a view echoed by Houston) because it does not focus properly on gender issues. In addition, by lumping together the unique characteristics and abilities of each individual (regardless of gender) as "all the same", it is easy to separate them into different groups. I think it is clear that this mechanical 'equalization' of education is out of step with the original purpose of education, which is to develop the individuality of each person (by eliminating discrimination and prejudice), while responding sensitively to gender issues.

Conversely, those who tried to promote gender-free education in Japan took the third meaning, i.e. to be free from gender bias. This is the perspective of eliminating gender discrimination from the perspective of human rights. From this perspective, the physical examination and changing of clothes in the same room for boys and girls is a big problem. This is because it clearly constitutes sexual harassment (Ito, 2009, 2014).

However, in Japanese society, a great deal of confusion has arisen because many people, including the mass media, have mistakenly taken gender freedom to mean the primary meaning. This misunderstanding of gender free has meant that the voices calling for the elimination of sex discrimination and freedom from gender bias have been taken to mean 'radicals', resulting in a lid being put on Japanese society's interest in gender issues, which had barely begun to move (thanks to Japan's gender policy, which was 20 years behind the international community, ended up creating further delays).

16.2 Backlash and Academic Research

In the context of this backlash against the trend towards gender equality, various discourses were developed in Japanese society. Interestingly, this discourse was often developed in an 'academic guise'. Among them, those who were hostile to the concept of 'gender' presented two 'enemies' and propagated through various media that 'gender studies in Japan are based on the wrong foundations, even from an academic perspective'.

One of the two 'enemies' was the work of Margaret Mead (Mead, 1928, 2001), who was to present the idea that gender is socially and culturally constructed. In her study of three social groups in New Guinea, she found that each of the three groups had distinctly different 'masculine' and 'feminine' temperaments (in her eyes, having grown up in the United States, one group was 'feminine' for both men and women, another was 'masculine' for both men and women, and the last one was seen as 'male-feminine and female-masculine'). From this she made it clear that the temperament of men and women is largely defined by culture and society (incidentally, Mead rather emphasized the decisive importance of biological sex differences and supported gender equality, but on the other hand was strongly opposed to the radical feminism of the 1960s).

It is clear that this important finding of Mead's (or rather, that women's and men's roles change depending on culture, as anyone who has travelled abroad and is a little careful will realise) was certainly one of the foundations for the discussion of gender as a social and cultural construct.

For the backlash detractors, perhaps because they feared that this Mead argument could easily lead people towards gender equality, they focused first on Mead's critique. The trigger was a post that spread online, saying that an anthropologist had debunked Mead's research with empirical studies.

Indeed, in response to the first study that made Mead's name—'Adolescence in Samoa' (the 'discovery' that there is no recognised equivalent of adolescence in Samoa as in Western societies), Australian anthropologist Freeman (Freeman, 1983) followed up her research and interviews with girls she interviewed at the time (Incidentally, Freeman's point was vigorously refuted by the North American Anthropological Association after its publication).

From this point, the argument that Mead is in academic denial spreads among the anti-gender free movements. From there, the argument that "gender researchers who talk about gender on the basis of the false Mead argument are telling a lie" is used to shut down the argument for gender equality (the author's text, which mentions Mead while making various assumptions, is published in a comprehensive high school Japanese language textbook). This led to a fierce attack from the Backlashists, who claimed that it was the first time that the word gender had appeared in a high school textbook. (See Ito, 2009).

The logic used here, while seemingly 'scientific' and 'academic' in appearance, was in fact a completely 'layman's' argument. In the first place, it would be 'erroneous', at least from a scholarly point of view, to conclude that 'all of her arguments are erroneous' just because some of Mead's research was wrong. This is because even if the details are wrong (and of course it is crucial to clarify the errors), the 'correctness' of the larger frame of the argument (if, of course, it is 'correct') is still reasonably assured.

More specifically, Freeman's studies used by backlash groups dealt with 'Samoa', as already mentioned. However, when discussing Mead's views from a gender perspective, what is most often used (as the author did) is, as already mentioned, the case of New Guinea, for example. Samoa and New Guinea are thousands of kilometres apart.

However, "Mead's doctrine is rejected. Gender theories that rely on Mead's argument are erroneous. Unfortunately, it is also true that the argument that 'it is not permissible to use Mead to explain gender' has 'passed' to a certain extent on the internet and in backlash discourse.

16.3 The Brenda Issue and the 'Money Vs Diamond Controversy'

The second target of the backlash faction was John Money (Money & Tucker, 1975), who was responsible for the spread of the use of gender concepts in Japan. It is undeniable that concepts such as 'gender roles' and 'gender identity', which he touched on in Money and P. Tucker's "Sexual Signatures on Beeing a Man or a Woman" (1975) played a major role in initiating discussions on gender in Japanese academia (although Money, a sexologist, also saw the gender concept as a purely 'social and cultural construct" = biological sex differences).

In terms of researchers who have made a semantic shift from the gender concept, which was originally a grammatical term, in the direction of 'socially and culturally constructed sex', Money can be positioned as one of the new founders of the gender concept, of whom there are several. (Incidentally, in academic terms, the gender concept is often used in the form of gender as a social and cultural construct distinct from biological sex difference, although it has now been 'watered down' and is sometimes used in newspaper and journal terms with a nuance closer to biological sex difference).

As in the case of Mead, Money's critique was also initiated in the context that 'if you reject Money's argument, you can also reject the concept of gender'. The occasion was the case of 'Brenda', who is closely associated with Money's research. For David Reimer, who was born as one of male identical twins and lost most of his penis due to a botched circumcision, Money performed a sex change operation and subsequently raised him as a girl named Brenda (Colapinto, 2000). With her twin brother having grown up as a man, Brenda's case was reputed to be an example of how a person's sense of self around gender is defined by environment rather than by birth (one of the grounds that gender is determined by environment, so to speak). However, David felt so uncomfortable with his female gender that he eventually changed his sex to male and later married a woman (he later committed suicide in his 30s).

It was Milton Diamond who was fundamentally critical of David, including Money's sometimes heavy-handed treatment of David. Through their interviews with David, Diamond and his colleagues also revealed that Money had been using considerable coercion against Brenda-David in order to reinforce his own theory.

David, on the other hand, was interviewed by Colapinto and published the incident in a magazine (Rolling Stones, Colapinto, 1998), and Colapinto subsequently compiled the interview into" As Nature Made Him "(in Japan as "The Boy Called Brenda"), A translation was published by Mumeisha Publishing in 2000. It was temporarily out of print, but was republished by Fusosha Publishing affiliated with Sankei Shimbun (conservative newspaper) in 2005 when the backlash spread.

The revelation of Manet's almost heavy-handed response to Brenda-David and Diamond's academic criticism were accepted as misunderstood (i.e. gender is biologically determined, not by environment) and spread in Japan as arguments for the rejection of the gender concept.

Here, as in the case of Mead, there are errors on two levels. The first is the argumentative approach of using a researcher's "one mistake" as an excuse to reject the academic achievements of a single scholar (Money) outright, without having done much research, and the second is the argumentative approach of "processing false information and using it to suit one's own argument" (Incidentally, in the post-writing of "A Boy Called Brenda" the translator's afterword quotes David's positive comments about the significance of the concept of gender, but in the Fusosha edition, this supposedly crucial statement has completely disappeared. It is clear from this that it was a political publication that wrote about fairness.

Thus, in Japan, Diamond came to be known as a leading theorist supporting "the conclusion that 'masculinity and femininity' are not acquired but inborn and innate" (Wikipedia in Japan "David Reimer" section, as of August 2011).

However, it is clear to see that this 'understanding' of Diamond's work is complete and utter bullshit. Perhaps it is Diamond himself who is most concerned about the 'appropriation' of Diamond's research in Japanese society, based on utter misunderstanding. In fact, Diamond repeatedly states that 'both birth and upbringing are closely intertwined', because that is his position. Naturally, Diamond also uses the concept of gender, which is deeply related to 'upbringing', quite accurately, as you will immediately understand when you read his papers. Rather than rejecting the concept of gender, it is clear at a glance that he uses the term as a key concept for the sexualisation of human beings (Diamond, 2011).

16.4 The Diamond Vs Butler Controversy

In this respect, it is the debate between Diamond and Judith Butler that will grab the attention of readers of this book. As is well known, Butler's view of gender departs from a very strong theoretical construction. That is, "this construct called 'sex' is as socially constructed as gender. In fact, perhaps sex is always gender" (Butler, 1990).

For Butler, who stands on a postmodern epistemology, sex, as a biological sex difference, is also constructed, as is gender, as a social construction. In a sense, this argument makes sense theoretically. This is because our perceptions are established by making sense of the world, mediated by language. In this sense, linking individual bodily diversity with some kind of concept is itself a kind of social construction.

What about our bodies in reality? Of course, the body is established according to the meanings we make of it in a certain culture. But at the same time, the body as a product of nature has also been repeatedly examined and scientifically verified as a subject of natural scientific research.

This is where the theoretical conflict between Diamond and Butler arises. In other words, the difference between Diamond, who starts from the physicality of human beings as living beings (of course, Diamond argues that there is a diversity that cannot be resolved by simple binary classifications such as male/female), and Butler, who from an epistemological standpoint rejects the simple binary theory of male/female and considers it a social construction. Butler, who considers this to be a social construction.

In their rejection of simple dualism, Diamond and Butler may be said to be similar. However, whereas Butler tries to organise gender from the side of human perception, Diamond tries to discuss it from a natural scientific perspective. What is more, it is probably true that Butler is unable to respond adequately to these questions from Diamond's side. Of course, this argument is not a simple one. I think it is possible to argue further from Butler's point of view (as in Butler's argument that the body is also a "matter" formed by speech act) (Butler, 1993). However, there is no doubt that Diamond's criticism of Butler is quite convincing.

16.5 Conclusion Nature Loves Diversity

Diamond's position is not at all the kind of 'masculinity and femininity are inborn, not acquired', as many Japanese backlash groups have strongly argued while 'misunderstanding' his arguments. Rather, the perspective of the diversity of human sexuality is Diamond's basic perspective. It almost overlaps with the arguments of many of those who are, so to speak, gender-free (although it is true that some of them think of gender as socio-environmental determinism = constructed).

There is a phrase that could be called Diamond's slogan. It is 'Nature loves variety'. He never tries to speak in terms of fixed gender binaries, including gender. Rather, his basic view of nature is that it is precisely nature that produces individual diversity (a viewpoint that Butler might even agree with somewhere).

How will future generations view the gender-free debate in Japan at the beginning of the twenty-first century? I think they will probably look at it as a ridiculous debate and discuss the 'incident' with an appalled gaze. This is because it clearly reflects the 'kinks' in Japanese society regarding academic debate and scientific recognition (acceptance of scientific debate and recognition in its immaturity). Moreover, the fact that such baseless arguments have expanded through the internet and other means, and have even gained a certain social power, is also problematic and a product of this era at the beginning of the twenty-first century.

References

Butler, J. (1990). *Gender trouble: Feminism and the subversion of identity*. Routledge.

Butler, J. (1993). *Bodies that matter, on the discursive limits of "sex"*. Routledge.

Colapinto, J. (1998) "The True Story of John/Joan", Rolling Stones, December 11.

Colapinto, J. (2000). *As nature made him: The boy was raised as a girl*. HarperCollins. (translated by Tomoyuki Murai and publishes by Mumeisha in 2000 titled Burenda to Yobareta Shounen (The Boy Called Brenda). With the same title and the same translator, this book was republished by Husosha in 2005).

Diamond, M. (2011). Discrimination, Psychosexual development and Japanese women in academia. In the Office for Promotion of Women's Researchers at Kyoto University (Ed.), *Seisa-Kagaku no Saizensen(frontier of sexual differences studies)*. The Office for Promotion of Women's Researchers at Kyoto University.

Freeman, D. (1983). *Margaret Mead and Samoa: The making and unmasking of an anthropological myth*. Harvard University Press.

Huston, B. (1985). Gender Freedom and the Subtleties of Sexist Education. *Educational Theory, 35*(4), 359–369 (Fall).

Huston, B. (1994). (Originally 1985) Should public education be gender free? In L. Stone (Ed.), *The education feminism reader*. Routledge.

Ito, K. (2009). *Danjo-Kyodo=sankaku Ga Toikakerumono-Gendainihonshakai to Jenda-poritikusu, Zouho-ban (What does the policy for gender equality ask ?—Gender politics in contemporary Japanese society, Revised and Enlarged Edition)*. Impact Sguppankai.

Ito, K. (2014). Emerging culture wars: Backlush against 'gender freedom'. In S. Sung & G. Pascal (Eds.), *Gender and welfare states in Eastern Asia Confucianism or gender Equality*. Macmillan.

Mead, M. (1928). *Coming of age in Samoa*. William Morrow and Co.

Mead, M. (2001). *(Originally 1935) Sex and temperament in three primitive societies*. Perennial.

Money, J., & Tucker, P. (1975). *Sexual signatures on being a man or a woman*. Little Brown & Co.

Ueda, M. (Ed.). (1999). *Jinken-rekishu-nenpyo(Chronological table of human rights)*. Yamakawashuppan.

Correction to: Fathers' Involvement in Childcare: A Historical View on Nature and Nurture in Edo and Meiji Japan

Atsuko Fujino

Correction to:
Chapter 11 in: J. Pérez-Jara, Í. Ongay (eds.), *Beyond Nature and Nurture*, Synthese Library 497, https://doi.org/10.1007/978-3-031-87715-5_11

The author had mistakenly added "(Overcoming the Nature Versus Nurture Debate)" before the actual chapter title, which has been corrected now.

The updated version of this chapter can be found at
https://doi.org/10.1007/978-3-031-87715-5_11

J. Pérez-Jara, Í. Ongay (eds.), *Beyond Nature and Nurture*, Synthese Library 497,
https://doi.org/10.1007/978-3-031-87715-5_17

Index

J. Pérez-Jara, Í. Ongay (eds.), *Beyond Nature and Nurture*, Synthese Library 497,
https://doi.org/10.1007/978-3-031-87715-5

MIX
Papier aus verantwortungsvollen Quellen
Paper from responsible sources
FSC® C105338

If you have any concerns about our products,
you can contact us on
ProductSafety@springernature.com

In case Publisher is established outside the EU,
the EU authorized representative is:
Springer Nature Customer Service Center GmbH
Europaplatz 3, 69115 Heidelberg, Germany

Printed by Libri Plureos GmbH
in Hamburg, Germany